## Student's Solutions Manual Part II

# CALCULUS

## Finney/Thomas

**Michael Schneider**
**Thomas L. Cochran**

**Belleville Area College**

**ADDISON-WESLEY PUBLISHING COMPANY**
Reading, Massachusetts • Menlo Park, California • New York
Don Mills, Ontario •Wokingham, England • Amsterdam • Bonn
Sydney • Singapore • Tokyo • Madrid • San Juan

*Reprinted with corrections, November 1991.*

Reproduced by Addison-Wesley from camera-ready copy supplied by the authors.



ISBN 0-201-51393-5

5 6 7 8 9 10 AL 9594939291

The authors would like to dedicate this book to their wives, Jane and Barbara, the two who have put up with so much for so long.

Michael Schneider
Tom Cochran
Melissa Slaim

The authors have attempted to make this manual as error free as possible but nobody's "perfeck". If you find errors, we would appreciate knowing them. You are more than welcome to write us at:

Belleville Area College
2500 Carlyle Road
Belleville, Illinois 62221

# TABLE OF CONTENTS

# CHAPTER 9

# INFINITE SERIES

## 9.1 SEQUENCES OF NUMBERS

1. $a_1 = 0, a_2 = -\frac{1}{4}, a_3 = -\frac{2}{9}, a_4 = -\frac{3}{16}$

3. $a_1 = 1, a_2 = -\frac{1}{3}, a_3 = \frac{1}{5}, a_4 = -\frac{1}{7}$

5. $1, \frac{3}{2}, \frac{7}{4}, \frac{15}{8}, \frac{63}{32}$

7. $2, 1, \frac{1}{2}, \frac{1}{4}, \frac{1}{8}, \frac{1}{16}$

9. 1, 1, 2, 3, 5, 8

11. $\lim_{n\to\infty} a_n = \lim_{n\to\infty} \left(2 + (0.1)^n\right) = 2 \Rightarrow$ converges

13. $\lim_{n\to\infty} a_n = \lim_{n\to\infty} \frac{1-2n}{1+2n} = -1 \Rightarrow$ converges

15. $\lim_{n\to\infty} a_n = \lim_{n\to\infty} \frac{n}{10} = \infty \Rightarrow$ diverges

17. $\lim_{n\to\infty} a_n = \lim_{n\to\infty} \frac{1-5n^4}{n^4+8n^3} = -5 \Rightarrow$ converges

19. $\lim_{n\to\infty} a_n = \lim_{n\to\infty} \frac{n^2-2n+1}{n-1} = \lim_{n\to\infty} n-1 = \infty \Rightarrow$ diverges

21. $\lim_{n\to\infty} a_n = \lim_{n\to\infty} 5 = 5 \Rightarrow$ converges

23. $\lim_{n\to\infty} a_n = \lim_{n\to\infty} 5^n = \infty \Rightarrow$ diverges

25. $\lim_{n\to\infty} a_n = \lim_{n\to\infty} \left(\frac{n+1}{2n}\right)\left(1-\frac{1}{n}\right) = \frac{1}{2} \Rightarrow$ converges

27. $\lim_{n\to\infty} a_n = \lim_{n\to\infty} \frac{(-1)^{(n+1)}}{2n-1} = 0$

29. $\lim_{n\to\infty} a_n = \lim_{n\to\infty} \frac{\sin n}{n} = 0$, by the Sandwich Theorem for Sequences

31. $\lim_{n\to\infty} a_n = \lim_{n\to\infty} \sqrt{\frac{2n}{n+1}} = \sqrt{\lim_{n\to\infty} \frac{2n}{n+1}} = \sqrt{2}$

33. $\lim_{n\to\infty} a_n = \lim_{n\to\infty} \tan^{-1} n = \frac{\pi}{2}$

35. $\lim_{n\to\infty} a_n = \lim_{n\to\infty} \frac{n}{2^n} = \lim_{n\to\infty} \frac{1}{2^n \ln 2} = 0$

37. $\lim_{n\to\infty} a_n = \lim_{n\to\infty} \frac{\ln(n+1)}{n} = 0$, due to the growth rates in secion 7.6

39. $\lim_{n\to\infty} a_n = \lim_{n\to\infty} 8^{1/n} = \lim_{n\to\infty} \exp\left(\frac{\ln 8}{n}\right) = e^0 = 1$

41. $\lim_{n\to\infty} a_n = \lim_{n\to\infty} \left(1 + \frac{7}{n}\right)^n = e^7$, due to table 9.1 part 5

43. $\lim_{n\to\infty} a_n = \lim_{n\to\infty} \frac{1}{(0.9)^n} = \lim_{n\to\infty} \left(\frac{10}{9}\right)^n = \infty \Rightarrow$ diverges

45. $\lim_{n\to\infty} a_n = \lim_{n\to\infty} \sqrt[n]{10\,n} = \lim_{n\to\infty} \exp\left(\frac{\ln(10\,n)}{n}\right) = \exp\left(\lim_{n\to\infty} \frac{\ln 10\,n}{n}\right) = \exp\left(\lim_{n\to\infty} \frac{1}{n}\right) = e^0 = 1$

47. $\lim_{n\to\infty} a_n = \lim_{n\to\infty} \left(\frac{3}{n}\right)^{1/n} = \frac{\lim_{n\to\infty} \sqrt[n]{3}}{\lim_{n\to\infty} \sqrt[n]{n}} = \frac{1}{1} = 1$, due to table 9.1, parts 2 and 3

49. $\lim_{n\to\infty} a_n = \lim_{n\to\infty} \frac{\ln n}{n^{1/n}} = \frac{\lim_{n\to\infty} \ln n}{\lim_{n\to\infty} \sqrt[n]{n}} = \frac{\infty}{1} = \infty$, due to table 9.1, part 2

51. $\lim_{n\to\infty} a_n = \lim_{n\to\infty} \left(\frac{1}{3}\right)^n = 0$, due to table 9.1, part 4

53. $\lim_{n\to\infty} a_n = \lim_{n\to\infty} \frac{1}{n!} = 0$, due to table 9.1, part 6

55. $\lim_{n\to\infty} a_n = \lim_{n\to\infty} \left(\frac{1}{n}\right)^{1/\ln(n)} = \lim_{n\to\infty} \exp\left(\frac{\ln(1/n)}{\ln(n)}\right) = \exp\left(\lim_{n\to\infty} \frac{\ln(1/n)}{\ln(n)}\right) = \exp\left(\lim_{n\to\infty} -1\right) = e^{-1}$

57. $\lim_{n\to\infty} a_n = \lim_{n\to\infty} \frac{n!}{10^{6n}} = \frac{1}{\lim_{n\to\infty} \frac{\left(10^6\right)^n}{n!}} = \infty \Rightarrow$ diverges

59. $\left|\sqrt[n]{0.5} - 1\right| < 10^{-3} \Rightarrow -\frac{1}{1000} < \left(\frac{1}{2}\right)^{1/n} - 1 < \frac{1}{1000} \Rightarrow n > \frac{\ln(1/2)}{\ln(999/1000)} \Rightarrow n > 692.8 \Rightarrow N = 693$

61. $(0.9)^n < 10^{-3} \Rightarrow n \ln(0.9) < -3 \ln 10 \Rightarrow n > 65.56 \Rightarrow N = 66$

62. $\frac{2^n}{n!} < 10^{-7} \Rightarrow n! > 10^7\, 2^n \Rightarrow n \geq 15 \Rightarrow N = 15$

64. $x_1 = 1$, $x_2 = \cos(1) \approx 0.540302305$, $x^3 \approx x_4 \approx 0.857553215$, $x_5 \approx 0.65428979 \Rightarrow x_{56} \approx 0.73908133$

## 9.2 INFINITE SERIES

1. $s_n = \dfrac{a\left(1 - r^n\right)}{(1 - r)} = \dfrac{2\left(1 - (1/3)^n\right)}{1 - 1/3} \Rightarrow \lim\limits_{n \to \infty} s_n = \dfrac{2}{1 - 1/3} = 3$

3. $s_n = \dfrac{a\left(1 - r^n\right)}{(1 - r)} = \dfrac{1 - (-1/2)^n}{1 - (-1/2)} \Rightarrow \lim\limits_{n \to \infty} s_n = \dfrac{1}{3/2} = \dfrac{2}{3}$

5. $\dfrac{1}{(n+1)(n+2)} = \dfrac{1}{n+1} - \dfrac{1}{n+2} \Rightarrow s_n = \left(\dfrac{1}{2} - \dfrac{1}{3}\right) + \left(\dfrac{1}{3} - \dfrac{1}{4}\right) + \ldots + \left(\dfrac{1}{n+1} - \dfrac{1}{n+2}\right) = \dfrac{1}{2} - \dfrac{1}{n+2} \Rightarrow$
$\lim\limits_{n \to \infty} s_n = \dfrac{1}{2}$

7. $1 + \dfrac{1}{4} + \dfrac{1}{16} + \ldots$ The sum of this geometric series is $\dfrac{1}{1 - 1/4} = \dfrac{4}{3}$.

9. $\dfrac{7}{4} + \dfrac{7}{16} + \dfrac{7}{64} + \ldots$ The sum of this geometric series is $\dfrac{7/4}{1 - 1/4} = \dfrac{7}{3}$.

11. $(5 + 1) + \left(\dfrac{5}{2} + \dfrac{1}{3}\right) + \left(\dfrac{5}{4} + \dfrac{1}{9}\right) + \left(\dfrac{5}{8} + \dfrac{1}{27}\right) + \ldots$ is the sum of two geometric series; the sum $\dfrac{5}{1 - 1/2} + \dfrac{1}{1 - 1/3} = 10 + \dfrac{3}{2} = \dfrac{23}{2}$.

13. $(1 + 1) + \left(\dfrac{1}{2} - \dfrac{1}{5}\right) + \left(\dfrac{1}{4} + \dfrac{1}{25}\right) + \left(\dfrac{1}{8} - \dfrac{1}{125}\right) + \ldots$ is the sum of two geometric series; the sum $\dfrac{1}{1 - 1/2} + \dfrac{1}{1 + 1/5} = 2 + \dfrac{5}{6} = \dfrac{17}{6}$.

15. $\dfrac{4}{(4n-3)(4n+1)} = \dfrac{1}{4n-3} - \dfrac{1}{4n+1} \Rightarrow s_n = \left(1 - \dfrac{1}{5}\right) + \left(\dfrac{1}{5} - \dfrac{1}{9}\right) + \left(\dfrac{1}{9} - \dfrac{1}{13}\right) + \ldots + \left(\dfrac{1}{4n-7} - \dfrac{1}{4n-3}\right)$
$+ \left(\dfrac{1}{4n-3} - \dfrac{1}{4n+1}\right) = 1 - \dfrac{1}{4n+1} \Rightarrow \lim\limits_{n \to \infty} s_n = \lim\limits_{n \to \infty} 1 - \dfrac{1}{4n+1} = 1$

17. $\dfrac{4}{(4n-3)(4n+1)} = \dfrac{1}{4n-3} - \dfrac{1}{4n+1} \Rightarrow s_n = \left(\dfrac{1}{9} - \dfrac{1}{13}\right) + \left(\dfrac{1}{13} - \dfrac{1}{17}\right) + \left(\dfrac{1}{17} - \dfrac{1}{21}\right) + \ldots +$
$\left(\dfrac{1}{4n-7} - \dfrac{1}{4n-3}\right) + \left(\dfrac{1}{4n-3} - \dfrac{1}{4n+1}\right) \Rightarrow \lim\limits_{n \to \infty} s_n = \lim\limits_{n \to \infty} \dfrac{1}{9} - \dfrac{1}{4n+1} = \dfrac{1}{9}$

19. A convergent geometric series with a sum of $\dfrac{1}{1 - 1/\sqrt{2}} = 2 + \sqrt{2}$.

21. A convergent geometric series with a sum of $\dfrac{3/2}{1 - (-1/2)} = 1$.

23. $\lim\limits_{n \to \infty} \cos(n\pi) \neq 0 \Rightarrow$ divergence.

25. A convergent geometric series with a sum of $\dfrac{1}{1 - 1/e^2} = \dfrac{e^2}{e^2 - 1}$.

27. $\lim\limits_{n \to \infty} a_n = \lim\limits_{n \to \infty} (-1)^{n+1} n \neq 0 \Rightarrow$ divergence.

29. The difference of two convergent geometric series with a sum of $\dfrac{1}{1 - 2/3} - \dfrac{1}{1 - 1/3} = 3 - \dfrac{3}{2} = \dfrac{3}{2}$.

31. $\lim\limits_{n \to \infty} a_n = \lim\limits_{n \to \infty} \dfrac{n!}{1000^n} = \dfrac{1}{\lim\limits_{n \to \infty} \dfrac{1000^n}{n!}} = \infty \neq 0 \Rightarrow$ divergence.

33. $\frac{1}{1+x} = \frac{1}{1-(-x)} \Rightarrow a = 1, r = -x$

35. distance $= 4 + 2\left[(4)\left(\frac{3}{4}\right) + (4)\left(\frac{3}{4}\right)^2 + \ldots\right] = 4 + 2\left[\frac{3}{1-3/4}\right] = 28$ m

37. $\frac{234}{10^3} + \frac{234}{10^6} + \frac{234}{10^9} + \ldots = \frac{234/10^3}{1-1/10^3} = \frac{26}{111}$

39. a) $\sum_{n=-2}^{\infty} \frac{1}{(n+4)(n+5)}$ b) $\sum_{n=0}^{\infty} \frac{1}{(n+2)(n+3)}$ c) $\sum_{n=5}^{\infty} \frac{1}{(n-3)(n-2)}$

41. area $= 2^2 + (\sqrt{2})^2 + (1)^2 + \left(\frac{1}{\sqrt{2}}\right)^2 + \ldots = 4 + 2 + 1 + \frac{1}{2} + \ldots = \frac{4}{1-1/2} = 8\text{ m}^2$

43. $\sum_{n=1}^{\infty} n$ and $\sum_{n=1}^{\infty} (-n)$

45. $\sum_{n=0}^{\infty} \left(\frac{1}{5}\right)^n$ converges to $\frac{5}{4}$ and $\sum_{n=0}^{\infty} \left(\frac{1}{2}\right)^n$ converges to 2,

but $\sum_{n=0}^{\infty} \frac{(1/5)^n}{(1/2)^n}$ converges to $\frac{5}{3} \neq \frac{5/4}{2} = \frac{5}{8}$.

## 9.3 SERIES WITH NONEGATIVE TERMS: COMPARISON AND INTEGRAL TESTS

1. converges, a geometric series with r = 1/10

3. converges, by the Comparison Test for Convergence, since $\frac{\sin^2 n}{2^n} \leq \frac{1}{2^n}$

5. converges, by the Comparison Test for Convergence, since $\frac{1 + \cos n}{n^2} \leq \frac{2}{n^2}$

7. diverges, by the Comparison Test for Convergence, since $\frac{1}{n} < \frac{\ln n}{n}$ for $n \geq 3$

9. converges, a geometric series with r = 2/3

11. diverges, by the Comparison Test for Convergence, since $\frac{1}{n+1} < \frac{1}{1+\ln n}$ and $\sum_{n=1}^{\infty} \frac{1}{n+1}$ diverges

by the Limit Comparison Test when compared with $\sum_{n=1}^{\infty} \frac{1}{n}$

13. diverges, $\lim_{n \to \infty} a_n = \lim_{n \to \infty} \frac{2^n}{n+1} = \lim_{n \to \infty} \frac{2^n \ln 2}{1} = \infty \neq 0$

15. converges, by the Limit Comparison Test when compared with $\sum_{n=1}^{\infty} \frac{1}{n^{3/2}}$

17. diverges, by the Limit Comparison Test when compared with $\sum_{n=1}^{\infty} \frac{1}{n}$

19. diverges, $\lim_{n \to \infty} a_n = \lim_{n \to \infty} \left(1 + \frac{1}{n}\right)^n = e \neq 0$

21. converges: $\sum_{n=1}^{\infty} \frac{1-n}{n\,2^n} = \sum_{n=1}^{\infty} \frac{1}{n\,2^n} + \sum_{n=1}^{\infty} \frac{-1}{2^n}$, the sum of two convergent series.

$\sum_{n=1}^{\infty} \frac{1}{n\,2^n}$ converges, by the Comparison Test for Convergence, since $\frac{1}{n\,2^n} < \frac{1}{2^n}$.

23. converges, by the Comparison Test for Convergence, since $\frac{1}{3^{n-1} + 1} < \frac{1}{3^{n-1}}$

25. There are $(13)(365)(24)(60)(60)(10^9)$ seconds in 13 billion years. $s_n \le 1 + \ln n$ where $n = (13)(365)(24)(60)(60)(10^9) \Rightarrow s_n \le 1 + \ln\left((13)(365)(24)(60)(60)(10^9)\right) = 1 + \ln(13) + \ln(365) + \ln(24) + 2\ln(60) + 9\ln(10) \approx 41.55$

27. If $\sum_{n=1}^{\infty} a_n$ converges and $\frac{a_n}{n} \le a_n$, then by the Comparison Test for Convergence,

$\sum_{n=1}^{\infty} \frac{a_n}{n}$ converges.

29. If $\{S_n\}$ is nonincreasing with lower bound M, then $\{-S_n\}$ is a nondecreasing sequence with upper bound $-M$. By Theorem 6, $\{-S_n\}$ converges, and hence, $\{S_n\}$ converges. If $\{S_n\}$ has no lower bound, then $\{-S_n\}$ has no upper bound and diverges. Hence, $\{S_n\}$ also diverges.

## 9.4 SERIES WITH NONNEGATIVE TERMS: RATIO AND ROOT TESTS

1. converges, by the Ratio Test for $\lim_{n\to\infty} \left|\frac{(n+1)^2}{2^{n+1}} \frac{2^n}{n^2}\right| = \frac{1}{2} < 1$

3. converges, by the Ratio Test for $\lim_{n\to\infty} \left|\frac{(n+1)^{10}}{10^{n+1}} \frac{10^n}{n^{10}}\right| = \frac{1}{10} < 1$

5. diverges, for $\lim_{n\to\infty} a_n = \lim_{n\to\infty} \left(\frac{n-2}{n}\right)^n = \lim_{n\to\infty} \left(1 + \frac{-2}{n}\right)^n = e^{-2} \neq 0$

7. diverges, by the Ratio Test for $\lim_{n\to\infty} \left|\frac{(n+1)!\, e^{-(n+1)}}{n!\, e^{-n}}\right| = \infty$

9. diverges, $\lim_{n\to\infty} a_n = \lim_{n\to\infty} \left(1 - \frac{3}{n}\right)^n = e^{-3} \neq 0$

11. converges, by the Comparison Test for Convergence, since $\frac{\ln n}{n^3} < \frac{n}{n^2} = \frac{1}{n^2}$ for $n \geq 2$

13. diverges, by the Comparison Test for Convergence, since $\frac{\ln n}{n} > \frac{1}{n}$ for $n \geq 3$

15. converges, by the Ratio Test for $\lim_{n\to\infty} \left|\frac{(n+2)(n+3)}{(n+1)!} \frac{n!}{(n+1)(n+2)}\right| = 0 < 1$

17. converges, by the Ratio Test for $\lim_{n\to\infty} \left|\frac{(n+4)!}{3!\,(n+1)!\,3^{n+1}} \frac{3!\,n!\,3^n}{(n+3)!}\right| = \frac{1}{3} < 1$

19. converges, by the Ratio Test for $\lim_{n\to\infty} \left|\frac{1}{(2n+3)!} \frac{(2n+1)!}{1}\right| = 0 < 1$

21. converges, by the Root Test for $\lim_{n\to\infty} \sqrt[n]{\frac{n}{(\ln n)^n}} = \lim_{n\to\infty} \frac{\sqrt[n]{n}}{\ln n} = \lim_{n\to\infty} \frac{1}{\ln n} = 0 < 1$

23. converges, by the Comparison Test for Convergence, since $\frac{n!}{(n+2)!} = \frac{1}{(n+1)(n+2)} < \frac{1}{n^2}$

25. converges, by the Ratio test for $\lim_{n\to\infty} \left|\frac{(n+1)!}{(2n+3)!} \frac{(2n+1)!}{n!}\right| = \lim_{n\to\infty} \frac{n+1}{(2n+3)(2n+2)} = 0 < 1$

27. converges, by the ratio Test for $\lim_{n\to\infty} \frac{\left(\frac{1+\sin n}{n}\right) a_n}{a_n} = 0 < 1$

29. diverges, the given series is $\sum_{n=1}^{\infty} \frac{3}{n}$

31. converges, by the Ratio Test for $\lim_{n\to\infty} \left|\frac{a_{n+1}}{a_n}\right| = \lim_{n\to\infty} \left|\frac{(1 + \ln n)\, a_n}{n\, a_n}\right| = \lim_{n\to\infty} \frac{1 + \ln n}{n} = 0 < 1$

33. converges, by the Ratio Test for $\lim_{n\to\infty} \left|\frac{2^{n+1}\,(n+1)!\,(n+1)!}{(2n+2)!}\;\frac{(2n)!}{2^n\,(n!)(n!)}\right| =$

$$\lim_{n\to\infty} \frac{2\,(n+1)^2}{(2n+2)(2n+1)} = \frac{1}{2} < 1$$

35. coverges, by the Comparison Test for Convergence, since $a_1 = 1 = \frac{12}{(1)(3)(2)^2}$, $a_2 = \frac{1\cdot 2}{3\cdot 4} =$

$\frac{12}{(2)(4)(3)^2}$, $a_3 = \frac{2\cdot 3}{4\cdot 5}\,\frac{1\cdot 2}{3\cdot 4} = \frac{12}{(3)(5)(4)^2}$, $a_4 = \frac{3\cdot 4}{5\cdot 6}\,\frac{2\cdot 3}{4\cdot 5}\,\frac{1\cdot 2}{3\cdot 4} = \frac{12}{(4)(6)(5)^2}, \ldots \Rightarrow$

$1 + \sum_{n=1}^{\infty} \frac{12}{(n+1)(n+3)(n+2)^2}$ represents the given series and $\frac{12}{(n+1)\,(n+3)\,(n+2)^2} < \frac{12}{n^4}$

## 9.5 ALTERNATING SERIES AND ABSOLUTE CONVERGENCE

1. converges absolutely, by the Absolute Convergence Theorem, since $\sum_{n=1}^{\infty} |a_n|$ is a convergent p–series

3. diverges, $\lim_{n\to\infty} a_n \neq 0$

5. converges, by the Alternating Series Theorem, since $f(x) = \frac{\sqrt{x} + 1}{x + 1} \Rightarrow f'(x) = \frac{1 - 2x - 2\sqrt{x}}{(x+1)^2} < 0 \Rightarrow$

f(x) is decreasing, $\lim_{n\to\infty} a_n = \lim_{n\to\infty} \frac{\sqrt{n} + 1}{n + 1} = 0$

7. converges absolutely, by the Absolute Convergence Theorem, since $\sum_{n=1}^{\infty} |a_n|$ is a convergent p–series

9. diverges, $\lim_{n\to\infty} a_n = \lim_{n\to\infty} \frac{3\sqrt{n+1}}{\sqrt{n} + 1} = \lim_{n\to\infty} \frac{3\sqrt{1 + 1/n}}{\sqrt{1} + 1/\sqrt{n}} = 3 \neq 0$

11. converges absolutely, by the Absolute Convergence Theorem, since $\sum_{n=1}^{\infty} |a_n|$ is a convergent geometric series

13. converges absolutely, by the Absolute Convergence Theorem, since $\sum_{n=1}^{\infty} |a_n|$ converges, by the Limit Comparison Test when compared with $\sum_{n=1}^{\infty} \frac{1}{n^2}$

15. converges conditionally, since $f(x) = \frac{1}{x+3} \Rightarrow f'(x) = \frac{-1}{(x+3)^2} < 0 \Rightarrow f(x)$ is decreasing and $\lim_{n\to\infty} \frac{1}{n+3} = 0 \Rightarrow$ the given series converges, by the Alternating Series Test, but $\sum_{n=1}^{\infty} \frac{1}{n+3}$ diverges, by the Limit Comparison Test when compared with $\sum_{n=1}^{\infty} \frac{1}{n}$

17. diverges, $\lim_{n\to\infty} a_n = \lim_{n\to\infty} \frac{3+n}{5+n} = 1 \neq 0$

19. converges conditionally, since $f(x) = \frac{1}{x^2} + \frac{1}{x} \Rightarrow f'(x) = -\left(\frac{1}{x^3} + \frac{1}{x}\right) < 0 \Rightarrow f(x)$ is decreasing and $\lim_{n\to\infty} a_n = \lim_{n\to\infty} \frac{1}{n^2} + \frac{1}{n} = 0 \Rightarrow$ the given series converges, by the Alternating Series Test, but

$\sum_{n=1}^{\infty} \frac{1+n}{n^2} = \sum_{n=1}^{\infty} \frac{1}{n^2} + \sum_{n=1}^{\infty} \frac{1}{n}$, the sum of a convergent and divergent series, diverges

21. converges absolutely, by the Ratio Test for $\lim_{n\to\infty} \left| \frac{(n+1)^2 \left(\frac{2}{3}\right)^{n+1}}{n^2 \left(\frac{2}{3}\right)^n} \right| = \frac{2}{3} < 1$

23. converges absolutely, by the Integral Test, since $\int_1^{\infty} \arctan x \left(\frac{1}{1+x^2}\right) dx =$

$$\lim_{t\to\infty} \int_1^t \arctan x \left(\frac{1}{1+x^2}\right) dx = \lim_{t\to\infty} \left[\frac{(\arctan t)^2}{2}\right]_1^t = \frac{3\pi^2}{32}$$

25. diverges; the given series is $\frac{1}{2}\left(\sum_{n=1}^{\infty}\frac{1}{n}\right)$

27. diverges, $\lim_{n\to\infty} a_n = \lim_{n\to\infty}\frac{(-1)^n n}{n+1} = \lim_{n\to\infty}(-1)^n \neq 0$

29. converges absolutely, by the Absolute Convergence Theorem, since $\left|\frac{-1}{n^2+2n+1}\right| < \frac{1}{n^2}$

31. converges absolutely, by the Ratio Test for $\lim_{n\to\infty}\left|\frac{(100)^{n+1}}{(n+1)!}\,\frac{n!}{100^n}\right| = \lim_{n\to\infty}\frac{100}{n+1} = 0 < 1$

33. converges absolutely, by the Absolute Convergence Theorem, since $\left|\frac{\cos n\pi}{n\sqrt{n}}\right| = \left|\frac{(-1)^{n+1}}{n^{3/2}}\right| = \frac{1}{n^{3/2}}$, a convergent p–series

35. converges conditionally, since $f(x) = \sqrt{x+1} - \sqrt{x} \Rightarrow f'(x) = \frac{1}{2}\left(\frac{1}{\sqrt{x+1}} - \frac{1}{\sqrt{x}}\right) < 0 \Rightarrow f(x)$ is decreasing and $\lim_{n\to\infty} a_n = \lim_{n\to\infty}\frac{1}{\sqrt{n}+\sqrt{n+1}} = 0 \Rightarrow$ the given series converges, by the Alternating Series Test, but $\sum_{n=1}^{\infty}\frac{1}{\sqrt{n}+\sqrt{n+1}} = \sum_{n=1}^{\infty}\sqrt{n+1} - \sum_{n=1}^{\infty}\sqrt{n}$, the difference of two diverging series

37. $|\text{error}| < \left|(-1)^6\frac{1}{5}\right| = 0.2$

39. $|\text{error}| < \left|(-1)^6\frac{(0.01)^5}{5}\right| = 2.\times 10^{-11}$

41. $\frac{1}{(2n)!} < \frac{5}{10^6} \Rightarrow (2n)! > \frac{10^6}{5} = 200000 \Rightarrow 2n = 10 \Rightarrow n = 5 \Rightarrow 1 - \frac{1}{2!} + \frac{1}{4!} - \frac{1}{6!} + \frac{1}{8!} \approx$ $0.540302579 \approx 0.54030$

43. a) $a_n \geq a_{n+1}$ fails, since $\frac{1}{3} < \frac{1}{2}$

b) $\left(\frac{1}{3}+\frac{1}{9}+\frac{1}{27}+\ldots\right) - \left(\frac{1}{2}+\frac{1}{4}+\frac{1}{8}+\ldots\right) = \frac{1/3}{1-1/3} - \frac{1/2}{1-1/2} = \frac{1}{2} - 1 = -\frac{1}{2}$

45. The unused terms are $\sum_{j=n+1}^{\infty}(-1)^{j+1}a_j = (-1)^{n+1}\left(a_{n+1}-a_{n+2}\right) + (-1)^{n+3}\left(a_{n+3}-a_{n+4}\right) + \ldots =$ $(-1)^{n+1}\left[\left(a_{n+1}-a_{n+2}\right) + \left(a_{n+3}-a_{n+4}\right) + \ldots\right]$. Each grouped term is positive, hence the remainder has the same sign as $(-1)^{n+1}$, which is the sign of the first unused term.

## 9.6 POWER SERIES

1. $\lim_{n\to\infty}\left|\frac{x^{n+1}}{x^n}\right| < 1 \Rightarrow -1 < x < 1$; when $x = -1$ we have $\sum_{n=1}^{\infty}(-1)^n$, a divergent series; when $x = 1$ we have $\sum_{n=1}^{\infty} 1$, a divergent series $\therefore$

a) $-1 < x < 1$ b) $-1 < x < 1$

3. $\lim_{n\to\infty}\left|\frac{(x+1)^{n+1}}{(x+1)^n}\right| < 1 \Rightarrow -2 < x < 0$; when $x = -2$ we have $\sum_{n=1}^{\infty} 1$, a divergent series; when $x = 0$ we have $\sum_{n=1}^{\infty}(-1)^n$, a divergent series $\therefore$

a) $-2 < x < 0$ b) $-2 < x < 0$

5. $\lim_{n\to\infty}\left|\frac{(x-2)^{n+1}}{10^{n+1}}\,\frac{10^n}{(x-2)^n}\right| < 1 \Rightarrow -8 < x < 12$; when $x = -8$ we have $\sum_{n=1}^{\infty}(-1)^n$, a divergent series; when $x = 12$ we have $\sum_{n=1}^{\infty} 1$, a divergent series $\therefore$

a) $-8 < x < 12$ b) $-8 < x < 12$

7. $\lim_{n\to\infty}\left|\frac{(n+1)\,x^{n+1}}{n+3}\,\frac{n+2}{n\,x^n}\right| < 1 \Rightarrow -1 < x < 1$; when $x = -1$ we have $\sum_{n=1}^{\infty}(-1)^n\frac{n}{n+2}$, a divergent series; when $x = 1$ we have $\sum_{n=1}^{\infty}\frac{n}{n+2}$, a divergent series $\therefore$

a) $-1 < x < 1$ b) $-1 < x < 1$

9. $\lim_{n\to\infty}\left|\frac{x^{n+1}}{(n+1)\sqrt{n+1}}\,\frac{n\sqrt{n}}{x^n}\right| < 1 \Rightarrow -1 < x < 1$; when $x = -1$ we have $\sum_{n=1}^{\infty}\frac{(-1)^n}{n^{3/2}}$, a convergent series; when $x = 1$ we have $\sum_{n=1}^{\infty}\frac{1}{n^{3/2}}$, a convergent series $\therefore$

α) $-1 \le x \le 1$ b) $-1 \le x \le 1$

11. $\lim_{n\to\infty}\left|\frac{x^{n+1}}{(n+1)!}\frac{n!}{x^n}\right| < 1 \Rightarrow |x| \lim_{n\to\infty} \frac{1}{n+1} < 1$ for all $x \Rightarrow$

a) For all x b) For all x

13. $\lim_{n\to\infty}\left|\frac{x^{2n+3}}{(n+1)!}\frac{n!}{x^{2n+1}}\right| < 1 \Rightarrow x^2 \lim_{n\to\infty} \left|\frac{1}{n+1}\right| < 1$ for all x

a) For all x b) For all x

15. $\lim_{n\to\infty}\left|\frac{x^{n+1}}{\sqrt{n^2+2n+4}}\frac{\sqrt{n^2+3}}{x^n}\right| < 1 \Rightarrow |x| \sqrt{\lim_{n\to\infty}\frac{n^2+3}{n^2+2n+4}} < 1 \Rightarrow$

$-1 < x < 1$; when $x = -1$ we have $\sum_{n=1}^{\infty}\frac{(-1)^n}{\sqrt{n^2+3}}$, a convergent series;

when $x = 1$ we have $\sum_{n=1}^{\infty}\frac{1}{\sqrt{n^2+3}}$, a divergent series $\therefore$

a) $-1 \le x < 1$ b) $-1 < x < 1$

17. $\lim_{n\to\infty}\left|\frac{(n+1)x^{n+1}}{n^2+2n+2}\frac{n^2+1}{nx^n}\right| < 1 \Rightarrow -1 < x < 1$; when $x = -1$ we have $\sum_{n=1}^{\infty}\frac{(-1)^n n}{n^2+1}$, a convergent

series; when $x = 1$ we have $\sum_{n=1}^{\infty}\frac{n}{n^2+1}$, a divergent series $\therefore$

a) $-1 \le x < 1$ b) $-1 < x < 1$

19. $\lim_{n\to\infty}\left|\frac{\sqrt{n+1}\,x^{n+1}}{3^{n+1}}\frac{3^n}{\sqrt{n}\,x^n}\right| < 1 \Rightarrow \frac{|x|}{3}\sqrt{\lim_{n\to\infty}\frac{n+1}{n}} < 1 \Rightarrow -3 < x < 3$; both series:

$\sum_{n=1}^{\infty}\sqrt{n}\,(-1)^n$, when $x = -3$ and $\sum_{n=1}^{\infty}\sqrt{n}$, when $x = 3$ diverge $\therefore$

a) $-3 < x < 3$ b) $-3 < x < 3$

21. $\lim_{t\to\infty}\left|\frac{\left(1+\frac{1}{n+1}\right)^{n+1}x^{n+1}}{\left(1+\frac{1}{n}\right)^n x^n}\right| < 1 \Rightarrow |x| \lim_{t\to\infty}\left(\frac{n(n+2)}{(n+1)^2}\right)^n\left(1+\frac{1}{n+1}\right) < 1 \Rightarrow -1 < x < 1;$

both series: $\sum_{n=1}^{\infty}(-1)^n\left(1+\frac{1}{n}\right)^n$, when $x = -1$ and $\sum_{n=1}^{\infty}\left(1+\frac{1}{n}\right)^n$, when $x = 1$ diverge $\therefore$

a) $-1 < x < 1$ b) $-1 < x < 1$

23. $\lim_{n\to\infty}\left|\frac{(n+1)^{n+1}x^{n+1}}{n^n x^n}\right| < 1 \Rightarrow |x|\left(\lim_{n\to\infty}\left(1+\frac{1}{n}\right)^n\right)\left(\lim_{n\to\infty}(n+1)\right) < 1 \Rightarrow$

$e|x|\lim_{n\to\infty}(n+1) < 1 \Rightarrow$

a) $x = 0$ b) $x = 0$

25. $\lim_{n\to\infty}\left|\frac{(x-3)^{n+1}}{2^{n+1}}\,\frac{2^n}{(x-3)^n}\right| < 1 \Rightarrow |x-3| < 2 \Rightarrow 1 < x < 5$; both series: $\sum_{n=1}^{\infty}(1)$,

when $x = 1$ and $\sum_{n=1}^{\infty}(-1)^n$, when $x = 5$ diverge $\therefore$ the interval of convergence is $1 < x < 5$;

the sum of this convergent geometric series is $\frac{1}{1+\frac{x-3}{2}} = \frac{2}{x-1}$;

$f(x) = 1 - \frac{1}{2}(x-3) + \frac{1}{4}(x-3)^2 + \ldots + \left(-\frac{1}{2}\right)^n(x-3)^n + \ldots = \frac{2}{x-1} \Rightarrow$

$f'(x) = -\frac{1}{2} + \frac{1}{2}(x-3) + \ldots + \left(-\frac{1}{2}\right)^n n(x-3)^{n-1} + \ldots$ is convergent when $1 < x < 5$, and diverges when $x = 1$ or $5$; its sum is $\frac{-2}{(x-1)^2}$, the derivative of $\frac{2}{x-1}$

27. a) $\ln|\sec x| + C = \int \tan x\,dx = \int x + \frac{x^3}{3} + \frac{2x^5}{15} + \frac{17x^7}{315} + \ldots\,dx = \frac{x^2}{2} + \frac{x^4}{12} + \frac{x^6}{45} + \frac{17x^8}{2520} + \ldots + C$, but $x = 0 \Rightarrow C = 0 \Rightarrow \ln|\sec x| = \frac{x^2}{2} + \frac{x^4}{12} + \frac{x^6}{45} + \frac{17x^8}{2520} + \ldots$, when $-\frac{\pi}{2} < x < \frac{\pi}{2}$

b) $\sec^2 x = \frac{d(\tan x)}{dx} = \frac{d\left(x + \frac{x^3}{3} + \frac{2x^5}{15} + \frac{17x^7}{315} + \ldots\right)}{dx} = 1 + x^2 + \frac{2x^4}{3} + \frac{17x^6}{45} + \ldots$, when $-\frac{\pi}{2} < x < \frac{\pi}{2}$

c) $\sec^2 x = (\sec x)(\sec x) = \left(1 + \frac{x^2}{2} + \frac{5x^4}{24} + \frac{61x^6}{720} + \ldots\right)\left(1 + \frac{x^2}{2} + \frac{5x^4}{24} + \frac{61x^6}{720} + \ldots\right) =$

$1 + \left(\frac{1}{2} + \frac{1}{2}\right)x^2 + \left(\frac{5}{24} + \frac{1}{4} + \frac{5}{24}\right)x^4 + \left(\frac{61}{720} + \frac{5}{48} + \frac{5}{48} + \frac{61}{720}\right)x^6 + \ldots = 1 + x^2 + \frac{2x^4}{3} + \frac{17x^6}{45} + \ldots$

## 9.7 TAYLOR AND MACLAURIN SERIES

1. $f(x) = \ln x$ $\quad f'(x) = \frac{1}{x}$ $\quad f''(x) = -\frac{1}{x^2}$ $\quad f'''(x) = \frac{2}{x^3}$

$f(1) = \ln 1 = 0$ $\quad f'(1) = 1$ $\quad f''(1) = -1$ $\quad f'''(1) = 2$

$P_0(x) = 0,\ P_1(x) = x - 1,\ P_2(x) = (x-1) - \frac{1}{2}(x-1)^2,\ P_3(x) = (x-1) - \frac{1}{2}(x-1)^2 + \frac{1}{3}(x-1)^3$

3. $f(x) = \frac{1}{x} + x^{-1}$ $\quad f'(x) = -x^{-2}$ $\quad f''(x) = 2x^{-3}$ $\quad f'''(x) = -6x^{-4}$

$f(2) = \frac{1}{2}$ $\quad f'(2) = -\frac{1}{4}$ $\quad f''(2) = \frac{1}{4}$ $\quad f'''(2) = -\frac{3}{8}$

$P_0(x) = \frac{1}{2},\ P_1(x) = \frac{1}{2} - \frac{1}{4}(x-2),\ P_2(x) = \frac{1}{2} - \frac{1}{4}(x-2) + \frac{1}{8}(x-2)^2,$

$P_3(x) = \frac{1}{2} - \frac{1}{4}(x-2) + \frac{1}{8}(x-2)^2 - \frac{1}{16}(x-2)^3$

5. $f(x) = \sin x \Rightarrow f'(x) = \cos x \Rightarrow f''(x) = -\sin x \Rightarrow f'''(x) = -\cos x \Rightarrow f(\pi/4) = \sin \pi/4 = \frac{\sqrt{2}}{2},$

$f'(\pi/4) = \cos \pi/4 = \frac{\sqrt{2}}{2},\ f''(\pi/4) = -\sin \pi/4 = -\frac{\sqrt{2}}{2},\ f'''(\pi/4) = -\cos \pi/4 = -\frac{\sqrt{2}}{2} \Rightarrow P_0 = \frac{\sqrt{2}}{2},$

$P_1(x) = \frac{\sqrt{2}}{2} + \frac{\sqrt{2}}{2}\left(x - \frac{\pi}{4}\right),\ P_2(x) = \frac{\sqrt{2}}{2} + \frac{\sqrt{2}}{2}\left(x - \frac{\pi}{4}\right) - \frac{\sqrt{2}}{4}\left(x - \frac{\pi}{4}\right)^2,$

$P_3(x) = \frac{\sqrt{2}}{2} + \frac{\sqrt{2}}{2}\left(x - \frac{\pi}{4}\right) - \frac{\sqrt{2}}{4}\left(x - \frac{\pi}{4}\right)^2 - \frac{\sqrt{2}}{12}\left(x - \frac{\pi}{4}\right)^3$

7. $f(x) = \sqrt{x} = x^{1/2} \Rightarrow f'(x) = (1/2)x^{-1/2} \Rightarrow f''(x) = (-1/4)x^{-3/2} \Rightarrow f'''(x) = (3/8)x^{-5/2} \Rightarrow$

$f(4) = \sqrt{4} = 4^{1/2} = 2,\ f'(4) = (1/2)4^{-1/2} = \frac{1}{4},\ f''(4) = (-1/4)4^{-3/2} = -\frac{1}{32},\ f'''(4) = (3/8)4^{-5/2} = \frac{3}{256} \Rightarrow$

$P_0(x) = 2, P_1(x) = 2 + \frac{1}{4}(x-4),\ P_2(x) = 2 + \frac{1}{4}(x-4) - \frac{1}{64}(x-4)^2,$

$P_3(x) = 2 + \frac{1}{4}(x-4) - \frac{1}{64}(x-4)^2 + \frac{1}{512}(x-4)^3$

9. $e^x = \sum_{n=0}^{\infty} \frac{x^n}{n!} \Rightarrow e^{-x} = \sum_{n=0}^{\infty} \frac{(-x)^n}{n!} = 1 - x + \frac{x^2}{2!} - \frac{x^3}{3!} + \frac{x^4}{4!} - \ldots$

11. $\sin x = \sum_{n=0}^{\infty} \frac{(-1)^n x^{2n+1}}{(2n+1)!} \Rightarrow \sin 3x = \sum_{n=0}^{\infty} \frac{(-1)^n (3x)^{2n+1}}{(2n+1)!} = 3x - \frac{(3x)^3}{3!} + \frac{(3x)^5}{5!} - \ldots$

13. $\cos(-x) = \cos(x) = \sum_{n=0}^{\infty} \frac{(-1)^n x^{2n}}{(2n)!} = 1 - \frac{x^2}{2!} + \frac{x^4}{4!} - \frac{x^6}{6!} + \ldots$, since cosine is an even function

15. $\cosh x = \frac{e^x + e^{-x}}{2} = \frac{1}{2}\left[\left(1 + x + \frac{x^2}{2!} + \frac{x^3}{3!} + \frac{x^4}{4!} + \ldots\right) + \left(1 - x + \frac{x^2}{2!} - \frac{x^3}{3!} + \frac{x^4}{4!} - \ldots\right)\right] =$

$1 + \frac{x^2}{2!} + \frac{x^4}{4!} + \frac{x^6}{6!} + \ldots$

17. $\frac{x^2}{2} - 1 + \cos x = \frac{x^2}{2} - 1 + \sum_{n=0}^{\infty} \frac{(-1)^n x^{2n}}{(2n)!} = \frac{x^2}{2} - 1 + 1 - \frac{x^2}{2!} + \frac{x^4}{4!} - \frac{x^6}{6!} + \ldots = \frac{x^4}{4!} - \frac{x^6}{6!} + \frac{x^8}{8!} - \frac{x^{10}}{10!} + \ldots$

19. $f(x) = \frac{1}{x+1} = (1+x)^{-1} \Rightarrow f'(x) = -(1+x)^{-2} \Rightarrow f''(x) = 2(1+x)^{-3} \Rightarrow f'''(x) = -6(1+x)^{-4} \Rightarrow$ $f(0) = 1$, $f'(0) = -(1)^{-2} = -1$, $f''(0) = 2(1)^{-3} = 2 \Rightarrow f'''(c) = -6(1+c)^{-4} = \frac{-6}{(1+c)^4}$, where c is between 0 and $x \Rightarrow P_2(x) = 1 - x + x^2 + R_2(x)$, where $R_2(x) = \frac{-x^3}{(1+c)^4}$

21. $f(x) = \ln(1+x) \Rightarrow f'(x) = (1+x)^{-1} \Rightarrow f''(x) = -(1+x)^{-2} \Rightarrow f'''(x) = 2(1+x)^{-3} \Rightarrow$ $f(0) = \ln(1) = 0$, $f'(0) = 1$, $f''(0) = -1$, $f'''(c) = 2(1+c)^{-3} = \frac{2}{(1+c)^3}$, where c is between 0 and $x \Rightarrow P_2(x) = x - \frac{x^2}{2} + R_2(x)$, where $R_2(x) = \frac{2}{(1+c)^3 \cdot 3!} x^3$

23. $f(x) = \sin x \Rightarrow f'(x) = \cos x \Rightarrow f''(x) = -\sin x \Rightarrow f'''(x) = -\cos x \Rightarrow f(0) = \sin 0 = 0, f'(0) = \cos 0 = 1$, $f''(0) = -\sin 0 = 0, f'''(c) = -\cos c$, where c is between 0 and $x \Rightarrow P_2(x) = x + R_2(x)$, where $R_2(x) = \frac{-\cos c}{3!} x^3$

25. If $e^x = \sum_{n=0}^{\infty} \frac{f^{(n)}(a)}{n!}(x-a)^n$ and $f(x) = e^x$, we have $f^{(n)}(a) = e^a$ for all n = 0, 1, 2, 3,...;

$e^x = e^a\left[\frac{(x-a)^0}{0!} + \frac{(x-a)^1}{1!} + \frac{(x-a)^2}{2!} + \ldots\right] = e^a\left[1 + (x-a) + \frac{(x-a)^2}{2!} + \ldots\right]$, at x = a

27. $\left|R_3(x)\right| = \left|\frac{-\cos c}{5!}(x-0)^5\right| < \left|\frac{x^5}{5!}\right| < 5 \times 10^{-4} \Rightarrow -5 \times 10^{-4} < \frac{x^5}{5!} < 5 \times 10^{-4} \Rightarrow$ $-\sqrt[5]{5!(5 \times 10^{-4})} < x < \sqrt[5]{5!(5 \times 10^{-4})} \Rightarrow -0.569679052 < x < 0.569679052$

29. $\sin x = x + R_1(x)$, when $|x| < 10^{-3} \Rightarrow \left|R_1(x)\right| = \left|\frac{-\cos c}{3!}x^3\right| < \left|\frac{(1)\,x^3}{3!}\right| < \frac{(10^{-3})^3}{3!} = 1.67 \times 10^{-10}$

From exercise 45 in section 9.5, $R_1(x)$ has the same sign as $-\frac{x^3}{3!}$. $x < \sin x \Rightarrow 0 < \sin x - x =$ $R_1(x)$, which has the same sign as $-\frac{x^3}{3!} \Rightarrow x < 0 \Rightarrow -10^{-3} < x < 0$.

31. $\left|R_2(x)\right| = \left|\frac{e^c x^3}{3!}\right| < \frac{3^{(0.1)}(0.1)^3}{3!} = 0.00018602$, where c is between 0 and x.

33. If we approximate sinh x with $x + \frac{x^3}{3!}$ and $|x| < 0.5$, then the $|\text{error}| < \left|\frac{(\cos c)\,x^5}{5!}\right| < \frac{(0.5)^5}{5!} =$ $\frac{1}{(32)(120)} = 0.000260416$, where c is between 0 and x.

35. sin x, when x = 0.1; the sum is $\sin(0.1) \approx 0.99833416$

37. $\sin x = x - \frac{x^3}{3!} + \frac{x^5}{5!} - \frac{x^7}{7!} + \ldots$; $\frac{d(\sin x)}{dx} = \frac{d\left(x - \frac{x^3}{3!} + \frac{x^5}{5!} - \frac{x^7}{7!} + \ldots\right)}{dx} = 1 - \frac{x^2}{2!} + \frac{x^4}{4!} - \frac{x^6}{6!} + \ldots = \cos x$;
$\frac{d(\cos x)}{dx} = \frac{d\left(1 - \frac{x^2}{2!} + \frac{x^4}{4!} - \frac{x^6}{6!} + \ldots\right)}{dx} = -x + \frac{x^3}{3!} - \frac{x^5}{5!} + \ldots = -\sin x$; $\frac{d\left(e^x\right)}{dx} =$
$\frac{d\left(1 + x + \frac{x^2}{2!} + \frac{x^3}{3!} + \frac{x^4}{4!} + \ldots\right)}{dx} = 1 + x + \frac{x^2}{2!} + \frac{x^3}{3!} + \frac{x^4}{4!} + \ldots$

39. $e^x \sin x = \left(1+x + \frac{x^2}{2!} + \frac{x^3}{3!} + \frac{x^4}{4!} + \frac{x^5}{5!} + \ldots\right)\left(x - \frac{x^3}{3!} + \frac{x^5}{5!} - \frac{x^7}{7!} + \ldots\right) = (1)\,x + (1)\,x^2 + \left(-\frac{1}{6} + \frac{1}{2}\right)x^3 +$
$\left(-\frac{1}{6} + \frac{1}{6}\right)x^4 + \left(\frac{1}{120} - \frac{1}{12} + \frac{1}{24}\right)x^5 + \left(\frac{1}{120} - \frac{1}{36} + \frac{1}{120}\right)x^6 + \ldots =$
$x + x^2 + \frac{1}{3}x^3 - \frac{1}{30}x^5 - \frac{1}{90}x^6 + \ldots$

41. $\sin x = x - \frac{x^3}{3!} + \frac{x^5}{5!} - \frac{x^7}{7!} + \ldots \Rightarrow \frac{\sin x}{x} = 1 - \frac{x^2}{3!} + \frac{x^4}{5!} - \frac{x^5}{7!} + \ldots$, $s_1 = 1$ and $s_2 = 1 - \frac{x^2}{6}$; If L is the sum of the series representing $\frac{\sin x}{x}$, then $L - s_1 = \frac{\sin x}{x} - 1 < 0$ and $L - s_2 = \frac{\sin x}{x} - \left(1 - \frac{x^2}{6}\right) > 0$, by the Alternating Series Estimation Theorem. $\therefore\ 1 - \frac{x^2}{6} < \frac{\sin x}{x} < 1$

43. a) $e^{-i\pi} = \cos(-\pi) + i\sin(-\pi) = -1 + i(0) = -1$

b) $e^{i\pi/4} = \cos\left(\frac{\pi}{4}\right) + i\sin\left(\frac{\pi}{4}\right) = \frac{1}{\sqrt{2}} + \frac{i}{\sqrt{2}} = \left(\frac{1}{\sqrt{2}}\right)(1 + i)$

c) $e^{-i\pi/2} = \cos\left(-\frac{\pi}{2}\right) + i\sin\left(-\frac{\pi}{2}\right) = 0 + i(-1) = -i$

45. $e^x = 1+x + \frac{x^2}{2!} + \frac{x^3}{3!} + \frac{x^4}{4!} + \ldots \Rightarrow e^{i\theta} = 1 + i\theta + \frac{(i\theta)^2}{2!} + \frac{(i\theta)^3}{3!} + \frac{(i\theta)^4}{4!} + \ldots$ and $e^{-i\theta} =$
$1 - i\theta + \frac{(-i\theta)^2}{2!} + \frac{(-i\theta)^3}{3!} + \frac{(-i\theta)^4}{4!} + \ldots = 1 - i\theta + \frac{(i\theta)^2}{2!} - \frac{(i\theta)^3}{3!} + \frac{(i\theta)^4}{4!} - \ldots$,

$$\frac{e^{i\theta} + e^{-i\theta}}{2} = \frac{\left(1 + i\theta + \frac{(i\theta)^2}{2!} + \frac{(i\theta)^3}{3!} + \frac{(i\theta)^4}{4!} + \ldots\right) + \left(1 - i\theta + \frac{(i\theta)^2}{2!} - \frac{(i\theta)^3}{3!} + \frac{(i\theta)^4}{4!} - \ldots\right)}{2} =$$

$1 - \frac{\theta^2}{2!} + \frac{\theta^4}{4!} - \frac{\theta^6}{6!} + \ldots = \cos\theta$;

$$\frac{e^{i\theta} - e^{-i\theta}}{2} = \frac{\left(1 + i\theta + \frac{(i\theta)^2}{2!} + \frac{(i\theta)^3}{3!} + \frac{(i\theta)^4}{4!} + \ldots\right) - \left(1 - i\theta + \frac{(i\theta)^2}{2!} - \frac{(i\theta)^3}{3!} + \frac{(i\theta)^4}{4!} - \ldots\right)}{2i} =$$

$\theta - \frac{\theta^3}{3!} + \frac{\theta^5}{5!} - \frac{\theta^7}{7!} + \ldots = \sin\theta$

47. $\frac{a-bi}{a^2+b^2}e^{(a+bi)x} + C_1 + iC_2 = \left(\frac{a-bi}{a^2+b^2}\right)e^{ax}(\cos bx + i\sin bx) + C_1 + iC_2 =$

$\frac{e^{ax}}{a^2+b^2}[a\cos bx + ia\sin bx - ib\cos bx + b\sin bx] + C_1 + iC_2 =$

$\frac{e^{ax}}{a^2+b^2}\left[(a\cos bx + b\sin bx) + (a\sin bx - b\cos bx)i\right] + C_1 + iC_2 =$

$\frac{e^{ax}[a\cos bx + b\sin bx]}{a^2+b^2} + C_1 + \frac{e^{ax}[a\sin bx - b\cos bx]}{a^2+b^2} + iC_2$, $e^{(a+bi)x} = e^{ax}e^{ibx} =$

$e^{ax}[\cos bx + i\sin bx] = e^{ax}\cos bx + ie^{ax}\sin bx$ and given that $\int e^{(a+bi)x}\,dx =$

$\frac{a-bi}{a^2+b^2}e^{(a+bi)x} + C_1 + iC_2$ we may conclude:

$\int e^{ax}\cos bx\,dx = \frac{e^{ax}[a\cos bx + b\sin bx]}{a^2+b^2} + C_1$ and

$\int e^{ax}\sin bx\,dx = \frac{e^{ax}[a\sin bx - b\cos bx]}{a^2+b^2} + iC_2$

## 9.8 FURTHER EXAMPLES OF TAYLOR SERIES

1. $f(x) = \cos x \Rightarrow f(\pi/3) = 0.5$, $f'(\pi/3) = -\frac{\sqrt{3}}{2}$, $f''(\pi/3) = -0.5$, $f'''(\pi/3) = \frac{\sqrt{3}}{2}$,

$f^{(4)}(\pi/3) = 0.5$; $\cos x = \frac{1}{2} - \frac{\sqrt{3}}{2}(x-\pi/3) - \frac{1}{4}(x-\pi/3)^2 + \frac{\sqrt{3}}{12}(x-\pi/3)^3 + \ldots$

3. $e^x = 1 + x + \frac{x^2}{2!} + \frac{x^3}{3!} + \ldots$

5. $f(x) = \cos x \Rightarrow f(22\pi) = 1$, $f'(22\pi) = 0$, $f''(22\pi) = -1$, $f'''(22\pi) = 0$, $f^{(4)}(22\pi) = 1$, $f^{(5)}(22\pi) = 0$,
$f^{(6)}(22\pi) = -1$; $\cos x = 1 - \frac{1}{2}(x-22\pi)^2 + \frac{1}{4!}(x-22\pi)^4 - \frac{1}{6!}(x-22\pi)^6 + \ldots$

7. $\int_0^{0.1}\frac{\sin x}{x}\,dx = \int_0^{0.1} 1 - \frac{x^2}{3!} + \frac{x^4}{5!} - \frac{x^6}{7!} + \ldots\,dx = \left[x - \frac{x^3}{3\cdot 3!} + \frac{x^5}{5\cdot 5!} - \frac{x^7}{7\cdot 7!} + \ldots\right]_0^{0.1} \approx$

$\left[x - \frac{x^3}{3\cdot 3!} + \frac{x^5}{5\cdot 5!}\right]_0^{0.1} \approx 0.099944461$

9. $\left(1+x^4\right)^{1/2} = (1)^{1/2} + \frac{1/2}{1}(1)^{-1/2}\left(x^4\right) + \frac{(1/2)(-1/2)}{2!}(1)^{-3/2}\left(x^4\right)^2 + \frac{(1/2)(-1/2)(-3/2)}{3!}(1)^{-5/2}\left(x^4\right)^3 +$

$\frac{(1/2)(-1/2)(-3/2)(-5/2)}{4!}(1)^{-7/2}\left(x^4\right)^4 + \ldots = 1 + \frac{x^4}{2} - \frac{x^8}{8} + \frac{x^{12}}{16} - \frac{5x^{16}}{128} + \ldots$;

$\int_0^{0.1} 1 + \frac{x^4}{2} - \frac{x^8}{8} + \frac{x^{12}}{16} - \frac{5x^{16}}{128} + \ldots\,dx = \left[x + \frac{x^5}{10} - \frac{x^9}{72} + \frac{x^{13}}{208} - \frac{5x^{17}}{2176} + \ldots\right]_0^{0.1} \approx 0.100001$

11. $\ln\left(\frac{1+x}{1-x}\right) = \ln(1+x) - \ln(1-x) = \left(x - \frac{x^2}{2} + \frac{x^3}{3} - \frac{x^4}{4} + \ldots\right) - \left(-x - \frac{x^2}{2} - \frac{x^3}{3} - \frac{x^4}{4} - \ldots\right) = 2\left(x + \frac{x^3}{3} + \frac{x^5}{5} + \ldots\right)$

13. $\tan^{-1}x = x - \frac{x^3}{3} + \frac{x^5}{5} - \frac{x^7}{7} + \frac{x^9}{9} - \ldots + \frac{(-1)^{n-1}x^{2n-1}}{2n-1} + \ldots$ and the $|\text{error}| = \left|\frac{(-1)^{n-1}x^{2n-1}}{2n-1}\right| = \frac{1}{2n-1}$, when $x = 1$; $\frac{1}{2n-1} < \frac{1}{10^3} \Rightarrow n > \frac{1001}{2} = 500.5 \Rightarrow$ the first term not used is 501st $\Rightarrow$ we must use 500 terms

15. $\tan^{-1}x = x - \frac{x^3}{3} + \frac{x^5}{5} - \frac{x^7}{7} + \frac{x^9}{9} - \ldots + \frac{(-1)^{n-1}x^{2n-1}}{2n-1} + \ldots$; when the series representing $48\tan^{-1}\left(\frac{1}{18}\right)$ has an error of magnitude less than $10^{-6}$, then the series representing $48\tan^{-1}\left(\frac{1}{18}\right) + 32\tan^{-1}\left(\frac{1}{57}\right) - 20\tan^{-1}\left(\frac{1}{239}\right)$ will also have an error of magnitude less than $10^{-6}$; $\frac{\left(\frac{1}{18}\right)^{2n-1}}{2n-1} < \frac{1}{10^6} \Rightarrow n \geq 3 \Rightarrow$ 3 terms

## PRACTICE EXERCISES

1. $\lim\limits_{n\to\infty} a_n = \lim\limits_{n\to\infty}\left(1 + \frac{(-1)^n}{n}\right) = 1$, converges to 1

3. $a_1 = \cos\left(\frac{\pi}{2}\right) = 0,\ a_2 = \cos(\pi) = -1,\ a_3 = \cos\left(\frac{3\pi}{2}\right) = 0,\ a_4 = \cos(2\pi) = 1, \ldots \Rightarrow$ the sequence diverges

5. $\lim\limits_{n\to\infty} a_n = \lim\limits_{n\to\infty} \frac{\ln n^2}{n} = 2\lim\limits_{n\to\infty} \frac{1/n}{1} = 0$, converges to 0

7. $\lim\limits_{n\to\infty} a_n = \lim\limits_{n\to\infty}\left(\frac{3^n}{n}\right)^{1/n} = \frac{3}{\lim\limits_{n\to\infty} n^{1/n}} = \frac{3}{1} = 3$, conveges to 3

9. $\lim\limits_{n\to\infty} a_n = \lim\limits_{n\to\infty} \frac{(-4)^n}{n!} = 0$, by table 9.1, converges to 0

11. $\lim\limits_{n\to\infty} a_n = \lim\limits_{n\to\infty} \frac{(n+1)!}{n!} = \lim\limits_{n\to\infty} n + 1 = \infty$, the sequence diverges

13. $\sum\limits_{n=1}^{\infty} \ln\left(\frac{n}{n+1}\right) = \sum\limits_{n=1}^{\infty} \ln(n) - \ln(n+1) \Rightarrow s_n = \left(\ln(1) - \ln(2)\right) + \left(\ln(2) - \ln(3)\right) + \left(\ln(3) - \ln(4)\right) + \ldots + \left(\ln(n-1) - \ln(n)\right) + \left(\ln(n) - \ln(n+1)\right) = \ln(1) - \ln(n+1) = -\ln(n+1) \Rightarrow$
$\lim\limits_{n\to\infty} s_n = -\infty$, $\Rightarrow$ the series diverges

15. $\sum_{n=0}^{\infty} e^{-n} = \sum_{n=0}^{\infty} \left(\frac{1}{e}\right)^n = 1 + \frac{1}{e} + \left(\frac{1}{e}\right)^2 + \left(\frac{1}{e}\right)^3 + \ldots = \frac{1}{1-\frac{1}{e}} = \frac{e}{e-1}$,

a convergent geometric series

17. diverges, a p–series where $p = \frac{1}{2}$

19. $f(x) = \frac{1}{x^{1/2}} \Rightarrow f'(x) = -\frac{1}{2x^{1/2}} < 0 \Rightarrow f(x)$ is decreasing, $\lim_{n \to \infty} a_n = \lim_{n \to \infty} \frac{(-1)^n}{\sqrt{n}} = 0$ and $a_n > a_{n+1} \Rightarrow$

$\sum_{n=1}^{\infty} \frac{(-1)^n}{\sqrt{n}}$ converges, by the Alternating Series Theorem; the series $\sum_{n=1}^{\infty} \frac{1}{\sqrt{n}}$ diverges $\Rightarrow$

the given series converges conditionally

21. the given series does not converge absolutlely, since $\frac{1}{\ln(n+1)} > \frac{1}{n+1}$ and $\sum_{n=1}^{\infty} \frac{1}{n+1}$ diverges;

$f(x) = \frac{1}{\ln(x+1)} \Rightarrow f'(x) = -\frac{1}{(\ln(x+1))^2(x+1)} < 0 \Rightarrow f(x)$ is decreasing, $\lim_{n \to \infty} a_n =$

$\lim_{n \to \infty} \frac{1}{\ln(n+1)} = 0$ and $a_n > a_{n+1}) \Rightarrow$ the given series converges conditionally, by

the Alternating Series Test

23. $\lim_{n \to \infty} \frac{\frac{1}{n\sqrt{n^2+1}}}{\frac{1}{n^2}} = \sqrt{\lim_{n \to \infty} \frac{n^2}{n^2+1}} = \sqrt{1} = 1 \Rightarrow$ converges absolutely, by the Limit Comparison Test

25. converges absolutely, by the Ratio Test, since $\lim_{n \to \infty} \left| \frac{n+2}{(n+1)!} \; \frac{n!}{n+1} \right| = \lim_{n \to \infty} \frac{n+2}{(n+1)^2} = 0$

27. converges absolutely, by the Ratio Test, since $\lim_{n \to \infty} \left| \frac{3^{n+1}}{(n+1)!} \; \frac{n!}{3^n} \right| = \lim_{n \to \infty} \left| \frac{3}{n+1} \right| = 0$

29. $\lim_{n \to \infty} \left| \frac{(x+2)^{n+1}}{3^{n+1}(n+1)} \; \frac{3^n n}{(x+2)^n} \right| < 1 \Rightarrow \frac{|x+2|}{3} \lim_{n \to \infty} \left| \frac{n}{n+1} \right| < 1 \Rightarrow -5 < x < 1$; when $x = -5$ we have

$\sum_{n=1}^{\infty} \frac{(-1)^n}{n}$, a convergent series; when $x = 1$ we have $\sum_{n=1}^{\infty} \frac{1}{n}$, a divergent series $\therefore$

a) $-5 \le x < 1$ b) $-5 < x < 1$

31. $\lim_{n\to\infty} \sqrt[n]{a_n} < 1 \Rightarrow \lim_{n\to\infty} \sqrt[n]{\frac{x^n}{n^n}} < 1 \Rightarrow \lim_{n\to\infty} \frac{|x|}{n} < 1 \Rightarrow |x|(0) < 1 \Rightarrow$

a) For all x b) For all x

33. $\lim_{n\to\infty} \left| \frac{(x-1)^{n+1}}{(n+1)^2} \frac{n^2}{(x-1)^n} \right| < 1 \Rightarrow |x-1| \lim_{n\to\infty} \left| \frac{n^2}{n^2+2n+1} \right| < 1 \Rightarrow 0 < x < 2$; when x = 0 we have

$\sum_{n=1}^{\infty} \frac{-1}{n^2}$, a convergent series; when x = 2 we have $\sum_{n=1}^{\infty} \frac{(-1)^{n-1}}{n}$, a convergent series $\therefore$

a) $0 \le x \le 2$ b) $0 \le x \le 2$

35. The given series is in the form $1 - x + x^2 - x^3 + \ldots + (-x)^n + \ldots = \frac{1}{1+x}$, where $x = \frac{1}{4}$.

The sum is $\frac{1}{1 + 1/4} = \frac{4}{5}$.

37. The given series is in the form $x - \frac{x^3}{3!} + \frac{x^5}{5!} - \ldots + (-1)^n \frac{x^{2n+1}}{(2n+1)!} + \ldots = \sin x$, where $x = \pi$.

The sum is $\sin \pi = 0$.

39. The given series is in the form $1 + x + \frac{x^2}{2!} + \frac{x^3}{3!} + \ldots + \frac{x^n}{n!} + \ldots = e^x$, where $x = \ln 2$.

The sum is $e^{\ln(2)} = 2$.

41. $f(x) = \sqrt{3 + x^2} = (3 + x^2)^{1/2} \Rightarrow f(-1) = 2,\ f'(-1) = -\frac{1}{2},\ f''(-1) = \frac{3}{8},\ f'''(-1) = \frac{9}{32}$;

$$\sqrt{3 + x^2} = 2 - \frac{(x+1)}{2 \cdot 1!} + \frac{3(x+1)^2}{2^3 \cdot 2!} + \frac{9(x+1)^3}{2^5 \cdot 3!} + \ldots$$

43. $$\left(x - \frac{x^3}{3!} + \frac{x^5}{5!} - \ldots\right)\left(2 - \frac{2x^2}{2!} + \frac{2x^4}{4!} - \ldots\right) = (2)\,x + (0)\,x^2 + \left(-\frac{2}{2!} - \frac{2}{3!}\right) x^3 + (0)\,x^4 +$$
$$\left(\frac{2}{4!} + \frac{2}{3!\,2!} + \frac{2}{5!}\right) x^5 + (0)\,x^6 + \left(-\frac{2}{6!} - \frac{2}{4!\,3!} - \frac{2}{5!\,2!} - \frac{2}{7!}\right) x^7 + \ldots = 2x - \frac{(2x)^3}{3!} + \frac{(2x)^5}{5!} - \frac{(2x)^7}{7!} + \ldots$$

45. $$\int_0^{1/2} \exp(-x^3)\,dx = \int_0^{1/2} 1 - x^3 + \frac{x^6}{2!} - \frac{x^9}{3!} + \frac{x^{12}}{4!} + \ldots dx =$$
$$\left[x - \frac{x^4}{4} + \frac{x^7}{7 \cdot 2!} - \frac{x^{10}}{10 \cdot 3!} + \frac{x^{13}}{13 \cdot 4!} - \ldots\right]_0^{1/2} \approx \frac{1}{2} - \frac{1}{2^4 \cdot 4} + \frac{1}{2^7 \cdot 7 \cdot 2!} - \frac{1}{2^{10} \cdot 10 \cdot 3!} \approx 0.484916759$$

47. a) the Maclaurin series is 0 + 0 + . . ., since $f^{(n)}(0) = 0$ for all n; this series conveges for all x; f(x) is zero only at x = 0

b) $f(x) = 0 + 0 + \ldots + 0 + R_n(x) \Rightarrow R_n(x) = \exp\left(-\frac{1}{x^2}\right)$

# CHAPTER 10

# PLANE CURVES AND POLAR COORDINATES

## 10.1 CONIC SECTIONS AND QUADRATIC EQUATIONS

1. $x = \frac{y^2}{8} \Rightarrow 4p = 8 \Rightarrow p = 2$. $\therefore$ Focus is (2,0), directrix is $x = -2$.

3. $y = -\frac{x^2}{6} \Rightarrow 4p = 6 \Rightarrow p = \frac{3}{2}$. $\therefore$ Focus is $(0, -\frac{3}{2})$, directrix is $y = \frac{3}{2}$.

5. $\frac{x^2}{4} - \frac{y^2}{9} = 1 \Rightarrow c = \sqrt{4+9} = \sqrt{13} \Rightarrow$ Foci are $(\pm\sqrt{13}, 0)$. $e = \frac{c}{a} = \frac{\sqrt{13}}{2} \Rightarrow \frac{a}{e} = \frac{2}{\frac{\sqrt{13}}{2}} = \frac{4}{\sqrt{13}} \Rightarrow$

   Directrices are $x = \pm\frac{4}{\sqrt{13}}$. Asymptotes are $y = \pm\frac{3}{2}x$.

7. $\frac{x^2}{2} + y^2 = 1 \Rightarrow c = \sqrt{2-1} = 1 \Rightarrow$ Foci are $(\pm 1, 0)$. $e = \frac{c}{a} = \frac{1}{\sqrt{2}} \Rightarrow$ Directrices are $x = \pm\frac{\sqrt{2}}{\frac{1}{\sqrt{2}}} = \pm 2$.

9. $16x^2 + 25y^2 = 400 \Rightarrow \frac{x^2}{25} + \frac{y^2}{16} = 1$

   $\Rightarrow c = \sqrt{a^2 - b^2} = \sqrt{25 - 16} = 3$.

   $e = \frac{c}{a} = \frac{3}{5}$

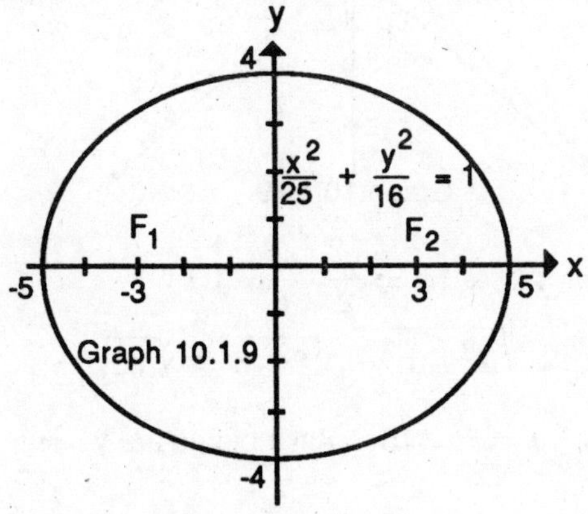

Graph 10.1.9

11. $2x^2 + y^2 = 2 \Rightarrow x^2 + \frac{y^2}{2} = 1$

   $\Rightarrow c = \sqrt{a^2 - b^2} = \sqrt{2 - 1} = 1$

   $e = \frac{c}{a} = \frac{1}{\sqrt{2}}$

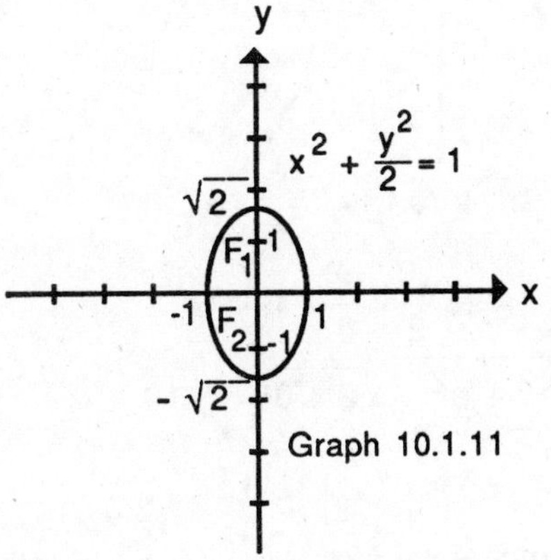

Graph 10.1.11

13. $3x^2 + 2y^2 = 6 \Rightarrow \frac{x^2}{2} + \frac{y^2}{3} = 1$

   $\Rightarrow c = \sqrt{a^2 - b^2} = \sqrt{3 - 2} = 1$

   $e = \frac{c}{a} = \frac{1}{\sqrt{3}}$

   (The graph is on the next page.)

15. $6x^2 + 9y^2 = 54 \Rightarrow \frac{x^2}{9} + \frac{y^2}{6} = 1$

   $c = \sqrt{a^2 - b^2} = \sqrt{9 - 6} = \sqrt{3}$

   $e = \frac{c}{a} = \frac{\sqrt{3}}{3}$

   (The graph is on the next page.)

13. (Graph)

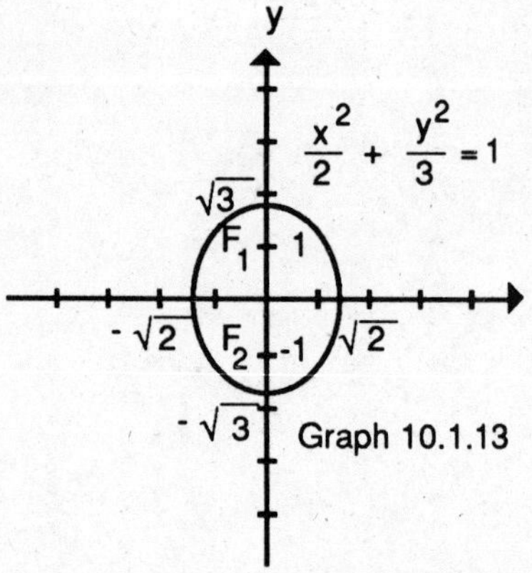


Graph 10.1.13

15. (Graph)

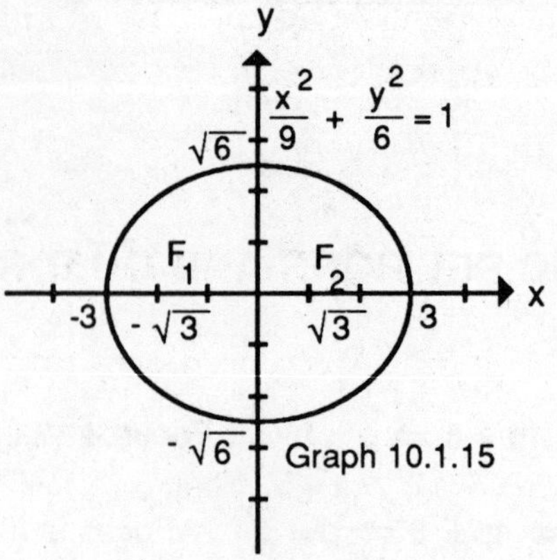


Graph 10.1.15

17. $x^2 - y^2 = 1 \Rightarrow c = \sqrt{a^2 + b^2} = \sqrt{1 + 1} = \sqrt{2}$

$e = \frac{c}{a} = \frac{\sqrt{2}}{1} = \sqrt{2}$

Asymptotes are $y = \pm x$

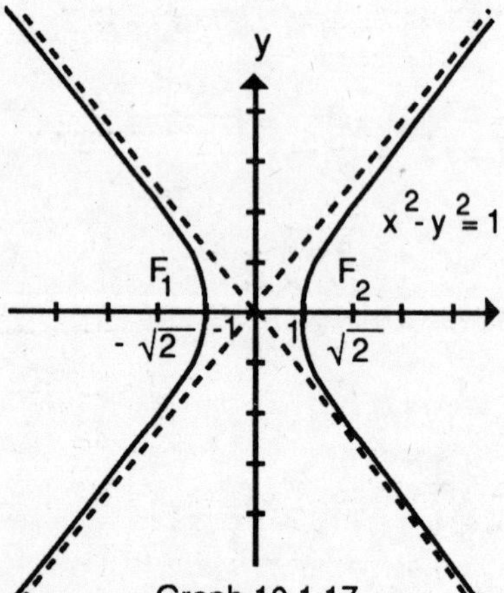


Graph 10.1.17

19. $y^2 - x^2 = 8 \Rightarrow \frac{y^2}{8} - \frac{x^2}{8} = 1$

$\Rightarrow c = \sqrt{a^2 + b^2} = \sqrt{8 + 8} = 4$

$e = \frac{c}{a} = \frac{4}{\sqrt{8}} = \sqrt{2}$ Asymptotes are $y = \pm x$.

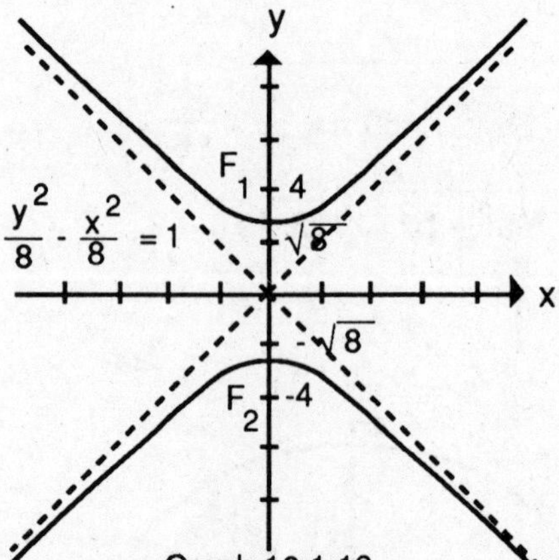


Graph 10.1.19

21. $8x^2 - y^2 = 16 \Rightarrow \frac{x^2}{2} - \frac{y^2}{8} = 1$

$\Rightarrow c = \sqrt{a^2 + b^2} = \sqrt{2 + 8} = \sqrt{10}$

$e = \frac{c}{a} = \frac{\sqrt{10}}{\sqrt{2}} = \sqrt{5}$

Asymptotes are $y = \pm 2x$.

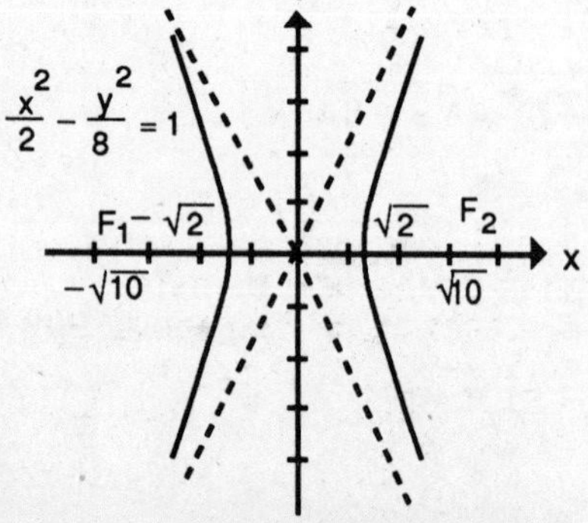


Graph 10.1.21

23. $8y^2 - 2x^2 = 16 \Rightarrow \frac{y^2}{2} - \frac{x^2}{8} = 1$

$\Rightarrow c = \sqrt{a^2 + b^2} = \sqrt{2 + 8} = \sqrt{10}$

$e = \frac{c}{a} = \frac{\sqrt{10}}{\sqrt{2}} = \sqrt{5}$ Asymptotes are $y = \pm \frac{x}{2}$.

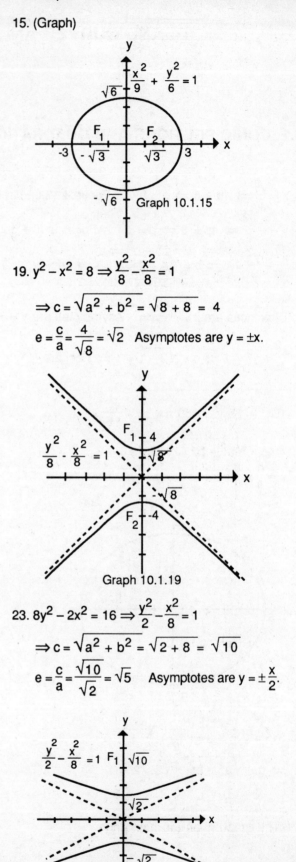

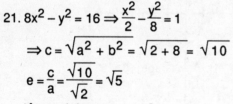

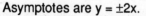

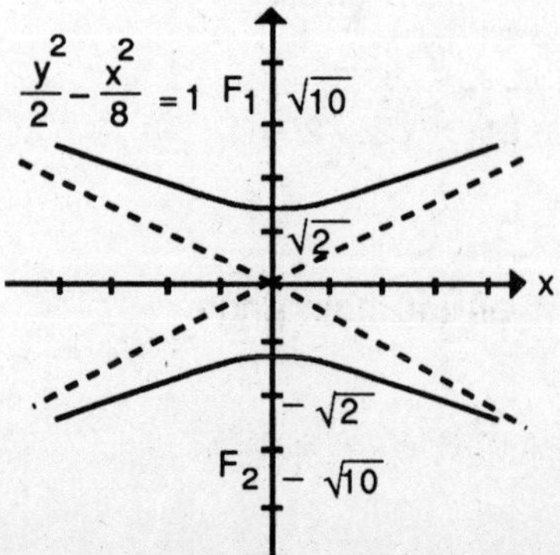


Graph 10.1.23

25. Volume of the Parabolic Solid: $V_1 = \int_0^{b/2} 2\pi x\left(h - \frac{4h}{b^2}x^2\right)dx = 2\pi h \int_0^{b/2}\left(x - \frac{4x^3}{b^2}\right)dx =$

$2\pi h\left[\frac{x^2}{2} - \frac{x^4}{b^2}\right]_0^{b/2} = \frac{\pi h b^2}{8}$ Volume of the Cone: $V_2 = \frac{1}{3}\pi\left(\frac{b}{2}\right)^2 h = \frac{1}{3}\pi\left(\frac{b^2}{4}\right)h = \frac{\pi h b^2}{12}$

$\therefore V_1 = \frac{3}{2}V_2$

27. $e = \frac{4}{5} \Rightarrow$ Let $c = 4$ and $a = 5$. Then

$c^2 = a^2 - b^2 \Rightarrow 16 = 25 - b^2 \Rightarrow$

$b^2 = 9 \Rightarrow b = 3 \quad \therefore \frac{x^2}{25} + \frac{y^2}{9} = 1$

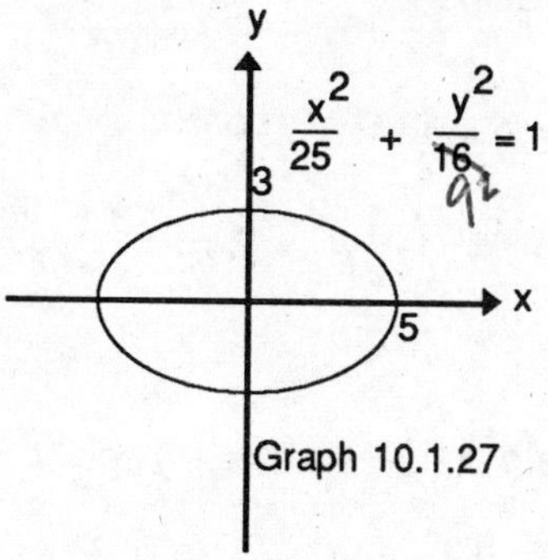


Graph 10.1.27

29. Let $y = \sqrt{1 - \frac{x^2}{4}}$ on the interval $0 \le x \le 2$. Then the area of the inscribed rectangle is given by

$$A(x) = 2x\left(2\sqrt{1 - \frac{x^2}{4}}\right) = 4x\sqrt{1 - \frac{x^2}{4}}$$

since the length is $2x$ and the height is $2y$.

Then $A'(x) = 4\sqrt{1 - \frac{x^2}{4}} - \frac{x^2}{\sqrt{1 - \frac{x^2}{4}}}$.

Let $A'(x) = 0 \Rightarrow 4\sqrt{1 - \frac{x^2}{4}} - \frac{x^2}{\sqrt{1 - \frac{x^2}{4}}} = 0 \Rightarrow$

29. (Continued)

$$4\left(1 - \frac{x^2}{4}\right) - x^2 = 0 \Rightarrow x^2 = 2 \Rightarrow x = \sqrt{2}$$

(Only the positive square root is in the interval.) Since $A(0) = 0$, $A(2) = 0$, $A(\sqrt{2}) = 4$ is the maximum area when the length is $2\sqrt{2}$ and the height is $\sqrt{2}$.

31. $x^2 - y^2 = 1 \Rightarrow x = \pm\sqrt{1 + y^2}$ on the interval $-3 \le y \le 3$. $\therefore$ Volume $= \int_{-3}^{3} \pi\left(\sqrt{1 + y^2}\right)^2 dy =$

$$2\int_0^3 \pi\left(\sqrt{1 + y^2}\right)^2 dy = 2\pi\int_0^3 (1 + y^2)dy = 2\pi\left[y + \frac{y^3}{3}\right]_0^3 = 24\pi$$

33. $\frac{dr_A}{dt} = \frac{dr_B}{dt} \Rightarrow \int \frac{dr_A}{dt} = \int \frac{dr_B}{dt} \Rightarrow r_A + C_1 = r_B + C_2 \Rightarrow r_A - r_B = C$, a constant $\Rightarrow$ The points, P(t), lie on a hyperbola with foci A and B.

## SECTION 10.2 THE GRAPHS OF QUADRATIC EQUATIONS IN X AND Y

1. $x^2 - y^2 - 1 = 0 \Rightarrow B^2 - 4AC = 0 - 4(1)(-1) = 4 > 0 \Rightarrow$ Hyperbola

3. $y^2 - 4x - 4 = 0 \Rightarrow B^2 - 4AC = 0^2 - 4(0)(1) = 0 \Rightarrow$ Parabola

5. $x^2 + 4y^2 - 4x - 8y + 4 = 0 \Rightarrow B^2 - 4AC = 0^2 - 4(1)(4) = -16 < 0 \Rightarrow$ Ellipse

7. $x^2 + 4xy + 4y^2 - 3x = 6 \Rightarrow B^2 - 4AC = 4^2 - 4(1)(4) = 0 \Rightarrow$ Parabola

9. $xy + y^2 - 3x = 5 \Rightarrow B^2 - 4AC = 1^2 - 4(0)(1) = 1 > 0 \Rightarrow$ Hyperbola

11. $x^2 - y^2 = 1 \Rightarrow B^2 - 4AC = 0^2 - 4(1)(-1) = 4 > 0 \Rightarrow$ Hyperbola

13. $x^2 - 3xy + 3y^2 + 6y = 7 \Rightarrow B^2 - 4AC = (-3)^2 - 4(1)(3) = -3 < 0 \Rightarrow$ Ellipse

15. $6x^2 + 3xy + 2y^2 + 17y + 2 = 0 \Rightarrow B^2 - 4AC = 3^2 - 4(6)(2) = -39 < 0 \Rightarrow$ Ellipse

17. $\cot 2\alpha = \frac{A-C}{B} = \frac{0}{1} = 0 \Rightarrow 2\alpha = \frac{\pi}{2} \Rightarrow \alpha = \frac{\pi}{4}$. $\therefore x = x'\cos\alpha - y'\sin\alpha,\ y = x'\sin\alpha + y'\cos\alpha \Rightarrow$
$x = x'\frac{\sqrt{2}}{2} - y'\frac{\sqrt{2}}{2},\ y = x'\frac{\sqrt{2}}{2} + y'\frac{\sqrt{2}}{2} \Rightarrow \left(\frac{\sqrt{2}}{2}x' - \frac{\sqrt{2}}{2}y'\right)\left(\frac{\sqrt{2}}{2}x' + \frac{\sqrt{2}}{2}y'\right) = 2 \Rightarrow \frac{1}{2}x'^2 - \frac{1}{2}y'^2 = 2 \Rightarrow x'^2 - y'^2 = 4$
$\Rightarrow$ Hyperbola

19. $\cot 2\alpha = \frac{A-C}{B} = \frac{3-1}{2\sqrt{3}} = \frac{1}{\sqrt{3}} \Rightarrow 2\alpha = \frac{\pi}{3} \Rightarrow \alpha = \frac{\pi}{6}$. $\therefore x = x'\cos\alpha - y'\sin\alpha,\ y = x'\sin\alpha + y'\cos\alpha \Rightarrow$
$x = \frac{\sqrt{3}}{2}x' - \frac{1}{2}y',\ y = \frac{1}{2}x' + \frac{\sqrt{3}}{2}y' \Rightarrow 3\left(\frac{\sqrt{3}}{2}x' - \frac{1}{2}y'\right)^2 + 2\sqrt{3}\left(\frac{\sqrt{3}}{2}x' - \frac{1}{2}y'\right)\left(\frac{1}{2}x' + \frac{\sqrt{3}}{2}y'\right) + \left(\frac{1}{2}x' + \frac{\sqrt{3}}{2}y'\right)^2 -$
$8\left(\frac{\sqrt{3}}{2}x' - \frac{1}{2}y'\right) + 8\sqrt{3}\left(\frac{1}{2}x' + \frac{\sqrt{3}}{2}y'\right) = 0 \Rightarrow 4x'^2 + 16y' = 0$, Parabola

21. $\cot 2\alpha = \frac{A-C}{B} = \frac{1-1}{-2} = 0 \Rightarrow 2\alpha = \frac{\pi}{2} \Rightarrow \alpha = \frac{\pi}{4}$. $\therefore x = x'\cos\alpha - y'\sin\alpha,\ y = x'\sin\alpha + y'\cos\alpha \Rightarrow$
$x = x'\frac{\sqrt{2}}{2} - y'\frac{\sqrt{2}}{2},\ y = x'\frac{\sqrt{2}}{2} + y'\frac{\sqrt{2}}{2} \Rightarrow \left(\frac{\sqrt{2}}{2}x' - \frac{\sqrt{2}}{2}y'\right)^2 - 2\left(\frac{\sqrt{2}}{2}x' - \frac{\sqrt{2}}{2}y'\right)\left(\frac{\sqrt{2}}{2}x' + \frac{\sqrt{2}}{2}y'\right) + \left(\frac{\sqrt{2}}{2}x' + \frac{\sqrt{2}}{2}y'\right)^2 = 2$
$\Rightarrow y'^2 = 1$, Parallel Horizontal Lines.

23. $\cot 2\alpha = \frac{A-C}{B} = \frac{\sqrt{2}-\sqrt{2}}{2\sqrt{2}} = 0 \Rightarrow 2\alpha = \frac{\pi}{2} \Rightarrow \alpha = \frac{\pi}{4}$. $\therefore x = x'\cos\alpha - y'\sin\alpha,\ y = x'\sin\alpha + y'\cos\alpha \Rightarrow$
$x = x'\frac{\sqrt{2}}{2} - y'\frac{\sqrt{2}}{2},\ y = x'\frac{\sqrt{2}}{2} + y'\frac{\sqrt{2}}{2} \Rightarrow \sqrt{2}\left(\frac{\sqrt{2}}{2}x' - \frac{\sqrt{2}}{2}y'\right)^2 + 2\sqrt{2}\left(\frac{\sqrt{2}}{2}x' - \frac{\sqrt{2}}{2}y'\right)\left(\frac{\sqrt{2}}{2}x' + \frac{\sqrt{2}}{2}y'\right) +$
$\sqrt{2}\left(\frac{\sqrt{2}}{2}x' + \frac{\sqrt{2}}{2}y'\right)^2 - 8\left(\frac{\sqrt{2}}{2}x' - \frac{\sqrt{2}}{2}y'\right) + 8\left(\frac{\sqrt{2}}{2}x' + \frac{\sqrt{2}}{2}y'\right) = 0 \Rightarrow 2\sqrt{2}\,x'^2 + 8\sqrt{2}\,y' = 0$, Parabola

25. $\cot 2\alpha = \frac{A-C}{B} = \frac{3-3}{2} = 0 \Rightarrow 2\alpha = \frac{\pi}{2} \Rightarrow \alpha = \frac{\pi}{4}$. $\therefore x = x'\cos\alpha - y'\sin\alpha,\ y = x'\sin\alpha + y'\cos\alpha \Rightarrow$
$x = x'\frac{\sqrt{2}}{2} - y'\frac{\sqrt{2}}{2},\ y = x'\frac{\sqrt{2}}{2} + y'\frac{\sqrt{2}}{2} \Rightarrow 3\left(\frac{\sqrt{2}}{2}x' - \frac{\sqrt{2}}{2}y'\right)^2 + 2\left(\frac{\sqrt{2}}{2}x' - \frac{\sqrt{2}}{2}y'\right)\left(\frac{\sqrt{2}}{2}x' + \frac{\sqrt{2}}{2}y'\right) + 3\left(\frac{\sqrt{2}}{2}x' + \frac{\sqrt{2}}{2}y'\right)^2$
$= 19 \Rightarrow 4x'^2 + 2y'^2 = 19$, Ellipse.

27. a) For $xy = 1$, $\alpha = \frac{\pi}{4} \Rightarrow x = x'\frac{\sqrt{2}}{2} - y'\frac{\sqrt{2}}{2}$, $y = x'\frac{\sqrt{2}}{2} + y'\frac{\sqrt{2}}{2} \Rightarrow \left(\frac{\sqrt{2}}{2}x' - \frac{\sqrt{2}}{2}y'\right)\left(\frac{\sqrt{2}}{2}x' + \frac{\sqrt{2}}{2}y'\right) = 1 \Rightarrow$ $x'^2 - y'^2 = 2$.

b) For $xy = a$, $\alpha = \frac{\pi}{4} \Rightarrow x = x'\frac{\sqrt{2}}{2} - y'\frac{\sqrt{2}}{2}$, $y = x'\frac{\sqrt{2}}{2} + y'\frac{\sqrt{2}}{2} \Rightarrow \left(\frac{\sqrt{2}}{2}x' - \frac{\sqrt{2}}{2}y'\right)\left(\frac{\sqrt{2}}{2}x' + \frac{\sqrt{2}}{2}y'\right) = 1 \Rightarrow$ $x'^2 - y'^2 = 2a$.

## SECTION 10.3 PARAMETRIC EQUATIONS FOR PLANE CURVES

1. $x = \cos t$, $y = \sin t$, $0 \le t \le \pi \Rightarrow$
$\cos^2 t + \sin^2 t = 1 \Rightarrow x^2 + y^2 = 1$

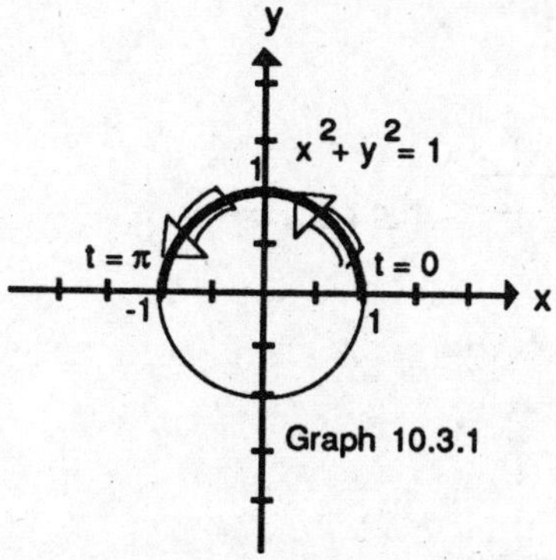

Graph 10.3.1

3. $x = \sin 2\pi t$, $y = \cos 2\pi t$, $0 \le t \le 1 \Rightarrow$
$\sin^2 2\pi t + \cos^2 2\pi t = 1 \Rightarrow x^2 + y^2 = 1$

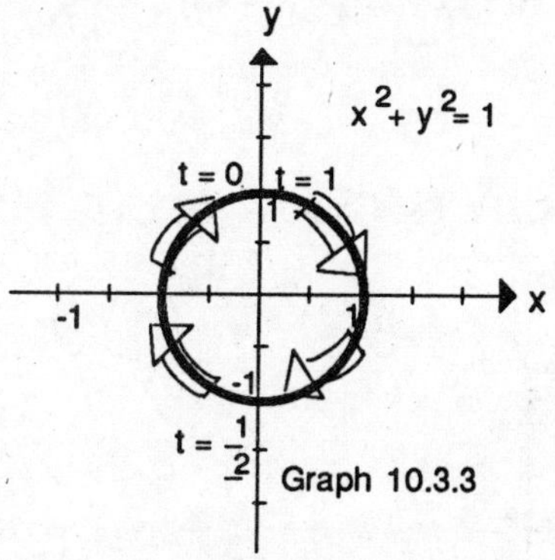

Graph 10.3.3

5. $x = 4\cos t$, $y = 2\sin t$, $0 \le t \le 2\pi \Rightarrow$
$\frac{16\cos^2 t}{16} + \frac{4\sin^2 t}{4} = 1 \Rightarrow \frac{x^2}{16} + \frac{y^2}{4} = 1$

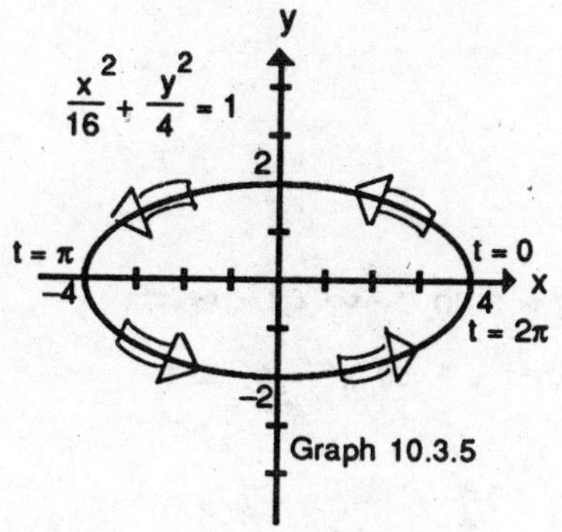

Graph 10.3.5

7. $x = 4\cos t$, $y = 5\sin t$, $0 \le t \le 2\pi \Rightarrow$
$\frac{16\cos^2 t}{16} + \frac{25\sin^2 t}{25} = 1 \Rightarrow \frac{x^2}{16} + \frac{y^2}{25} = 1$

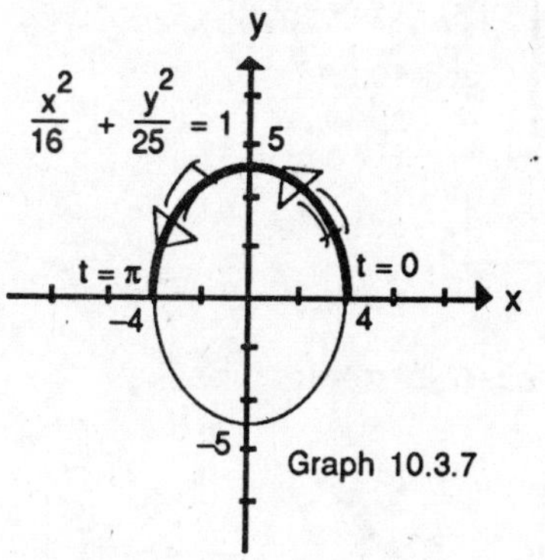

Graph 10.3.7

9. $x = 3t$, $y = 9t^2$, $-\infty < t < \infty \Rightarrow$
$y = x^2$

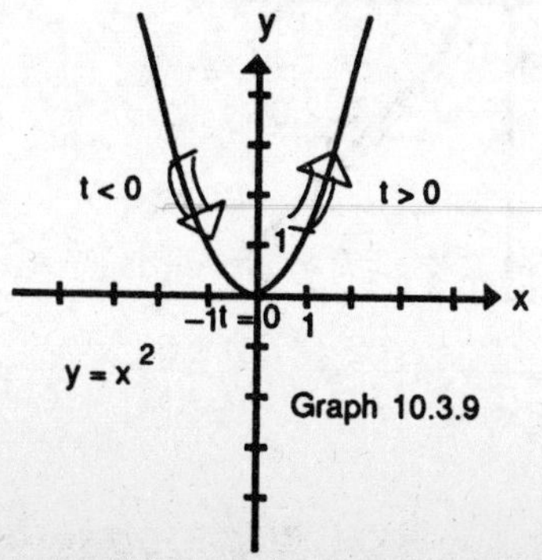

Graph 10.3.9

11. $x = t$, $y = \sqrt{t}$, $t \ge 0$
$x = y^2$

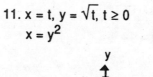

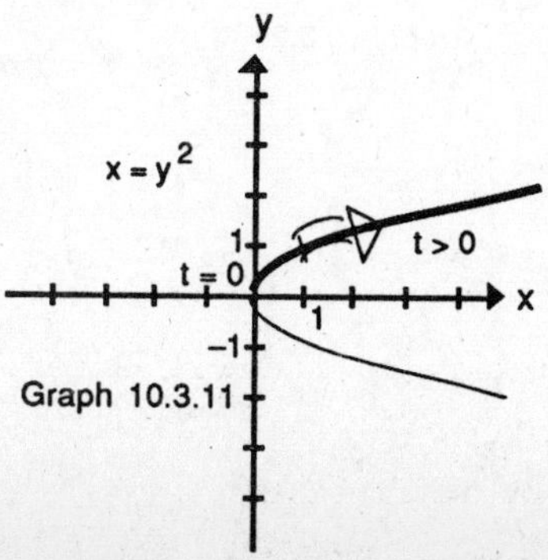

Graph 10.3.11

13. $x = -\sec t$, $y = \tan t$, $-\frac{\pi}{2} < t < \frac{\pi}{2} \Rightarrow$

$\sec^2 t - \tan^2 t = 1 \Rightarrow x^2 - y^2 = 1$

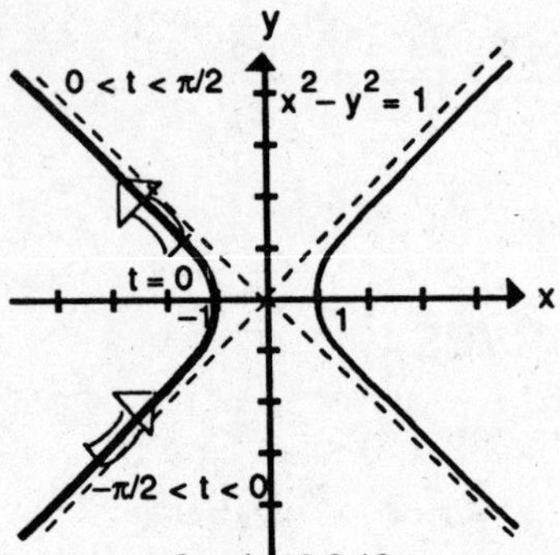


Graph 10.3.13

15. $x = 2t - 5$, $y = 4t - 7$, $-\infty < t < \infty \Rightarrow$

$x + 5 = 2t \Rightarrow 2(x + 5) = 4t \Rightarrow$
$y = 2(x + 5) - 7 \Rightarrow y = 2x + 3$

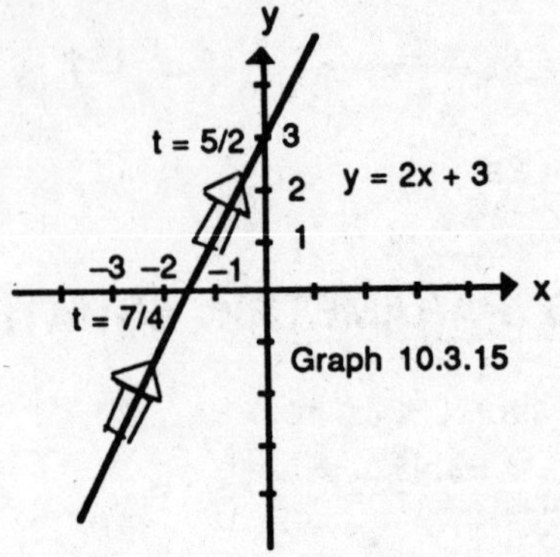


Graph 10.3.15

17. $x = t$, $y = 1 - t$, $0 \le t \le 1 \Rightarrow$

$y = 1 - x$

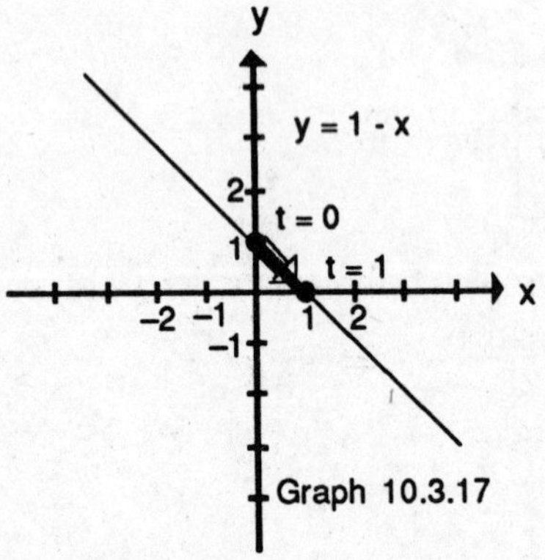


Graph 10.3.17

19. $x = t$, $y = \sqrt{1 - t^2}$, $-1 \le t \le 1 \Rightarrow$

$y = \sqrt{1 - x^2}$

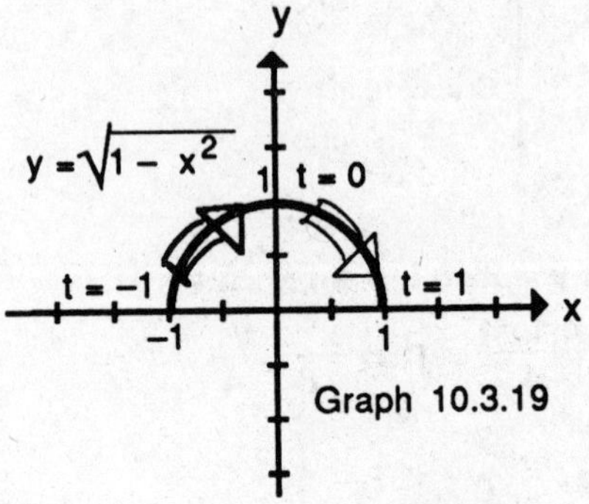


Graph 10.3.19

21. $x = t^2$, $y = \sqrt{t^4 + 1}$, $t \ge 0 \Rightarrow$

$y = \sqrt{x^2 + 1}$, $x \ge 0$

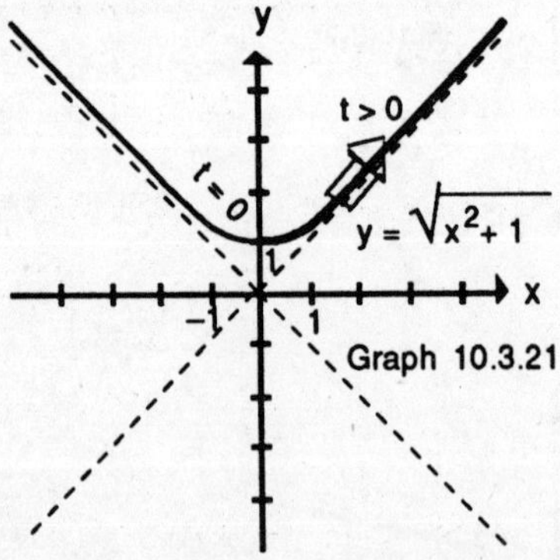


Graph 10.3.21

23. $x = \cosh t$, $y = \sinh t$, $-\infty < t < \infty \Rightarrow$

$\cosh^2 t - \sinh^2 t = 1 \Rightarrow x^2 - y^2 = 1$

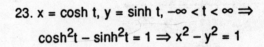

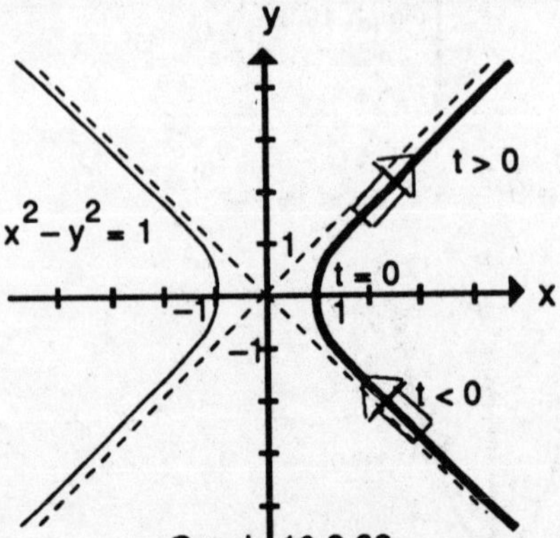


Graph 10.3.23

25. a) $x = a\cos t,\ y = -a\sin t,\ 0 \le t \le 2\pi$  b) $x = a\cos t,\ y = a\sin t,\ 0 \le t \le 2\pi$

c) $x = a\cos t,\ y = -a\sin t,\ 0 \le t \le 4\pi$  d) $x = a\cos t,\ y = a\sin t,\ 0 \le t \le 4\pi$

27.

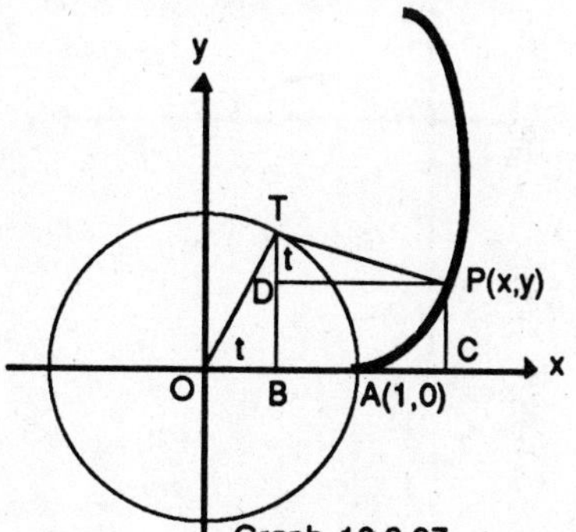

Graph 10.3.27

$\angle PTB = \angle TOB = t$. PT = arc(AT) = t since PT = length of the unwound string = length of arc(AT).

$\therefore\ x = OB + BC = OB + DP = \cos t + t\sin t$.

$y = PC = TB - TD = \sin t - t\cos t$

29. $d = \sqrt{(x-2)^2 + (y-\frac{1}{2})^2} \Rightarrow d^2 = (x-2)^2 + (y-\frac{1}{2})^2 = (t-2)^2 + (t^2-\frac{1}{2})^2 \Rightarrow d^2 = t^4 - 4t + \frac{17}{4}$

$\frac{d(d^2)}{dt} = 4t^3 - 4$. Let $4t^3 - 4 = 0 \Rightarrow t = 1$. The second derivative is always positve for $t \ne 0 \Rightarrow$ $t = 1$ gives a local minimum which is an absolute minimum since it is the only extremum. $\therefore$ the closest point on the parabola is (1,1).

31. a) $x = x_0 + (x_1 - x_0)t,\ y = y_0 + (y_1 - y_0)t \Rightarrow t = \frac{x - x_0}{x_1 - x_0} \Rightarrow y = y_0 + (y_1 - y_0)\left(\frac{x - x_0}{x_1 - x_0}\right) \Rightarrow$

$y - y_0 = \left(\frac{y_1 - y_0}{x_1 - x_0}\right)(x - x_0)$ which is the equation of the line through the points.

b) $x = x_1 t,\ y = y_1 t$ (Answer not unique)  c) $x = -1 + t,\ y = t$ or $x = -t,\ y = 1 - t$

33. a)

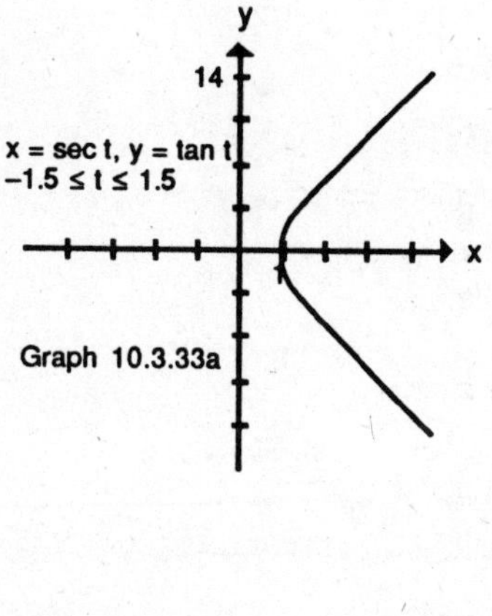

Graph 10.3.33a

b)

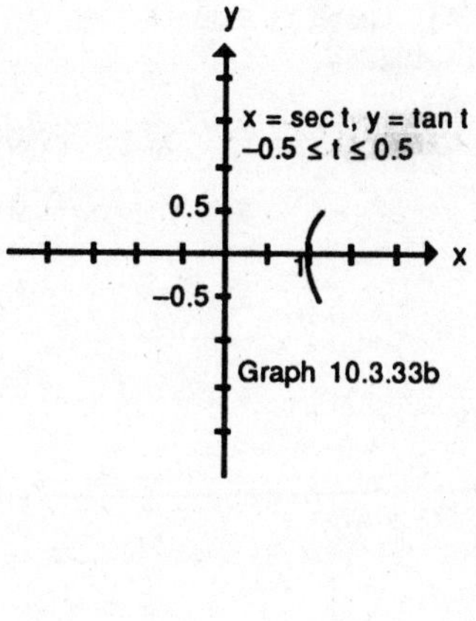

Graph 10.3.33b

c)

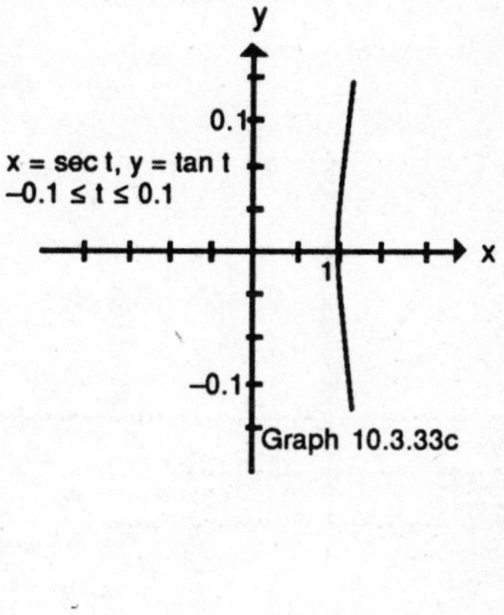

Graph 10.3.33c

35. a)

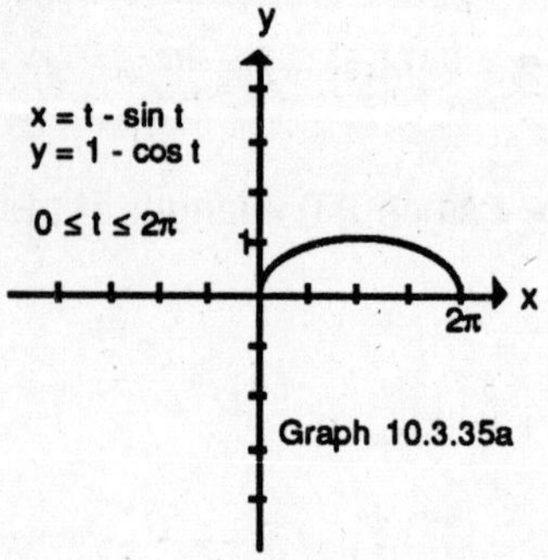


Graph 10.3.35a

b)

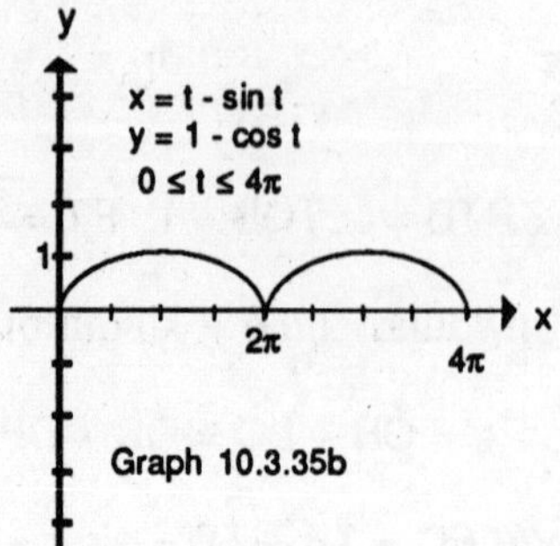


Graph 10.3.35b

c)

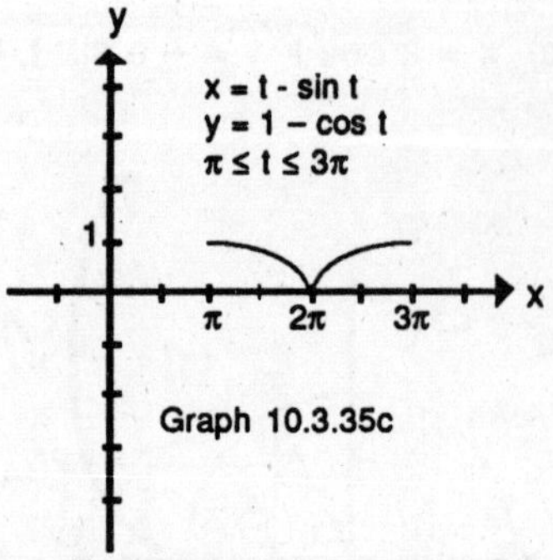


Graph 10.3.35c

37. a)

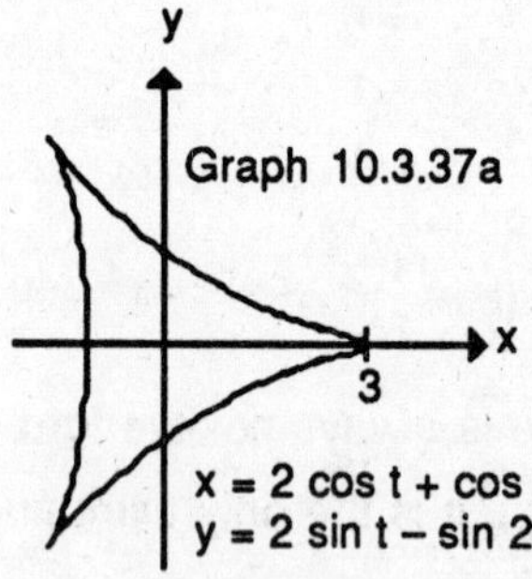


Graph 10.3.37a

b)

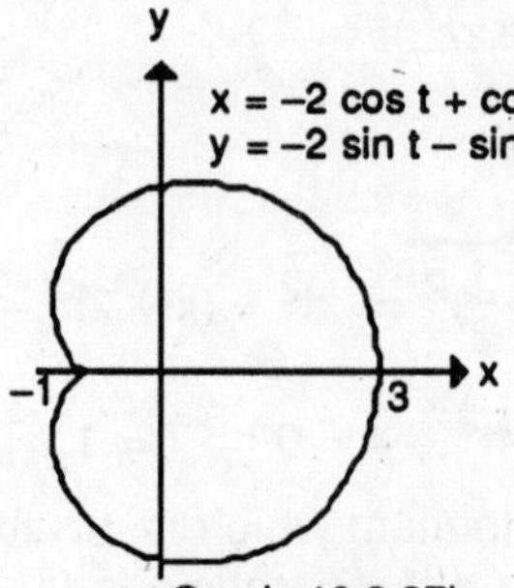


Graph 10.3.37b

39. a)

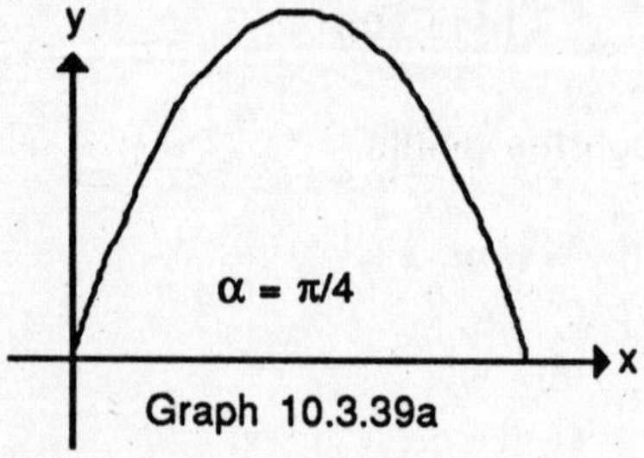


Graph 10.3.39a

b)

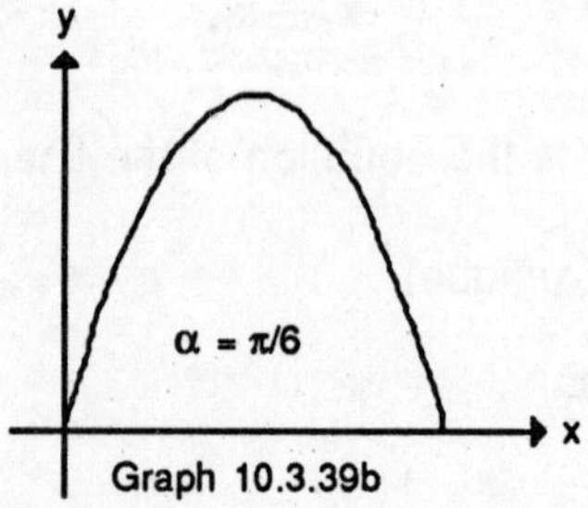


Graph 10.3.39b

c)

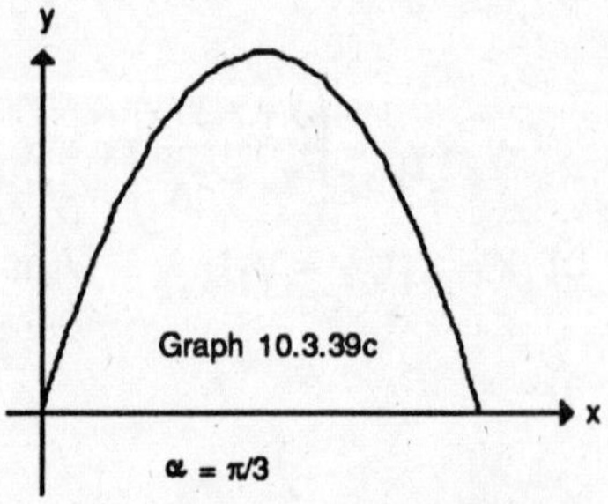


Graph 10.3.39c

d)

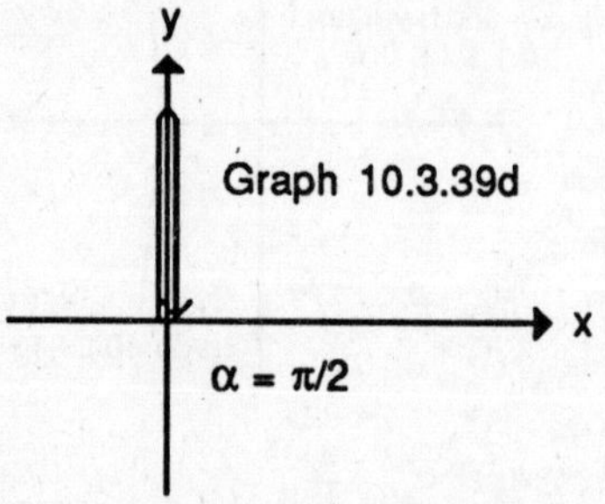


Graph 10.3.39d

## SECTION 10.4 THE CALCULUS OF PARAMETRIC EQUATIONS

1. $t = \frac{\pi}{4} \Rightarrow x = 2\cos\frac{\pi}{4} = \sqrt{2}$, $y = 2\sin\frac{\pi}{4} = \sqrt{2}$. $\frac{dx}{dt} = -2\sin t$, $\frac{dy}{dt} = 2\cos t \Rightarrow \frac{dy}{dx} = \frac{2\cos t}{-2\sin t} = -\cot t$.
$\therefore \frac{dy}{dx}\left(\frac{\pi}{4}\right) = -\cot\frac{\pi}{4} = -1$. $\therefore$ Tangent line is $y - \sqrt{2} = -1(x - \sqrt{2}) \Rightarrow y = -x + 2\sqrt{2}$.
$\frac{dy'}{dt} = \csc^2 t \Rightarrow \frac{d^2y}{dx^2} = \frac{\csc^2 t}{-2\sin t} = -\frac{1}{2\sin^3 t} \Rightarrow \frac{d^2y}{dx^2}\left(\frac{\pi}{4}\right) = -\sqrt{2}$

3. $t = \frac{\pi}{4} \Rightarrow x = 4\sin\frac{\pi}{4} = 2\sqrt{2}$, $y = 2\cos\frac{\pi}{4}$. $\frac{dx}{dt} = 4\cos t$, $\frac{dy}{dt} = -2\sin t \Rightarrow \frac{dy}{dx} = \frac{-2\sin t}{4\cos t} = -\frac{1}{2}\tan t \Rightarrow$
$\frac{dy}{dx}\left(\frac{\pi}{4}\right) = -\frac{1}{2}\tan\frac{\pi}{4} = -\frac{1}{2}$. $\therefore$ Tangent line is $y - \sqrt{2} = -\frac{1}{2}(x - 2\sqrt{2}) \Rightarrow y = -\frac{1}{2}x + 2\sqrt{2}$.
$\frac{dy'}{dt} = -\frac{1}{2}\sec^2 t \Rightarrow \frac{d^2y}{dx^2} = \frac{-\frac{1}{2}\sec^2 t}{4\cos t} = -\frac{1}{8\cos^3 t} \Rightarrow \frac{d^2y}{dx^2}\left(\frac{\pi}{4}\right) = -\frac{\sqrt{2}}{4}$

5. $t = \frac{1}{4} \Rightarrow x = \frac{1}{4}$, $y = \frac{1}{2}$. $\frac{dx}{dt} = 1$, $\frac{dy}{dt} = \frac{1}{2\sqrt{t}} \Rightarrow \frac{dy}{dx} = \frac{1}{2\sqrt{t}} \Rightarrow \frac{dy}{dx}\left(\frac{1}{4}\right) = \frac{1}{2\sqrt{\frac{1}{4}}} = 1$.
$\therefore$ Tangent line is $y - \frac{1}{2} = 1\left(x - \frac{1}{4}\right) \Rightarrow y = x + \frac{1}{4}$. $\frac{dy'}{dt} = -\frac{1}{4}t^{-3/2} \Rightarrow \frac{d^2y}{dx^2} = -\frac{1}{4}t^{-3/2} \Rightarrow \frac{d^2y}{dx^2}\left(\frac{1}{4}\right) = -2$

7. $t = \frac{\pi}{6} \Rightarrow x = \sec\frac{\pi}{6} = \frac{2}{\sqrt{3}}$, $y = \tan\frac{\pi}{6} = \frac{1}{\sqrt{3}}$. $\frac{dx}{dt} = \sec t\tan t$, $\frac{dy}{dt} = \sec^2 t \Rightarrow \frac{dy}{dx} = \frac{\sec^2 t}{\sec t\tan t} = \csc t \Rightarrow$
$\frac{dy}{dx}\left(\frac{\pi}{6}\right) = \csc\frac{\pi}{6} = 2$. $\therefore$ Tangent line is $y - \frac{1}{\sqrt{3}} = 2\left(x - \frac{2}{\sqrt{3}}\right) \Rightarrow y = 2x - \sqrt{3}$. $\frac{dy'}{dt} = -\csc t\cot t \Rightarrow$
$\frac{d^2y}{dx^2} = \frac{-\csc t\cot t}{\sec t\tan t} = -\cot^3 t \Rightarrow \frac{d^2y}{dx^2}\left(\frac{\pi}{6}\right) = -3\sqrt{3}$

9. $t = -1 \Rightarrow x = 5$, $y = 1$. $\frac{dx}{dt} = 4t$, $\frac{dy}{dt} = 4t^3 \Rightarrow \frac{dy}{dx} = \frac{4t^3}{4t} = t^2 \Rightarrow \frac{dy}{dx}(-1) = (-1)^2 = 1$. $\therefore$ Tangent line is
$y - 1 = 1(x - 5) \Rightarrow y = x - 4$. $\frac{dy'}{dt} = 2t \Rightarrow \frac{d^2y}{dx^2} = \frac{2t}{4t} = \frac{1}{2} \Rightarrow \frac{d^2y}{dx^2}(-1) = \frac{1}{2}$

11. $t = \frac{\pi}{3} \Rightarrow x = \frac{\pi}{3} - \sin\frac{\pi}{3} = \frac{\pi}{3} - \frac{\sqrt{3}}{2}$, $y = 1 - \cos\frac{\pi}{3} = 1 - \frac{1}{2} = \frac{1}{2}$. $\frac{dx}{dt} = 1 - \cos t$, $\frac{dy}{dt} = \sin t \Rightarrow \frac{dy}{dx} = \frac{\sin t}{1 - \cos t}$
$\Rightarrow \frac{dy}{dx}\left(\frac{\pi}{3}\right) = \frac{\sin\frac{\pi}{3}}{1 - \cos\frac{\pi}{3}} = \frac{\frac{\sqrt{3}}{2}}{\frac{1}{2}} = \sqrt{3}$. $\therefore$ Tangent line is $y - \frac{1}{2} = \sqrt{3}\left(x - \frac{\pi}{3} + \frac{\sqrt{3}}{2}\right) \Rightarrow y = \sqrt{3}x - \frac{\pi\sqrt{3}}{3} + 2$.
$\frac{dy'}{dt} = \frac{(1 - \cos t)\cos t - \sin t(\sin t)}{(1 - \cos t)^2} = \frac{-1}{1 - \cos t} \Rightarrow \frac{d^2y}{dx^2} = \frac{\frac{-1}{1 - \cos t}}{1 - \cos t} = \frac{-1}{(1 - \cos t)^2} \Rightarrow \frac{d^2y}{dx^2}\left(\frac{\pi}{3}\right) = -4$

13. $\frac{dx}{dt} = -\sin t,\ \frac{dy}{dt} = 1 + \cos t \Rightarrow \sqrt{\left(\frac{dx}{dt}\right)^2 + \left(\frac{dy}{dt}\right)^2} = \sqrt{(-\sin t)^2 + (1 + \cos t)^2} = \sqrt{2 + 2\cos t}$.

$\therefore$ Length $= \int_0^{\pi} \sqrt{2 + 2\cos t}\, dt = \sqrt{2} \int_0^{\pi} \sqrt{\frac{1 - \cos t}{1 - \cos t}(1 + \cos t)}\, dt = \sqrt{2} \int_0^{\pi} \sqrt{\frac{\sin^2 t}{1 - \cos t}}\, dt =$

$\sqrt{2} \int_0^{\pi} \frac{\sin t}{\sqrt{1 - \cos t}} dt$ (since $\sin t \geq 0$ on $[0,\pi]$) $= \sqrt{2} \int_0^{2} u^{-1/2}\, du = \sqrt{2}\left[2u^{1/2}\right]_0^2 = 4$.

(Let $u = 1 - \cos t \Rightarrow du = \sin t\, dt;\ t = 0 \Rightarrow u = 0,\ t = \pi \Rightarrow u = 2$)

15. $\frac{dx}{dt} = t,\ \frac{dy}{dt} = (2t + 1)^{1/2} \Rightarrow \sqrt{\left(\frac{dx}{dt}\right)^2 + \left(\frac{dy}{dt}\right)^2} = \sqrt{t^2 + \left((2t + 1)^{1/2}\right)^2} = |t + 1| = t + 1$ since $0 \leq t \leq 4$.

$\therefore$ Length $= \int_0^{4} (t + 1)\, dt = \left[\frac{t^2}{2} + t\right]_0^4 = 12$.

17. $\frac{dx}{dt} = 8t\cos t,\ \frac{dy}{dt} = 8t\sin t \Rightarrow \sqrt{\left(\frac{dx}{dt}\right)^2 + \left(\frac{dy}{dt}\right)^2} = \sqrt{(8t\cos t)^2 + (8t\sin t)^2} = \sqrt{64t^2\cos^2 t + 64t^2\sin^2 t} = |8t|$

$= 8t$ since $0 \leq t \leq \frac{\pi}{2}$. $\therefore$ Length $= \int_0^{\pi/2} 8t\, dt = \left[4t^2\right]_0^{\pi/2} = \pi^2$.

19. $\frac{dx}{dt} = -\sin t,\ \frac{dy}{dt} = \cos t \Rightarrow \sqrt{\left(\frac{dx}{dt}\right)^2 + \left(\frac{dy}{dt}\right)^2} = \sqrt{(-\sin t)^2 + (\cos t)^2} = 1$.

$\therefore$ Area $= \int_0^{2\pi} 2\pi(2 + \sin t)1\, dt = 2\pi[2t - \cos t]_0^{2\pi} = 8\pi^2$

21. $\frac{dx}{dt} = 1,\ \frac{dy}{dt} = t + \sqrt{2} \Rightarrow \sqrt{\left(\frac{dx}{dt}\right)^2 + \left(\frac{dy}{dt}\right)^2} = \sqrt{1^2 + (t + \sqrt{2})^2} = \sqrt{t^2 + 2\sqrt{2}\,t + 3}$.

$\therefore$ Area $= \int_{-\sqrt{2}}^{\sqrt{2}} 2\pi(t + \sqrt{2})\sqrt{t^2 + 2\sqrt{2}\,t + 3}\, dt = \int_1^9 \pi\sqrt{u}\, du = \left[\frac{2}{3}\pi u^{3/2}\right]_1^9 = \frac{52}{3}\pi$

(Let $u = t^2 + 2\sqrt{2}\,t + 3 \Rightarrow du = 2t + 2\sqrt{2};\ t = -\sqrt{2} \Rightarrow u = 1,\ t = \sqrt{2} \Rightarrow u = 9$)

23. $\frac{dx}{dt} = 2,\ \frac{dy}{dt} = 1 \Rightarrow \sqrt{\left(\frac{dx}{dt}\right)^2 + \left(\frac{dy}{dt}\right)^2} = \sqrt{2^2 + 1^2} = \sqrt{5}$. $\therefore$ Area $= \int_0^1 2\pi(t + 1)\sqrt{5}\, dt =$

$2\pi\sqrt{5}\left[\frac{t^2}{2} + t\right]_0^1 = 3\pi\sqrt{5}$. The slant height is $\sqrt{5} \Rightarrow$ Area $= \pi(1 + 2)\sqrt{5} = 3\pi\sqrt{5}$.

25. a) $\frac{dx}{dt} = -2\sin 2t,\ \frac{dy}{dt} = 2\cos 2t \Rightarrow \sqrt{\left(\frac{dx}{dt}\right)^2 + \left(\frac{dy}{dt}\right)^2} = \sqrt{(-2\sin 2t)^2 + (2\cos 2t)^2} = 2.$

$\therefore$ Length $= \int_0^{\pi/2} 2\,dt = [2t]_0^{\pi/2} = \pi.$

b) $\frac{dx}{dt} = \pi\cos \pi t,\ \frac{dy}{dt} = -\pi\sin \pi t \Rightarrow \sqrt{\left(\frac{dx}{dt}\right)^2 + \left(\frac{dy}{dt}\right)^2} = \sqrt{(\pi\cos \pi t)^2 + (-\pi\sin \pi t)^2} = \pi.$

$\therefore$ Length $= \int_{-1/2}^{1/2} \pi\,dt = [\pi t]_{-1/2}^{1/2} = \pi.$

27. $\frac{dx}{dt} = \cos t,\ \frac{dy}{dt} = 2\cos 2t \Rightarrow \frac{dy}{dx} = \frac{2\cos 2t}{\cos t} = \frac{2(2\cos^2 t - 1)}{\cos t}$. Let $\frac{2(2\cos^2 t - 1)}{\cos t} = 0 \Rightarrow 2\cos^2 t - 1 = 0$

$\Rightarrow \cos t = \pm\frac{1}{\sqrt{2}} \Rightarrow t = \frac{\pi}{4}, \frac{3\pi}{4}, \frac{5\pi}{4}, \frac{7\pi}{4}$. In the 1st quadrant, $t = \frac{\pi}{4} \Rightarrow x = \sin\frac{\pi}{4} = \frac{\sqrt{2}}{2},\ y = \sin 2\left(\frac{\pi}{4}\right) = 1$

$\Rightarrow \left(\frac{\sqrt{2}}{2}, 1\right)$ is the point in the 1st quadrant where the tangent line is horizontal.

$x = 0, y = 0 \Rightarrow \sin t = 0 \Rightarrow t = 0$ or $t = \pi$ and $\sin 2t = 0 \Rightarrow t = 0, \frac{\pi}{2}, \pi, \frac{3\pi}{2}$. $\therefore$ $t = 0$ and $t = \pi$ give the tangent lines at the origin. $\frac{dy}{dx}(0) = 2 \Rightarrow y = 2x$ and $\frac{dy}{dx}(\pi) = -2 \Rightarrow y = -2x$.

29.

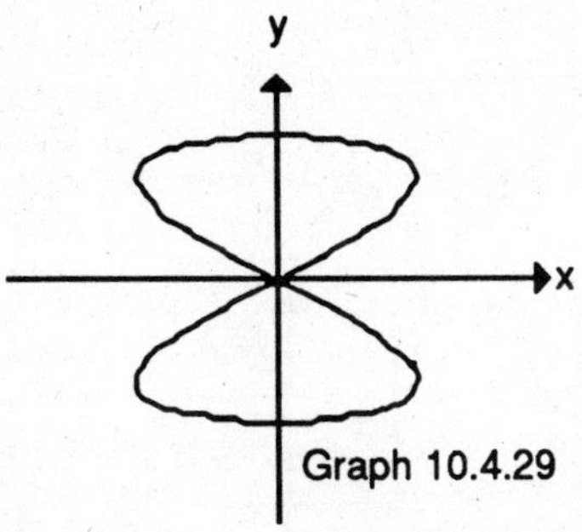


Graph 10.4.29

31.

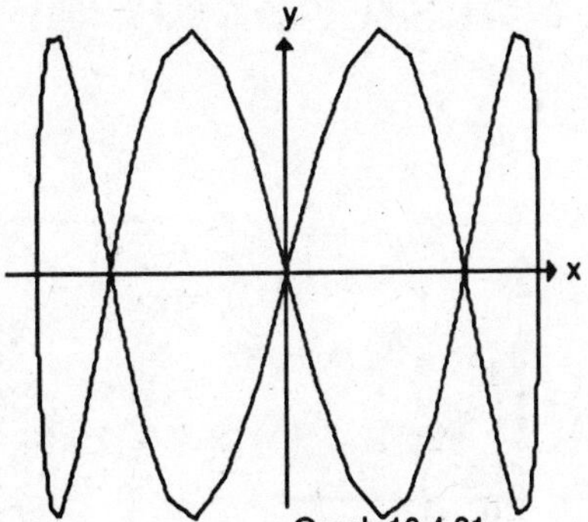


Graph 10.4.31

33.

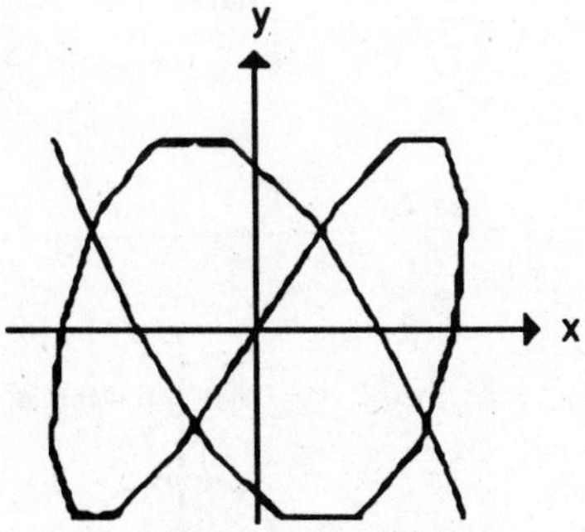


Graph 10.4.33

35.

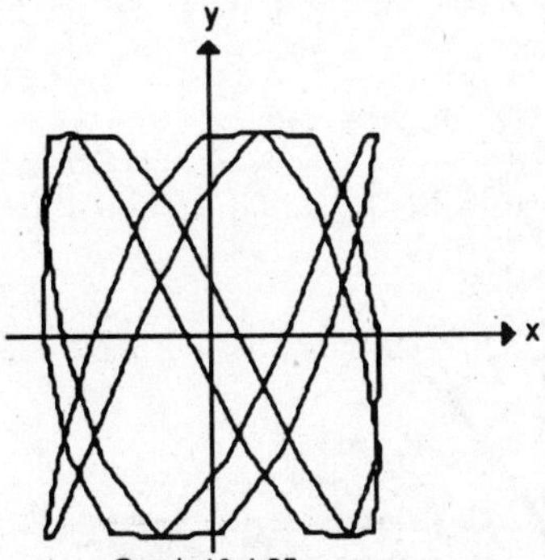


Graph 10.4.35

## SECTION 10.5 POLAR COORDINATES

1. a, c; b, d; e, k; g, j; h, f; i, l; m, o; n, p

3. a) $\left(2, \frac{\pi}{2} + 2n\pi\right)$ and $\left(-2, \frac{\pi}{2} + (2n+1)\pi\right)$, n an integer

   b) $(2, 2n\pi)$ and $(-2, (2n+1)\pi)$, n an integer

   c) $\left(2, \frac{3\pi}{2} + 2n\pi\right)$ and $\left(-2, \frac{3\pi}{2} + (2n+1)\pi\right)$, n an integer

   d) $(2, (2n+1)\pi)$ and $(-2, 2n\pi)$, n an integer

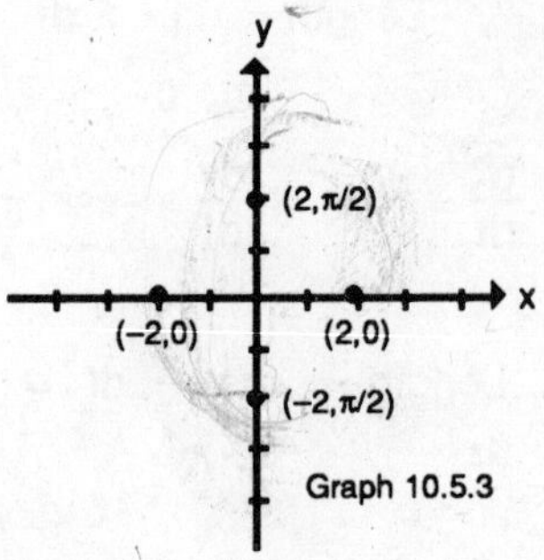

Graph 10.5.3

5. a) $x = \sqrt{2}\cos\frac{\pi}{4} = 1,\ y = \sqrt{2}\sin\frac{\pi}{4} = 1 \Rightarrow (1,1)$

   b) $x = 1\cos 0 = 1,\ y = 1\sin 0 = 1 \Rightarrow (1,0)$

   c) $x = 0\cos\frac{\pi}{2} = 0,\ y = 0\sin\frac{\pi}{2} = 0 \Rightarrow (0,0)$

   d) $x = -\sqrt{2}\cos\frac{\pi}{4} = -1,\ y = -\sqrt{2}\sin\frac{\pi}{2} = -1 \Rightarrow (-1,-1)$

   e) $x = -3\cos\frac{5\pi}{6} = \frac{3\sqrt{3}}{2},\ y = -3\sin\frac{5\pi}{6} = -\frac{3}{2} \Rightarrow \left(\frac{3\sqrt{3}}{2}, -\frac{3}{2}\right)$

   f) $x = 5\cos(\tan^{-1}\frac{4}{3}) = 3,\ y = 5\sin(\tan^{-1}\frac{4}{3}) = 4 \Rightarrow (3,4)$

   g) $x = -1\cos 7\pi = 1,\ y = -1\sin 7\pi = 0 \Rightarrow (1,0)$

   h) $x = 2\sqrt{3}\cos\frac{2\pi}{3} = -\sqrt{3},\ y = 2\sqrt{3}\sin\frac{2\pi}{3} = 3 \Rightarrow (-\sqrt{3}, 3)$

7.

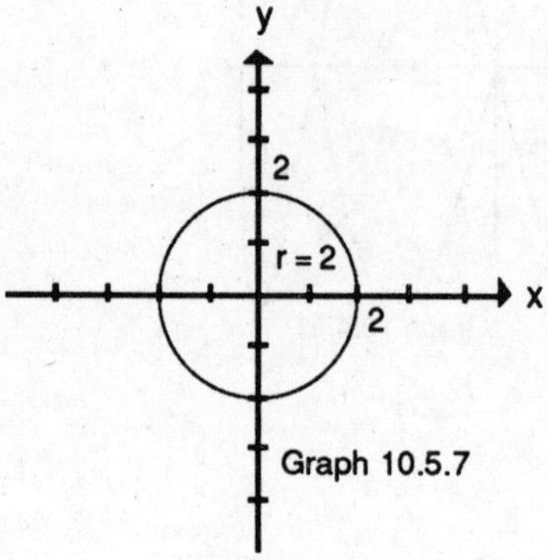

Graph 10.5.7

9.

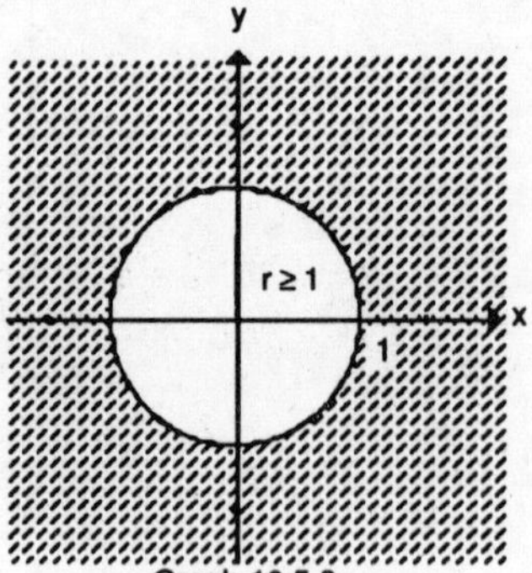

Graph 10.5.9

11.

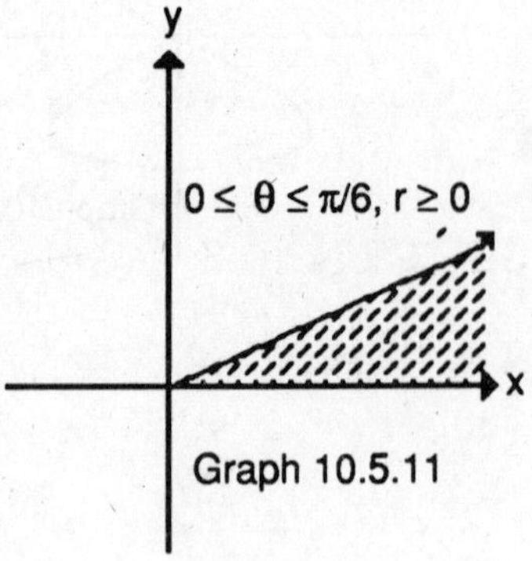

Graph 10.5.11

13.

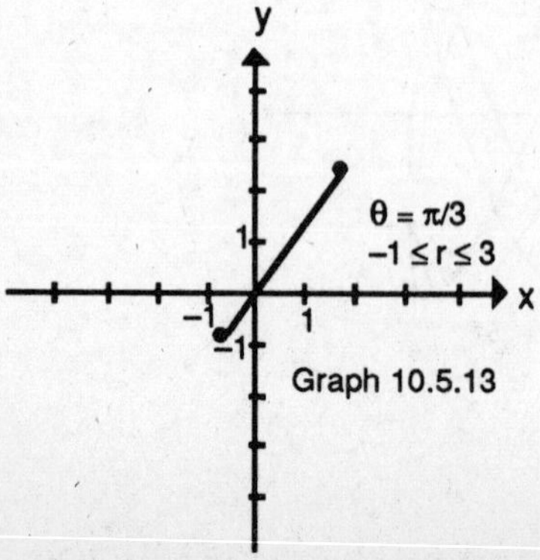

Graph 10.5.13

15.

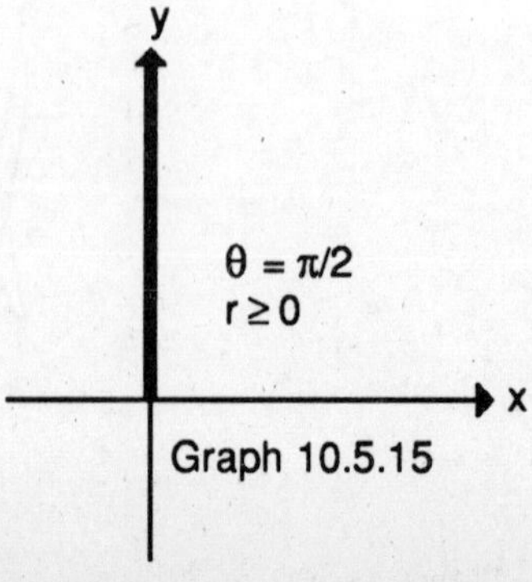

Graph 10.5.15

17.

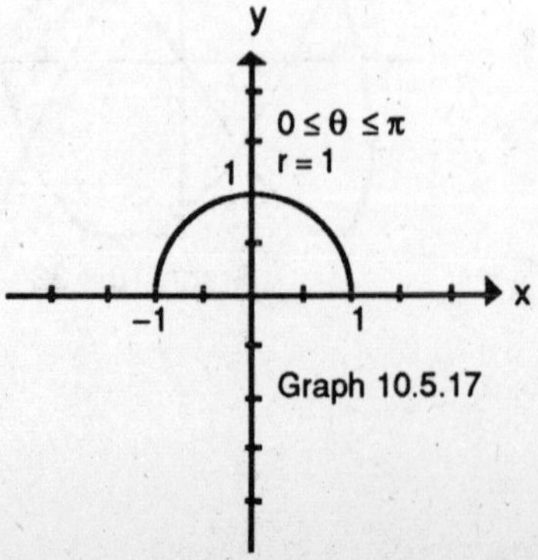

Graph 10.5.17

19.

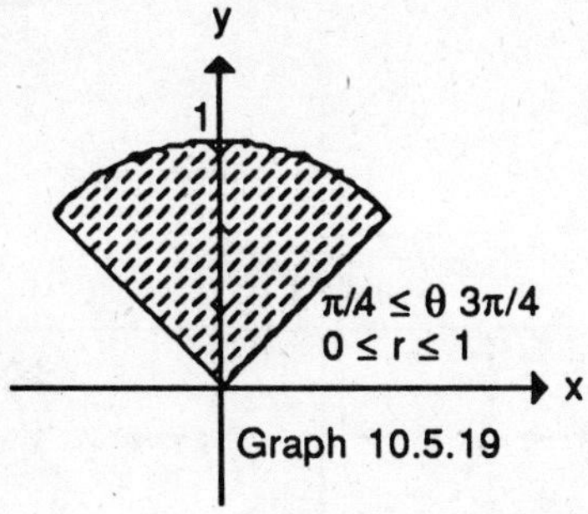

Graph 10.5.19

21.

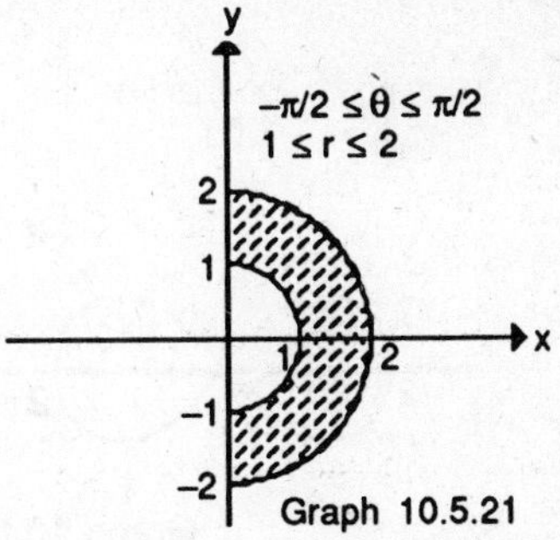

Graph 10.5.21

23. $r\cos\theta = 2 \Rightarrow x = 2$, vertical line through (2,0).

25. $r\sin\theta = 4 \Rightarrow y = 4$, horizontal line through (0,4).

27. $r\sin\theta = 0 \Rightarrow y = 0$, the x–axis.

29. $r\cos\theta + r\sin\theta = 1 \Rightarrow x + y = 1$,
line, $m = -1$, $b = 1$

31. $r^2 = 1 \Rightarrow x^2 + y^2 = 1$, circle, C = (0,0), r = 1

33. $r = \dfrac{5}{\sin\theta - 2\cos\theta} \Rightarrow r\sin\theta - 2r\cos\theta = 5 \Rightarrow$
$y - 2x = 5$, line, $m = 2$, $b = 5$

35. $x = 7 \Rightarrow r\cos\theta = 7$

37. $x = y \Rightarrow r\cos\theta = r\sin\theta \Rightarrow \theta = \dfrac{\pi}{4}$

39. $x^2 + y^2 = 4 \Rightarrow r^2 = 4$ or $r = 2$ or $r = -2$

41. $\dfrac{x^2}{9} + \dfrac{y^2}{4} = 1 \Rightarrow 4x^2 + 9y^2 = 36 \Rightarrow$
$4r^2\cos^2\theta + 9r^2\sin^2\theta = 36$

43. $y^2 = 4x \Rightarrow r^2\sin^2\theta = 4r\cos\theta$

## SECTION 10.6 GRAPHING IN POLAR COORDINATES

1.

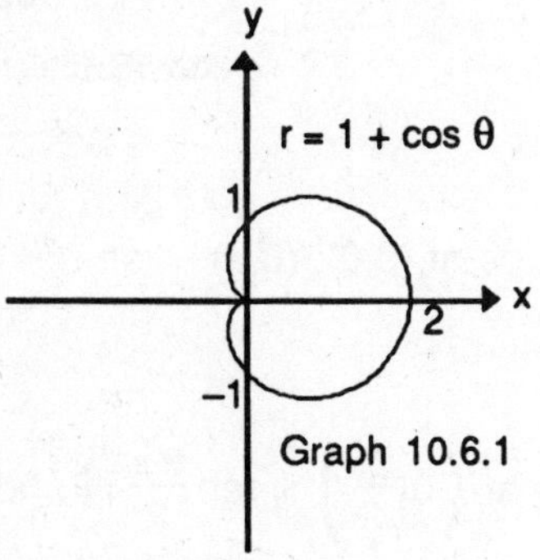

Graph 10.6.1

3.

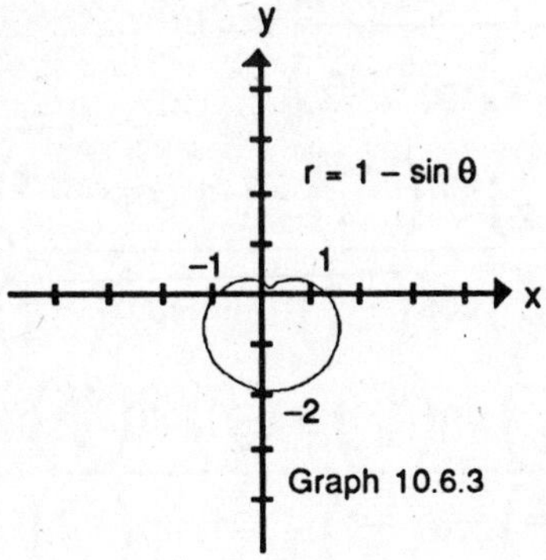

Graph 10.6.3

5.

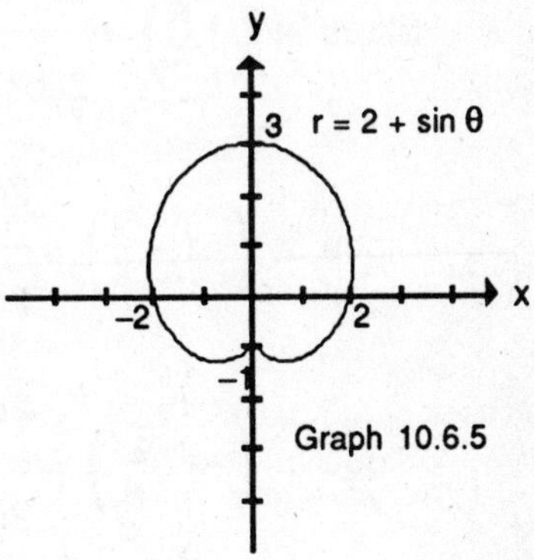

Graph 10.6.5

7.

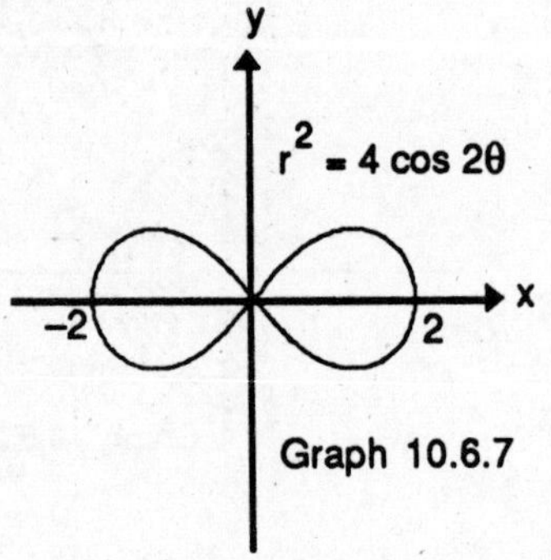


Graph 10.6.7

9.

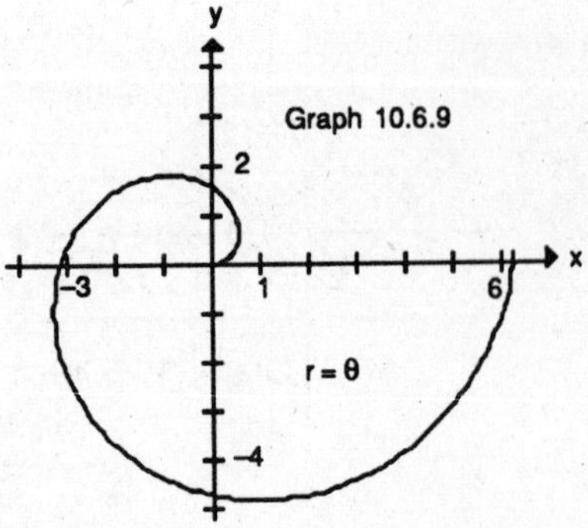


Graph 10.6.9

11. $\theta = \frac{\pi}{2} \Rightarrow r = -1 \Rightarrow \left(-1, \frac{\pi}{2}\right)$, $\theta = -\frac{\pi}{2} \Rightarrow r = -1 \Rightarrow \left(-1, -\frac{\pi}{2}\right)$

$r' = \frac{dr}{d\theta} = -\sin\theta$. Slope $= \frac{r'\sin\theta + r\cos\theta}{r'\cos\theta - r\sin\theta} = \frac{-\sin^2\theta + r\cos\theta}{-\sin\theta\cos\theta - r\sin\theta} \Rightarrow$

Slope at $\left(-1, \frac{\pi}{2}\right) = \frac{-\sin^2\left(\frac{\pi}{2}\right) + (-1)\cos\frac{\pi}{2}}{-\sin\frac{\pi}{2}\cos\frac{\pi}{2} - (-1)\sin\frac{\pi}{2}} = -1$. Slope at $\left(-1, -\frac{\pi}{2}\right) =$

$$\frac{-\sin^2\left(-\frac{\pi}{2}\right) + (-1)\cos\left(-\frac{\pi}{2}\right)}{-\sin\left(-\frac{\pi}{2}\right)\cos\left(-\frac{\pi}{2}\right) - (-1)\sin\left(-\frac{\pi}{2}\right)} = 1$$

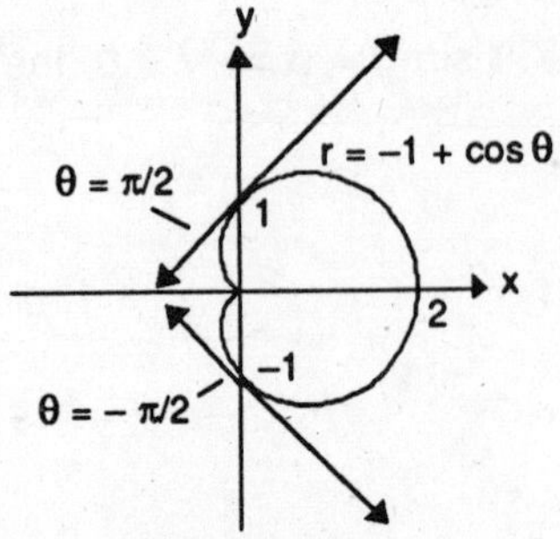


Graph 10.6.11

13. $\theta = \frac{\pi}{4} \Rightarrow r = 1 \Rightarrow \left(1, \frac{\pi}{4}\right)$. $\theta = -\frac{\pi}{4} \Rightarrow r = -1 \Rightarrow \left(-1, -\frac{\pi}{4}\right)$

$\theta = \frac{3\pi}{4} \Rightarrow r = -1 \Rightarrow \left(-1, \frac{3\pi}{4}\right)$. $\theta = -\frac{3\pi}{4} \Rightarrow r = 1 \Rightarrow \left(1, -\frac{3\pi}{4}\right)$

$r = 0 \Rightarrow \theta = 0, \frac{\pi}{2}, \pi, \frac{3\pi}{2}$. $r' = \frac{dr}{d\theta} = 2\cos 2\theta \Rightarrow$

Slope $= \frac{r'\sin\theta + r\cos\theta}{r'\cos\theta - r\sin\theta} = \frac{2\cos 2\theta\sin\theta + r\cos\theta}{2\cos 2\theta\cos\theta - r\sin\theta} \Rightarrow$

Slope at $\left(1, \frac{\pi}{4}\right) = \frac{2\cos\left(\frac{\pi}{2}\right)\sin\frac{\pi}{4} + (1)\cos\frac{\pi}{4}}{2\cos\left(\frac{\pi}{2}\right)\cos\frac{\pi}{4} - (1)\sin\frac{\pi}{4}} = -1$

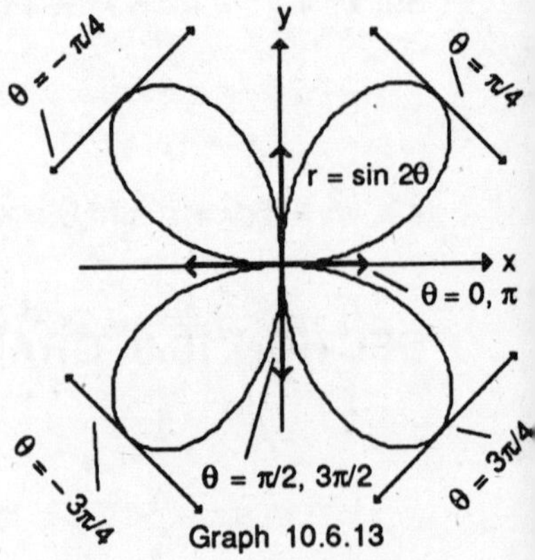


Graph 10.6.13

Slope at $\left(-1, -\frac{\pi}{4}\right) = \frac{2\cos\left(-\frac{\pi}{2}\right)\sin\left(-\frac{\pi}{4}\right) + (-1)\cos\left(-\frac{\pi}{4}\right)}{2\cos\left(-\frac{\pi}{2}\right)\cos\left(-\frac{\pi}{4}\right) - (-1)\sin\left(-\frac{\pi}{4}\right)} = 1$ Slope at $(0,0)$, $(0,\pi) = \tan 0 = 0$

Slope at $\left(-1, \frac{3\pi}{4}\right) = \frac{2\cos\left(\frac{3\pi}{2}\right)\sin\left(\frac{3\pi}{4}\right) + (-1)\cos\left(\frac{3\pi}{4}\right)}{2\cos\left(\frac{3\pi}{2}\right)\cos\left(\frac{3\pi}{4}\right) - (-1)\sin\left(\frac{3\pi}{4}\right)} = 1$ Slope at $\left(0, \frac{3\pi}{2}\right) = \tan\frac{3\pi}{2}$ is undefined

Slope at $\left(1, -\frac{3\pi}{4}\right) = \frac{2\cos\left(-\frac{3\pi}{2}\right)\sin\left(-\frac{3\pi}{4}\right) + (1)\cos\left(-\frac{3\pi}{4}\right)}{2\cos\left(-\frac{3\pi}{2}\right)\cos\left(-\frac{3\pi}{4}\right) - (1)\sin\left(-\frac{3\pi}{4}\right)} = -1$ Slope at $\left(0, \frac{\pi}{2}\right) = \tan\frac{\pi}{2}$ is undefined

15.

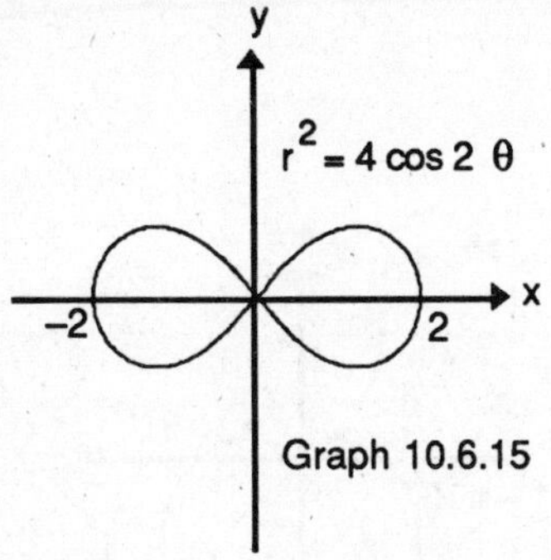

Graph 10.6.15

17. a)

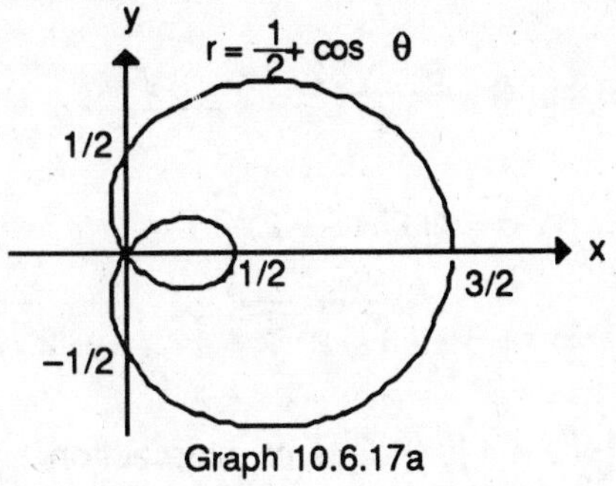

Graph 10.6.17a

17. b)

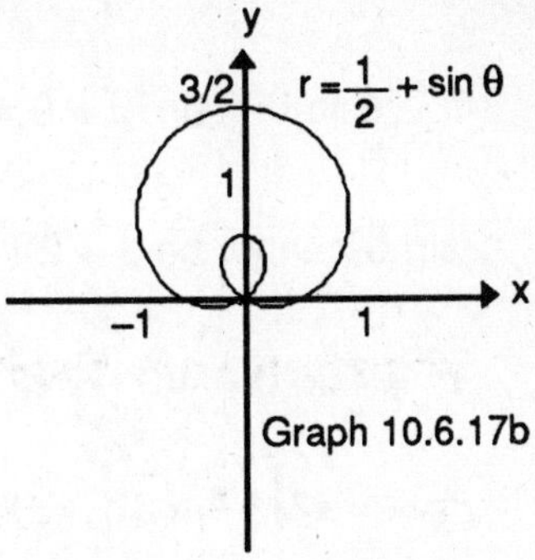

Graph 10.6.17b

19. a)

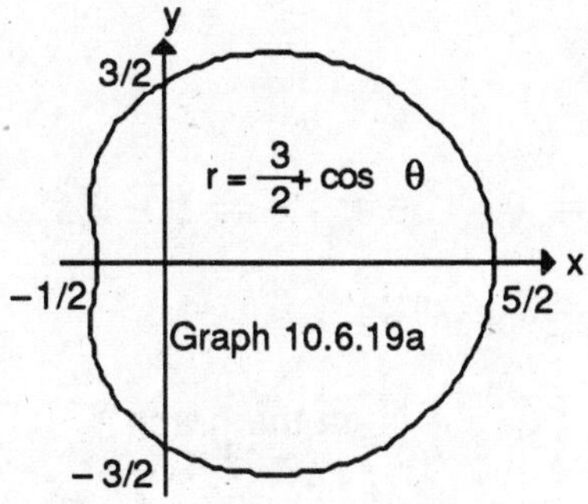

Graph 10.6.19a

19. b)

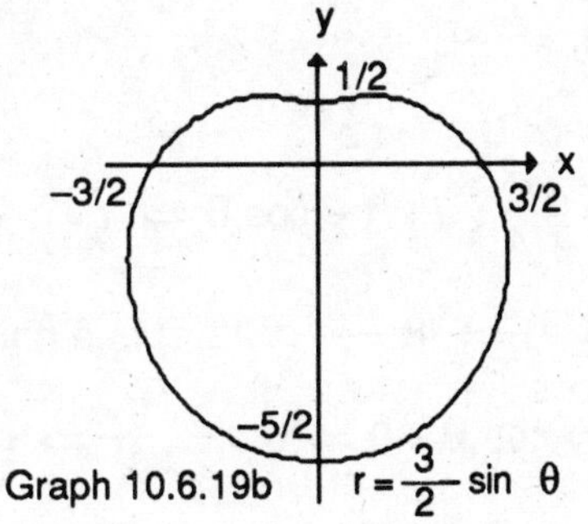

Graph 10.6.19b

21.

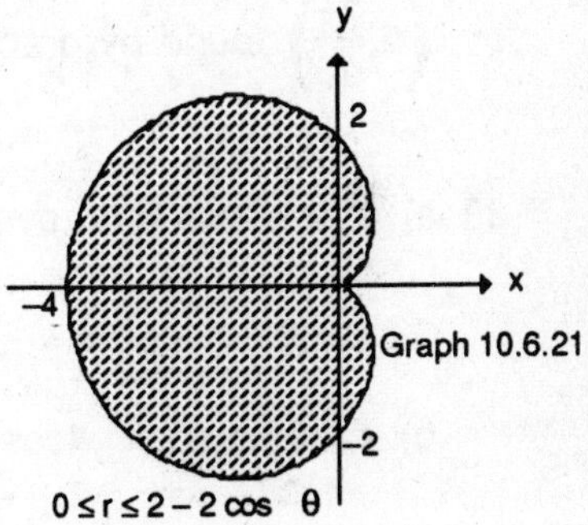

Graph 10.6.21

23. $\left(2, \frac{3\pi}{4}\right)$ is the same point as $\left(-2, -\frac{\pi}{4}\right)$. $r = 2\sin 2\left(-\frac{\pi}{4}\right) = 2\sin\left(-\frac{\pi}{2}\right) = -2 \Rightarrow \left(-2, -\frac{\pi}{4}\right)$ is on the graph $\Rightarrow \left(2, \frac{3\pi}{4}\right)$ is on the graph.

25. $1 + \cos\theta = 1 - \cos\theta \Rightarrow \cos\theta = 0 \Rightarrow$

$\theta = \frac{\pi}{2}, \frac{3\pi}{2} \Rightarrow r = 1$. Points of intersection are $\left(1, \frac{\pi}{2}\right)$ and $\left(1, \frac{3\pi}{2}\right)$. The point of intersection, (0,0), is found by graphing.

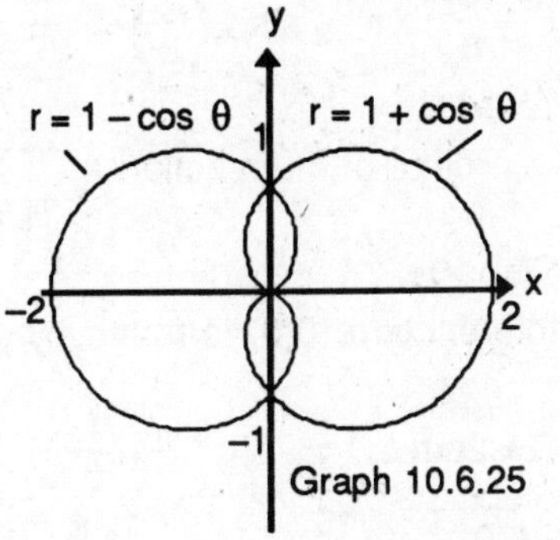

Graph 10.6.25

27. $(1 - \sin\theta)^2 = 4\sin\theta \Rightarrow 1 - 2\sin\theta + \sin^2\theta = 4\sin\theta \Rightarrow$

$1 - 6\sin\theta + \sin^2\theta = 0 \Rightarrow \sin\theta = \dfrac{6 \pm \sqrt{32}}{2} = 3 \pm 2\sqrt{2}.$

$\sin\theta$ cannot be $3 + 2\sqrt{2}$. $\therefore\ \theta = \sin^{-1}(3 - 2\sqrt{2}) \Rightarrow$

$r^2 = 4\sin\theta = 4(3 - 2\sqrt{2}) \Rightarrow r = \pm 2\sqrt{3 - 2\sqrt{2}} = \pm 2\sqrt{1 - 2\sqrt{2} + 2}$

$\Rightarrow r = \pm 2\left|1 - \sqrt{2}\right| = \pm 2\left(\sqrt{2} - 1\right)$. Points of intersection are

$\left(\pm 2\left(\sqrt{2} - 1\right), \sin^{-1}\left(3 - 2\sqrt{2}\right)\right)$. Points of intersection (0,0) and

$\left(2, \frac{3\pi}{2}\right)$ found by graphing.

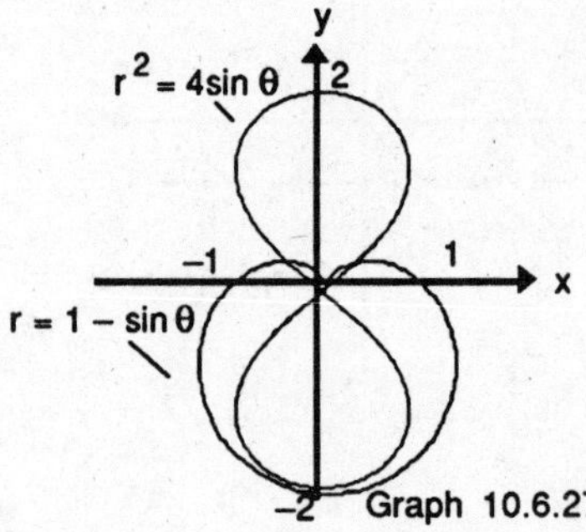

Graph 10.6.27

29. a) $r^2 = -4\cos\theta \Rightarrow \cos\theta = -\dfrac{r^2}{4}$. $r = 1 - \cos\theta \Rightarrow r = 1 - \left(-\dfrac{r^2}{4}\right) \Rightarrow 0 = r^2 - 4r + 4 \Rightarrow (r - 2)^2 = 0 \Rightarrow$

$r = 2$. $\therefore\ \cos\theta = -\dfrac{2^2}{4} = -1 \Rightarrow \theta = \pi\ \therefore\ (2,\pi)$ is a point of intersection.

b) $r = 0 \Rightarrow 0^2 = -4\cos\theta \Rightarrow \cos\theta = 0 \Rightarrow \theta = \frac{\pi}{2}, \frac{3\pi}{2} \Rightarrow \left(0, \frac{\pi}{2}\right)$ or $\left(0, \frac{3\pi}{2}\right)$ is on the graph.

$r = 0 \Rightarrow 0 = 1 - \cos\theta \Rightarrow \cos\theta = 1 \Rightarrow \theta = 0 \Rightarrow (0,0)$ is on the graph. Since $(0,0) = \left(0, \frac{\pi}{2}\right)$, the graphs intersect at the origin.

31. $r^2 = \sin 2\theta$ and $r^2 = \cos 2\theta$ are generated completely for $0 \le \theta \le \frac{\pi}{2}$. Then $\sin 2\theta = \cos 2\theta$ yields $2\theta = \frac{\pi}{4}$ as the only solution on that interval $\Rightarrow$

$\theta = \frac{\pi}{8} \Rightarrow r^2 = \sin 2\left(\frac{\pi}{8}\right) = \dfrac{1}{\sqrt{2}} \Rightarrow r = \pm\dfrac{1}{\sqrt[4]{2}}$.

$\therefore$ Points of intersection are $\left(\pm\dfrac{1}{\sqrt[4]{2}}, \dfrac{\pi}{8}\right)$. The point of intersection (0,0) is found by graphing.

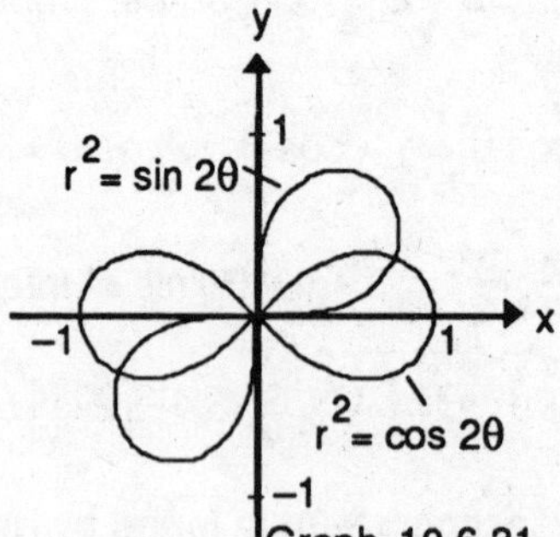

Graph 10.6.31

33. $1 = 2\sin 2\theta \Rightarrow \sin 2\theta = \frac{1}{2} \Rightarrow 2\theta = \frac{\pi}{6}, \frac{5\pi}{6}, \frac{13\pi}{6}, \frac{17\pi}{6}$

$\Rightarrow \theta = \frac{\pi}{12}, \frac{5\pi}{12}, \frac{13\pi}{12}, \frac{17\pi}{12}$. Points of intersection are

$\left(1, \frac{\pi}{12}\right)$, $\left(1, \frac{5\pi}{12}\right)$, $\left(1, \frac{13\pi}{12}\right)$, and $\left(1, \frac{17\pi}{12}\right)$.

Points of intersection $\left(1, \frac{7\pi}{12}\right)$, $\left(1, \frac{11\pi}{12}\right)$, $\left(1, \frac{19\pi}{12}\right)$, and $\left(1, \frac{23\pi}{12}\right)$ found by graphing and symmetry.

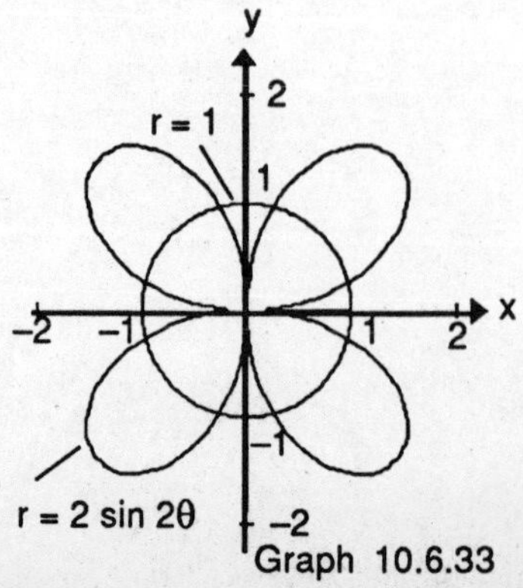

Graph 10.6.33

35.

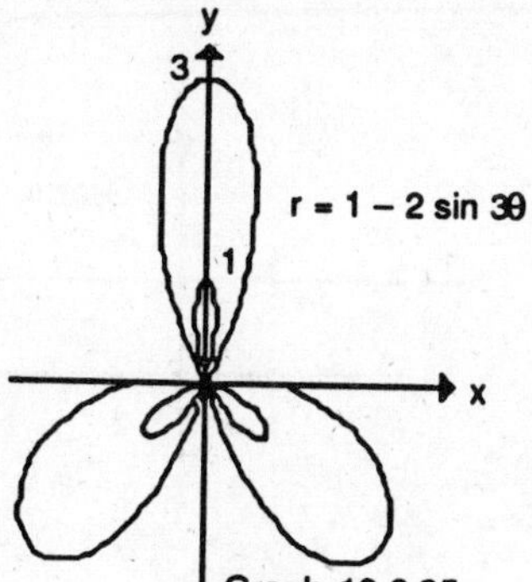

Graph 10.6.35

37. a)

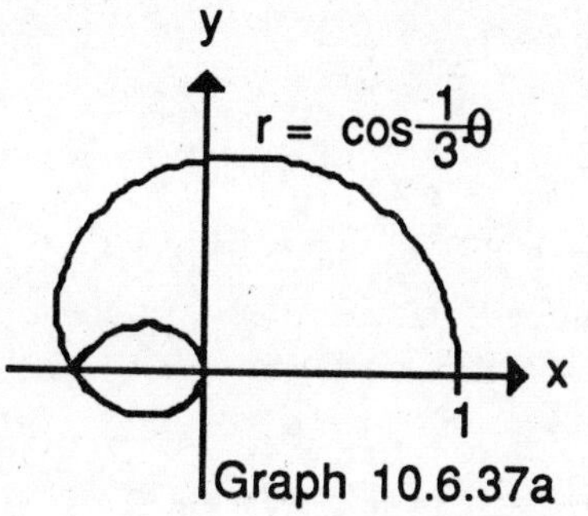

Graph 10.6.37a

37. b)

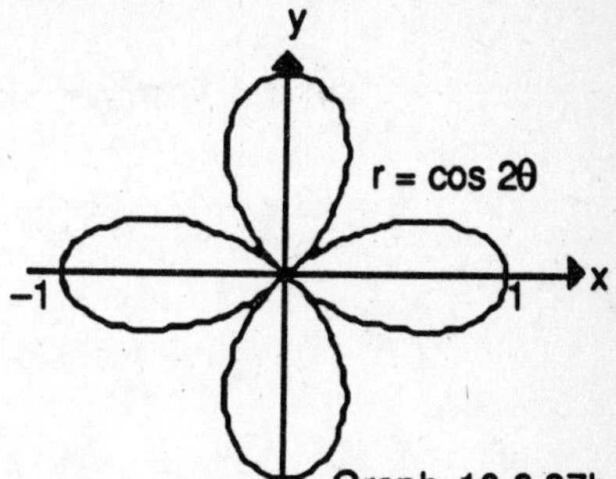

Graph 10.6.37b

37. c)

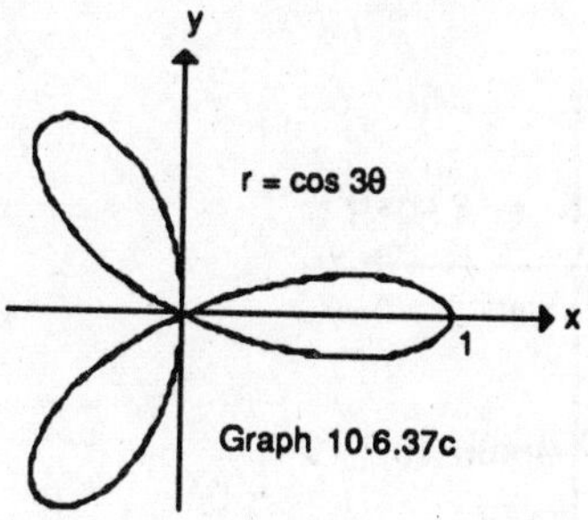

Graph 10.6.37c

37. d)

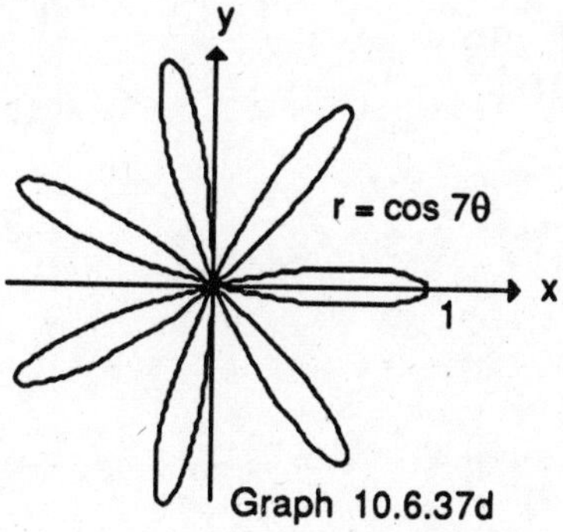

Graph 10.6.37d

## SECTION 10.7 POLAR EQUATIONS OF CONIC SECTIONS

1. $r\cos\left(\theta - \frac{\pi}{6}\right) = 5 \Rightarrow r\left(\cos\theta\cos\frac{\pi}{6} + \sin\theta\sin\frac{\pi}{6}\right) = 5 \Rightarrow \frac{\sqrt{3}}{2} r\cos\theta + \frac{1}{2} r\sin\theta = 5 \Rightarrow$
$\frac{\sqrt{3}}{2}x + \frac{1}{2}y = 5 \Rightarrow \sqrt{3}\,x + y = 10.$

3. $r\cos\left(\theta - \frac{4\pi}{3}\right) = 3 \Rightarrow r\left(\cos\theta\cos\frac{4\pi}{3} + \sin\theta\sin\frac{4\pi}{3}\right) = 3 \Rightarrow -\frac{1}{2} r\cos\theta - \frac{\sqrt{3}}{2} r\sin\theta = 3 \Rightarrow$
$-\frac{1}{2}x - \frac{\sqrt{3}}{2}y = 3 \Rightarrow x + \sqrt{3}\,y = -6$

5. $r\cos\left(\theta - \frac{\pi}{4}\right) = \sqrt{2} \Rightarrow r\left(\cos\theta\cos\frac{\pi}{4} + \sin\theta\sin\frac{\pi}{4}\right) = \sqrt{2} \Rightarrow$

$\frac{1}{\sqrt{2}} r\cos\theta + \frac{1}{\sqrt{2}} r\sin\theta = \sqrt{2} \Rightarrow$
$\frac{1}{\sqrt{2}}x + \frac{1}{\sqrt{2}}y = \sqrt{2} \Rightarrow x + y = 2.$

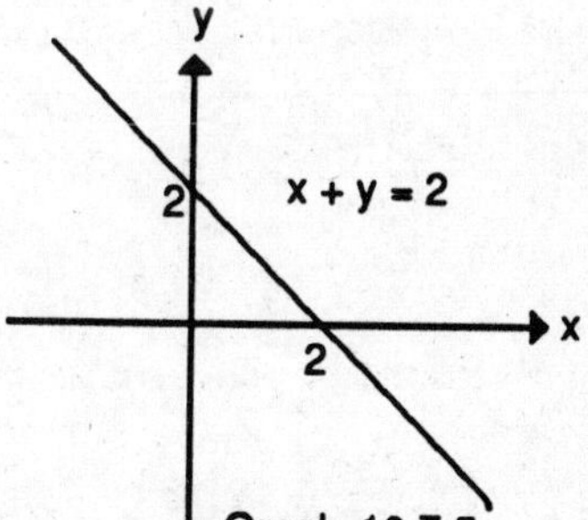

Graph 10.7.5

7. $r\cos\left(\theta - \frac{3\pi}{2}\right) = 1 \Rightarrow r\left(\cos\theta\cos\frac{3\pi}{2} + \sin\theta\sin\frac{3\pi}{2}\right) = 1 \Rightarrow$

$-r\sin\theta = 1 \Rightarrow y = -1.$

$x = \sqrt{3}\,y = 4$

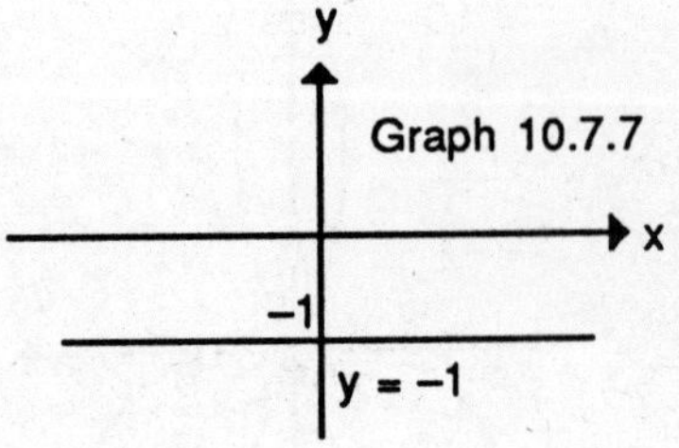


9. $r = 2(4)\cos\theta = 8\cos\theta$

11. $r = 2\sqrt{2}\sin\theta$

13.

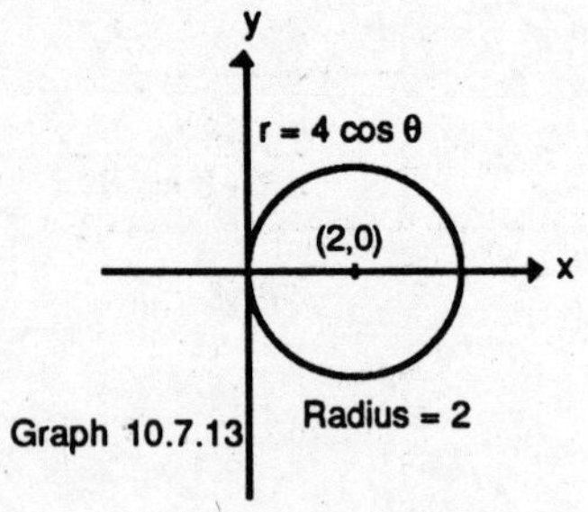


15.

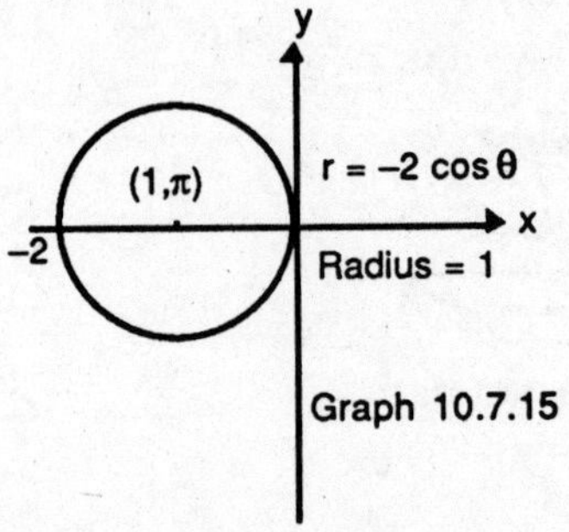


17. $e = 1,\ x = 2 \Rightarrow k = 2 \Rightarrow$

$r = \dfrac{2(1)}{1 + (1)\cos\theta}$

$= \dfrac{2}{1 + \cos\theta}$

19. $e = 2,\ x = 4 \Rightarrow k = 4 \Rightarrow$

$r = \dfrac{4(2)}{1 + (2)\cos\theta} = \dfrac{8}{1 + 2\cos\theta}$

21. $e = \frac{1}{2},\ x = 1 \Rightarrow k = 1 \Rightarrow r = \dfrac{\frac{1}{2}(1)}{1 + (1)\cos\theta} = \dfrac{1}{2 + \cos\theta}$

23. $e = \frac{1}{5},\ y = -10 \Rightarrow k = 10 \Rightarrow r = \dfrac{\frac{1}{5}(10)}{1 - \frac{1}{5}\sin\theta} = \dfrac{2}{1 - \frac{1}{5}\sin\theta} = \dfrac{10}{5 - \sin\theta}$

25. $r = \dfrac{1}{1 + \cos\theta} \Rightarrow e = 1,\ k = 1 \Rightarrow x = 1$

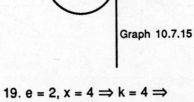

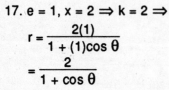

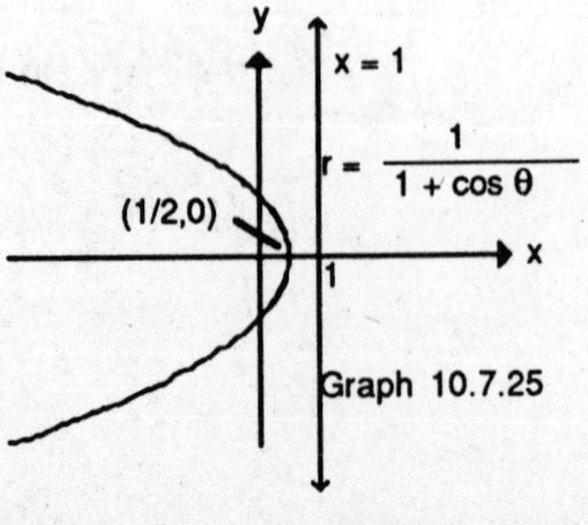

27. $r = \dfrac{25}{10 - 5\cos\theta} \Rightarrow r = \dfrac{\frac{25}{10}}{1 - \frac{5}{10}\cos\theta} = \dfrac{\frac{5}{2}}{1 - \frac{1}{2}\cos\theta} \Rightarrow$

$e = \frac{1}{2}$, $k = 5$. $a(1 - e^2) = ke \Rightarrow a\left(1 - \left(\frac{1}{2}\right)^2\right) = \frac{5}{2} \Rightarrow$
$\frac{3}{4}a = \frac{5}{2} \Rightarrow a = \frac{10}{3}$. $a - ae = \frac{10}{3} - \left(\frac{10}{3}\right)\frac{1}{2} = \frac{5}{3}$

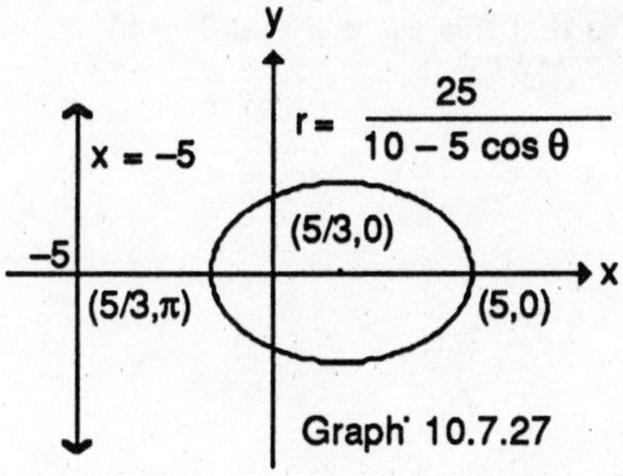

Graph 10.7.27

29. $r = \dfrac{400}{16 + 8\sin\theta} \Rightarrow r = \dfrac{\frac{400}{16}}{1 + \frac{8}{16}\sin\theta} \Rightarrow r = \dfrac{25}{1 + \frac{1}{2}\sin\theta} \Rightarrow$

$e = \frac{1}{2}$, $k = 50$. $a(1 - e^2) = ke \Rightarrow a\left(1 - \left(\frac{1}{2}\right)^2\right) = 25 \Rightarrow$
$\frac{3}{4}a = 25 \Rightarrow a = \frac{100}{3}$. $a - ae = \frac{100}{3} - \frac{100}{3}\left(\frac{1}{2}\right) = \frac{50}{3}$.

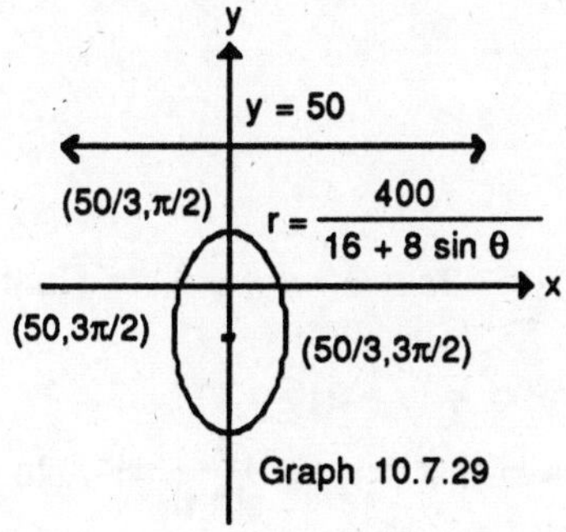

Graph 10.7.29

31. $r = \dfrac{8}{2 - 2\sin\theta} \Rightarrow r = \dfrac{4}{1 - \sin\theta} \Rightarrow e = 1, k = 4$

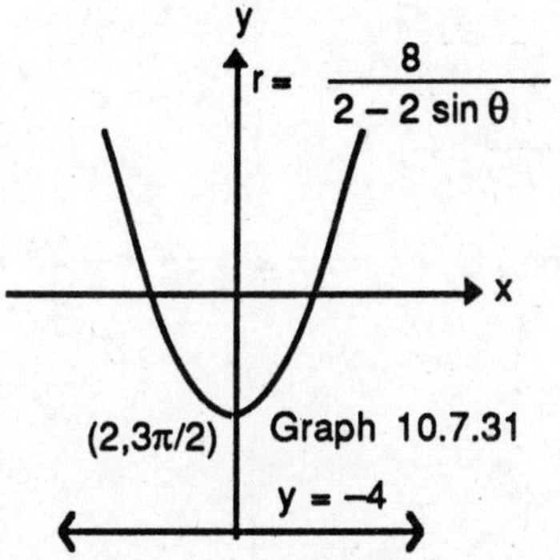

Graph 10.7.31

33.

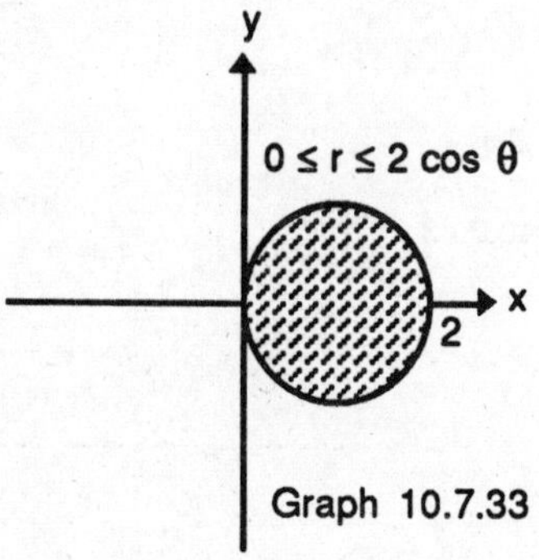

Graph 10.7.33

35. a) Perihelion = $a - ae = a(1 - e)$

Aphelion = $ea + a = a(1 + e)$

b)

| Planet | Perihelion | Aphelion |
|---|---|---|
| Mercury | 0.3075 AU | 0.4667 AU |
| Venus | 0.7184 AU | 0.7282 AU |
| Earth | 0.9833 AU | 1.0167 AU |
| Mars | 1.3817 AU | 1.6663 AU |
| Jupiter | 4.9512 AU | 5.4548 AU |
| Saturn | 9.0210 AU | 10.0570 AU |
| Uranus | 18.2977 AU | 20.0623 AU |
| Neptune | 29.8135 AU | 30.3065 AU |
| Pluto | 29.6549 AU | 49.2251 AU |

37. a) $r = 2\sin\theta \Rightarrow a = 1 \Rightarrow$ Center = (0,1)

$\Rightarrow x^2 + (y-1)^2 = 1.$

$r = \csc\theta \Rightarrow r = \dfrac{1}{\sin\theta} \Rightarrow r\sin\theta = 1$

$\Rightarrow y = 1$

b)

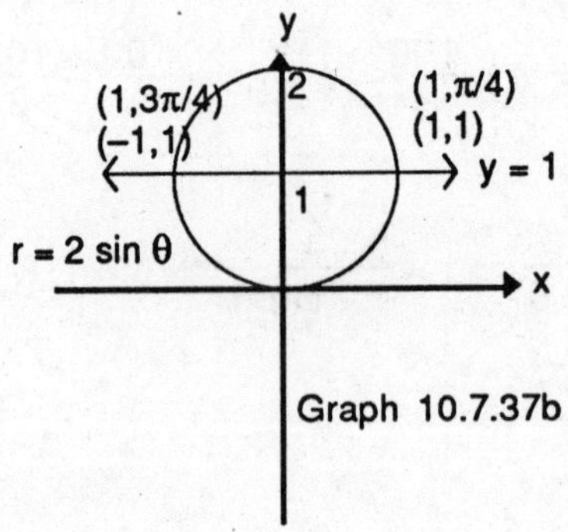


Graph 10.7.37b

39. $r\cos\theta = 4 \Rightarrow x = 4 \Rightarrow k = 4$. Parabola $\Rightarrow e = 1$. $\therefore\ r = \dfrac{4}{1 + \cos\theta}$

41.

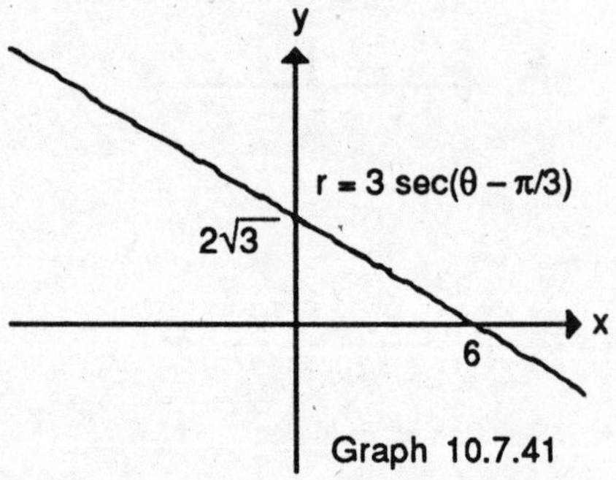


Graph 10.7.41

43.

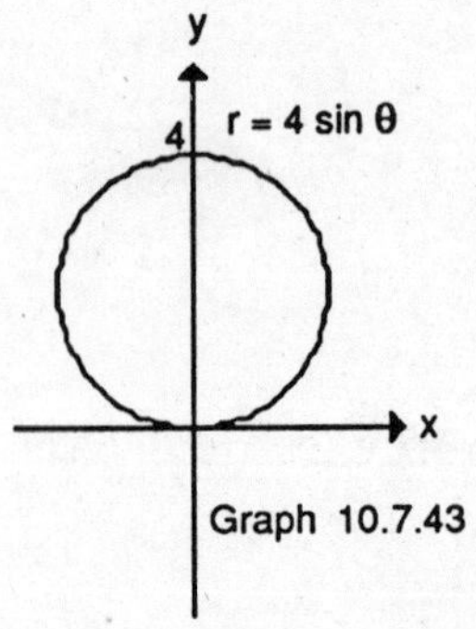


Graph 10.7.43

45.

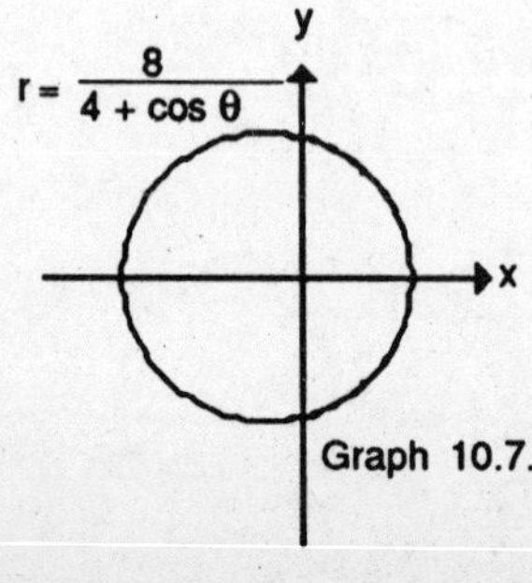


Graph 10.7.45

47.

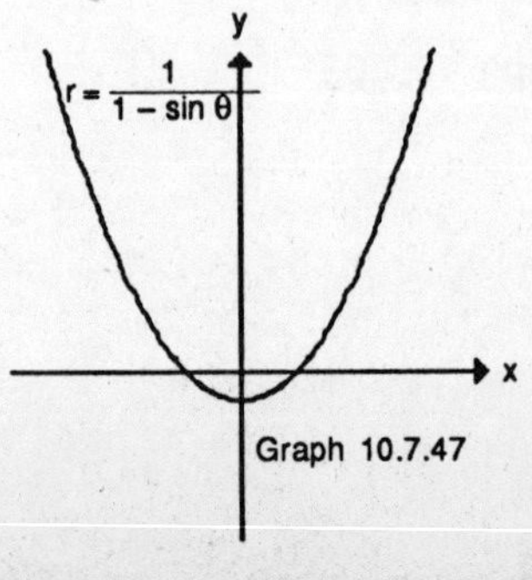


Graph 10.7.47

45.

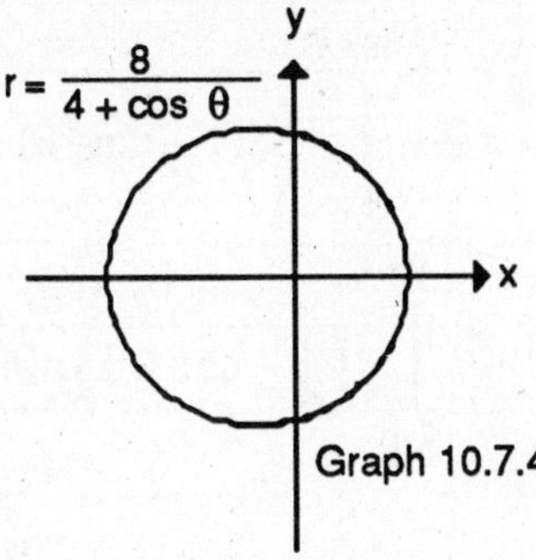


Graph 10.7.45

47.

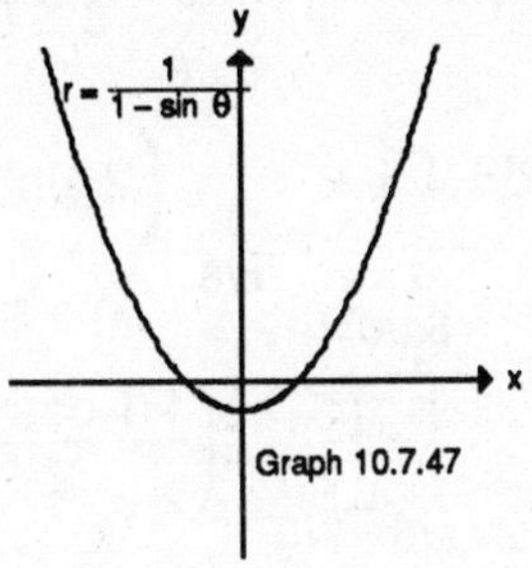


Graph 10.7.47

49.

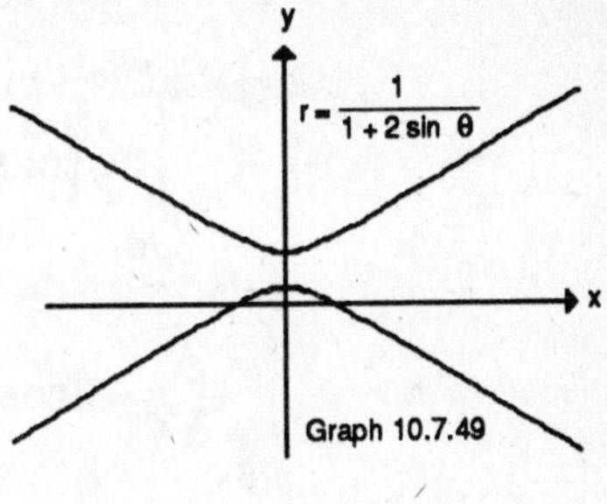


Graph 10.7.49

## SECTION 10.8 INTEGRATION IN POLAR COORDINATES

1. $A = \displaystyle\int_0^{\pi/4} \frac{1}{2} r^2\, d\theta = \int_0^{\pi/4} \frac{1}{2}\cos^2\theta\, d\theta = \int_0^{\pi/4} \frac{1}{2}\left(\frac{1 + \cos 2\theta}{2}\right) d\theta = \frac{1}{4}\int_0^{\pi/4} (1 + \cos 2\theta)\, d\theta$

$= \dfrac{1}{4}\left[\theta + \dfrac{\sin 2\theta}{2}\right]_0^{\pi/4} = \dfrac{\pi}{16} + \dfrac{1}{8}$

3. $A = \displaystyle\int_0^{2\pi} \frac{1}{2}(4 + 2\cos\theta)^2 d\theta = \int_0^{2\pi} \frac{1}{2}(16 + 16\cos\theta + 4\cos^2\theta)\, d\theta$

$= \displaystyle\int_0^{2\pi} \left(8 + 8\cos\theta + 2\left(\frac{1 + \cos 2\theta}{2}\right)\right) d\theta = \int_0^{2\pi} (9 + 8\cos\theta + \cos 2\theta)\, d\theta =$

$\left[9\theta + 8\sin\theta + \dfrac{1}{2}\sin 2\theta\right]_0^{2\pi} = 18\pi$

5. $A = 2\displaystyle\int_0^{\pi/4} \frac{1}{2}\cos^2 2\theta\, d\theta = \int_0^{\pi/4} \frac{1 + \cos 4\theta}{2}\, d\theta = \frac{1}{2}\left[\theta + \frac{\sin 4\theta}{4}\right]_0^{\pi/4} = \frac{\pi}{8}$

7. $A = 2\displaystyle\int_{-\pi/4}^{\pi/4} \frac{1}{2}\left(2a^2\cos 2\theta\right) d\theta = 2a^2 \int_{-\pi/4}^{\pi/4} \cos 2\theta\, d\theta = 2a^2\left[\frac{\sin 2\theta}{2}\right]_{-\pi/4}^{\pi/4} = 2a^2$

9. $A = 6\displaystyle\int_0^{\pi/6} \frac{1}{2}(2\sin 3\theta)\, d\theta = 6\int_0^{\pi/6} \sin 3\theta\, d\theta = 6\left[-\frac{\cos 3\theta}{3}\right]_0^{\pi/6} = 2$

11. $r = 1,\ r = 2\sin\theta \Rightarrow 2\sin\theta = 1 \Rightarrow \sin\theta = \frac{1}{2} \Rightarrow \theta = \frac{\pi}{6}, \frac{5\pi}{6}$

$$A = \pi(1)^2 - \int_{\pi/6}^{5\pi/6} \frac{1}{2}\left((2\sin\theta)^2 - 1^2\right)d\theta = \pi - \int_{\pi/6}^{5\pi/6}\left(2\sin^2\theta - \frac{1}{2}\right)d\theta = \pi - \int_{\pi/6}^{5\pi/6}\left(2(1-\cos^2\theta) - \frac{1}{2}\right)d\theta$$

$$= \pi - \int_{\pi/6}^{5\pi/6}\left(\frac{3}{2} - 2\cos^2\theta\right)d\theta = \pi - \int_{\pi/6}^{5\pi/6}\left(\frac{3}{2} - 2\left(\frac{1+\cos 2\theta}{2}\right)\right)d\theta = \pi - \int_{\pi/6}^{5\pi/6}\left(\frac{1}{2} - \cos 2\theta\right)d\theta =$$

$$\pi - \left[\frac{1}{2}\theta - \frac{\sin 2\theta}{2}\right]_{\pi/6}^{5\pi/6} = \frac{4\pi - 3\sqrt{3}}{6}$$

13. $r = 2(1-\cos\theta),\ r = 2(1+\cos\theta) \Rightarrow 1 - \cos\theta = 1 + \cos\theta \Rightarrow \cos\theta = 0 \Rightarrow \theta = \frac{\pi}{2}, \frac{3\pi}{2}$. The graph gives the point of intersection (0,0).

$$A = 2\int_0^{\pi/2} \frac{1}{2}\left(2(1-\cos\theta)\right)^2 d\theta + 2\int_{\pi/2}^{\pi} \frac{1}{2}\left(2(1+\cos\theta)\right)^2 d\theta =$$

$$\int_0^{\pi/2} 4(1 - 2\cos\theta + \cos^2\theta)d\theta + \int_{\pi/2}^{\pi} 4(1 + 2\cos\theta + \cos^2\theta)d\theta =$$

$$\int_0^{\pi/2}\left(4 - 8\cos\theta + 4\left(\frac{1+\cos 2\theta}{2}\right)\right)d\theta + \int_{\pi/2}^{\pi}\left(4 + 8\cos\theta + 4\left(\frac{1+\cos 2\theta}{2}\right)\right)d\theta =$$

$$\int_0^{\pi/2} (6 - 8\cos\theta + 2\cos 2\theta)d\theta + \int_{\pi/2}^{\pi} (6 + 8\cos\theta + 2\cos 2\theta)d\theta =$$

$$\left[6\theta - 8\sin\theta + \sin 2\theta\right]_0^{\pi/2} + \left[6\theta + 8\sin\theta + \sin 2\theta\right]_{\pi/2}^{\pi} = 6\pi - 16$$

15. $r = 3a\cos\theta,\ r = a(1+\cos\theta) \Rightarrow 3a\cos\theta = a(1+\cos\theta) \Rightarrow 3\cos\theta = 1 + \cos\theta \Rightarrow \cos\theta = \frac{1}{2} \Rightarrow$

$$\theta = \frac{\pi}{3} \text{ or } -\frac{\pi}{3}.\ \therefore\ A = 2\int_0^{\pi/3} \frac{1}{2}\left((3a\cos\theta)^2 - (a(1+\cos\theta))^2\right)d\theta =$$

$$\int_0^{\pi/3}\left(9a^2\cos^2\theta - a^2 - 2a^2\cos\theta - a^2\cos^2\theta\right)d\theta = \int_0^{\pi/3}\left(8a^2\cos^2\theta - 2a^2\cos\theta - a^2\right)d\theta =$$

$$\int_0^{\pi/3}\left(4a^2(1+\cos 2\theta) - 2a^2\cos\theta - a^2\right)d\theta = \int_0^{\pi/3}\left(3a^2 + 4a^2\cos 2\theta - 2a^2\cos\theta\right)d\theta =$$

$$\left[3a^2\theta + 2a^2\sin 2\theta - 2a^2\sin\theta\right]_0^{\pi/3} = \pi a^2$$

17. a) $A = 2\int_0^{2\pi/3} \frac{1}{2}(2\cos\theta + 1)^2 d\theta = \int_0^{2\pi/3} (4\cos^2\theta + 4\cos\theta + 1)d\theta =$

$\int_0^{2\pi/3} \Big(2(1+\cos 2\theta) + 4\cos\theta + 1\Big)d\theta = \int_0^{2\pi/3} (3 + 2\cos 2\theta + 4\cos\theta)d\theta =$

$\Big[3\theta + \sin 2\theta + 4\sin\theta\Big]_0^{2\pi/3} = 2\pi + \frac{3\sqrt{3}}{2}$

b) $A = \left(2\pi + \frac{3\sqrt{3}}{2}\right) - \left(\pi - \frac{3\sqrt{3}}{2}\right) = \pi + 3\sqrt{3}$ (From 17 a) above and Example 2, page 702, in the text.)

19. $r = \theta^2, 0 \le \theta \le \sqrt{5} \Rightarrow \frac{dr}{d\theta} = 2\theta$. $\therefore$ Length $= \int_0^{\sqrt{5}} \sqrt{(\theta^2)^2 + (2\theta)^2}\, d\theta = \int_0^{\sqrt{5}} \sqrt{\theta^4 + 4\theta^2}\, d\theta =$

$\int_0^{\sqrt{5}} |\theta|\sqrt{\theta^2 + 4}\, d\theta = \int_0^{\sqrt{5}} \theta\sqrt{\theta^2 + 4}\, d\theta = \int_4^9 \frac{1}{2}\sqrt{u}\, du = \frac{1}{2}\Big[\frac{2}{3}u^{3/2}\Big]_4^9 = \frac{19}{3}$

Let $u = \theta^2 + 4 \Rightarrow \frac{1}{2}du = \theta d\theta$; $\theta = 0 \Rightarrow u = 4$, $\theta = \sqrt{5} \Rightarrow u = 9$

21. $r = \sec\theta, 0 \le \theta \le \frac{\pi}{4} \Rightarrow \frac{dr}{d\theta} = \sec\theta\tan\theta$. $\therefore$ Length $= \int_0^{\pi/4} \sqrt{\sec^2\theta + (\sec\theta\tan\theta)^2}\, d\theta =$

$\int_0^{\pi/4} \sqrt{\sec^2\theta(1 + \tan^2\theta)}\, d\theta = \int_0^{\pi/4} |\sec\theta|\sqrt{1 + \tan^2\theta}\, d\theta = \int_0^{\pi/4} |\sec\theta|\sqrt{\sec^2\theta}\, d\theta =$

$\int_0^{\pi/4} \sec^2\theta\, d\theta = \Big[\tan\theta\Big]_0^{\pi/4} = 1$

23. $r = 1 + \cos\theta \Rightarrow \frac{dr}{d\theta} = -\sin\theta$. $\therefore$ Length $= \int_0^{2\pi} \sqrt{(1+\cos\theta)^2 + (-\sin\theta)^2}\, d\theta =$

$2\int_0^{\pi} \sqrt{1 + 2\cos\theta + \cos^2\theta + \sin^2\theta}\, d\theta = 2\int_0^{\pi} \sqrt{2 + 2\cos\theta}\, d\theta =$

$2\int_0^{\pi} \sqrt{\frac{4(1+\cos\theta)}{2}}\, d\theta = 4\int_0^{\pi} \sqrt{\frac{1+\cos\theta}{2}}\, d\theta = 4\int_0^{\pi} \cos\frac{1}{2}\theta\, d\theta = 4\Big[2\sin\frac{1}{2}\theta\Big]_0^{\pi} = 8$

25. $r = \sqrt{\cos 2\theta}\,,\ 0 \le \theta \le \frac{\pi}{4} \Rightarrow \frac{dr}{d\theta} = \frac{1}{2}(\cos 2\theta)^{-1/2}(-\sin 2\theta)(2) = \frac{-\sin 2\theta}{\sqrt{\cos 2\theta}}$

$$\therefore \text{ Surface Area} = \int_0^{\pi/4} 2\pi r \cos\theta \sqrt{\left(\sqrt{\cos 2\theta}\right)^2 + \left(\frac{-\sin 2\theta}{\sqrt{\cos 2\theta}}\right)^2}\, d\theta =$$

$$\int_0^{\pi/4} 2\pi\sqrt{\cos 2\theta}\cos\theta \sqrt{\cos 2\theta + \frac{\sin^2 2\theta}{\cos 2\theta}}\, d\theta = \int_0^{\pi/4} 2\pi\sqrt{\cos 2\theta}\cos\theta\sqrt{\frac{1}{\cos 2\theta}}\, d\theta =$$

$$\int_0^{\pi/4} 2\pi\cos\theta\, d\theta = \Big[2\pi \sin\theta\Big]_0^{\pi/4} = \pi\sqrt{2}$$

27. $r^2 = \cos 2\theta \Rightarrow r = \pm\sqrt{\cos 2\theta}$. Use $r = \sqrt{\cos 2\theta}$ on $\left[0, \frac{\pi}{4}\right]$. Then $\frac{dr}{d\theta} = \frac{1}{2}(\cos 2\theta)^{-1/2}(-\sin 2\theta)(2) =$

$$\frac{-\sin 2\theta}{\sqrt{\cos 2\theta}} \quad \therefore \text{ Surface Area} = \int_0^{\pi/4} 2\pi\sqrt{\cos 2\theta}\sin\theta\sqrt{\left(\sqrt{\cos 2\theta}\right)^2 + \left(\frac{-\sin 2\theta}{\sqrt{\cos 2\theta}}\right)^2}\, d\theta =$$

$$\int_0^{\pi/4} 2\pi\sqrt{\cos 2\theta}\sin\theta\sqrt{\cos 2\theta + \frac{\sin^2 2\theta}{\cos 2\theta}}\, d\theta = \int_0^{\pi/4} 2\pi\sqrt{\cos 2\theta}\sin\theta\sqrt{\frac{1}{\cos 2\theta}}\, d\theta =$$

$$\int_0^{\pi/4} 2\pi\sin\theta\, d\theta = \Big[-2\pi\cos\theta\Big]_0^{\pi/4} = \pi(2 - \sqrt{2})$$

29. $$\bar{x} = \frac{\frac{2}{3}\int_0^{2\pi} r^3\cos\theta\, d\theta}{\int_0^{2\pi} r^2\, d\theta} = \frac{\frac{2}{3}\int_0^{2\pi} (a(1+\cos\theta))^3\cos\theta\, d\theta}{\int_0^{2\pi} (a(1+\cos\theta))^2\, d\theta} =$$

$$\frac{\frac{2}{3}a^3\int_0^{2\pi}(1 + 3\cos\theta + 3\cos^2\theta + \cos^3\theta)\cos\theta\, d\theta}{a^2\int_0^{2\pi}(1 + 2\cos\theta + \cos^2\theta)d\theta} =$$

29. (Continued)

$$\frac{\frac{2}{3}a\int_0^{2\pi}\left(\cos\theta+3\left(\frac{1+\cos 2\theta}{2}\right)+3(1-\sin^2\theta)\cos\theta+\left(\frac{1+\cos 2\theta}{2}\right)^2\right)d\theta}{\int_0^{2\pi}\left(1+2\cos\theta+\left(\frac{1+\cos 2\theta}{2}\right)\right)d\theta}=$$

(After much work using the identity $\cos A=\frac{1+\cos 2A}{2}$)

$$\frac{a\int_0^{2\pi}\left(\frac{15}{12}+\frac{8}{3}\cos\theta+\frac{4}{3}\cos 2\theta-2\cos\theta\sin^2\theta+\frac{1}{12}\cos 4\theta\right)d\theta}{\int_0^{2\pi}\left(\frac{3}{2}+2\cos\theta+\frac{1}{2}\cos 2\theta\right)d\theta}=$$

$$\frac{a\left[\frac{15}{12}\theta+\frac{8}{3}\sin\theta+\frac{2}{3}\sin 2\theta-\frac{2}{3}\sin^3\theta+\frac{1}{48}\sin 4\theta\right]_0^{2\pi}}{\left[\frac{3}{2}\theta+2\sin\theta+\frac{1}{4}\sin 2\theta\right]_0^{2\pi}}=\frac{a\left(\frac{15}{6}\pi\right)}{3\pi}=\frac{5}{6}a$$

$$\overline{y}=\frac{\frac{2}{3}\int_0^{2\pi}r^3\sin\theta\,d\theta}{\int_0^{2\pi}r^2\,d\theta}=\frac{\frac{2}{3}\int_0^{2\pi}(a(1+\cos\theta))^3\sin\theta\,d\theta}{3\pi}=\frac{\frac{2}{3}\int_{2a}^{2a}-\frac{1}{a}u^3du}{3\pi}=\frac{0}{3\pi}=0$$ Centroid is $\left(\frac{5}{6}a,0\right)$

Let $u=a(1+\cos\theta)\Rightarrow-\frac{1}{a}du=\sin\theta\,d\theta$; $\theta=0\Rightarrow u=2a$, $\theta=2\pi\Rightarrow u=2a$

## PRACTICE EXERCISES

1. $x = \frac{y^2}{8} \Rightarrow 4p = 8 \Rightarrow p = 2$

   $\therefore$ Focus: (2,0); Directrix: $x = -2$

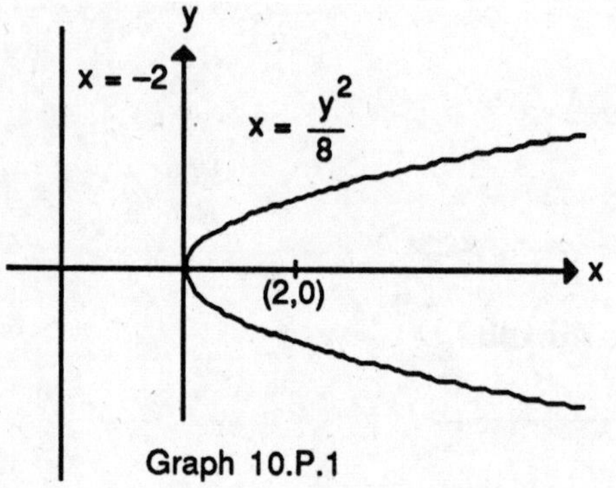


Graph 10.P.1

3. $16x^2 + 7y^2 = 112 \Rightarrow \frac{x^2}{7} + \frac{y^2}{16} = 1 \Rightarrow$

   $c^2 = 16 - 7 = 9 \Rightarrow c = 3.\ e = \frac{c}{a} = \frac{3}{4} \Rightarrow$

   $y = \frac{a}{e} = \frac{4}{\frac{3}{4}} = \frac{16}{3}$ and $y = -\frac{16}{3}$

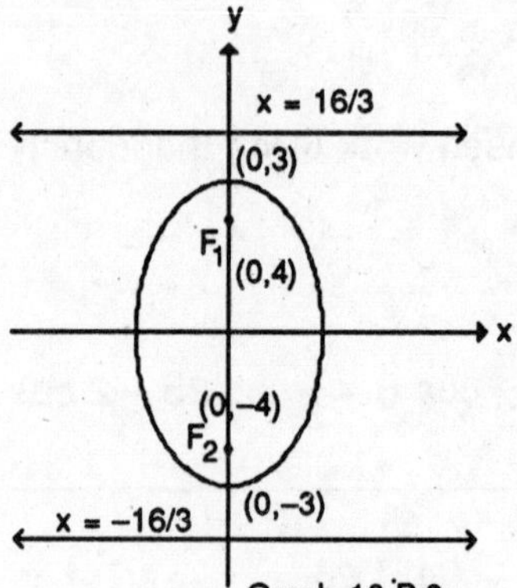


Graph 10.P.3

5. $3x^2 - y^2 = 3 \Rightarrow x^2 - \frac{y^2}{3} = 1 \Rightarrow$

   $c^2 = 1 + 3 = 4 \Rightarrow c = 2.\ e = \frac{c}{a} = \frac{2}{1} = 2 \Rightarrow$

   $x = \frac{a}{e} = \frac{1}{2}$ and $x = -\frac{1}{2}$

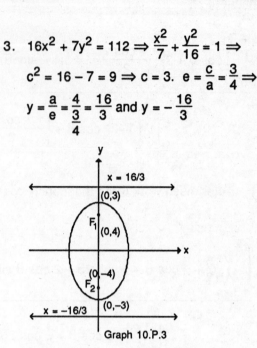

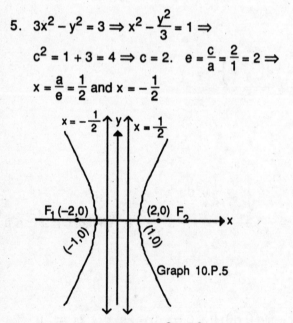

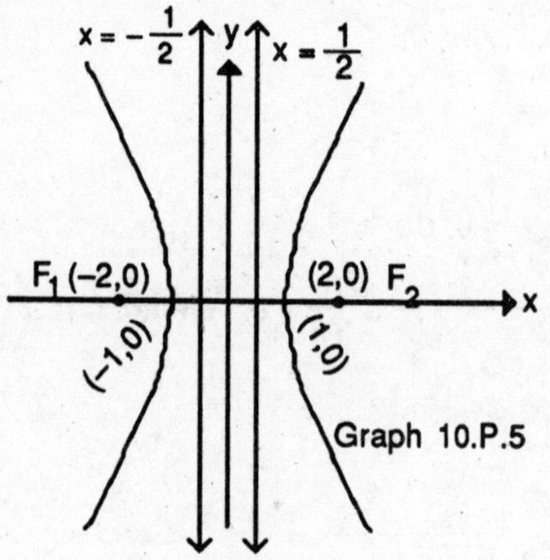


Graph 10.P.5

7. $B^2 - 4AC = 1 - 4(1)(1) = -3 < 0 \Rightarrow$ Ellipse

9. $B^2 - 4AC = 4^2 - 4(1)(4) = 0 \Rightarrow$ Parabola

11. $B^2 - 4AC = 1^2 - 4(2)(2) = -15 < 0 \Rightarrow$ Ellipse.

    $\cot 2\alpha = \frac{A - C}{B} = 0 \Rightarrow 2\alpha = \frac{\pi}{2} \Rightarrow \alpha = \frac{\pi}{4}$

    $x = \frac{\sqrt{2}}{2}x' - \frac{\sqrt{2}}{2}y',\ y = \frac{\sqrt{2}}{2}x' + \frac{\sqrt{2}}{2}y' \Rightarrow$

    $2\left(\frac{\sqrt{2}}{2}x' - \frac{\sqrt{2}}{2}y'\right)^2 + \left(\frac{\sqrt{2}}{2}x' - \frac{\sqrt{2}}{2}y'\right)\left(\frac{\sqrt{2}}{2}x' + \frac{\sqrt{2}}{2}y'\right) +$

    $2\left(\frac{\sqrt{2}}{2}x' + \frac{\sqrt{2}}{2}y'\right)^2 - 15 = 0 \Rightarrow$

    $5x'^2 + 3y'^2 - 30 = 0.$

13. $xy = 2 \Rightarrow \alpha = \frac{\pi}{4}$ and $x'^2 - y'^2 = 4$ (See Exercise 17, Section 10.2). $a = 1, b = 1 \Rightarrow c = \sqrt{2} \Rightarrow e = \frac{c}{a} = \sqrt{2}$

15. a) Around the x–axis: $9x^2 + 4y^2 = 36 \Rightarrow y^2 = 9 - \frac{9}{4}x^2 \Rightarrow y = \pm\sqrt{9 - \frac{9}{4}x^2}$, use the positive root.

$$V = 2\int_0^2 \pi\left(\sqrt{9 - \frac{9}{4}x^2}\right)^2 dx = 2\int_0^2 \pi\left(9 - \frac{9}{4}x^2\right) dx = 2\pi\left[9x - \frac{3}{4}x^3\right]_0^2 = 24\pi$$

15. b) Around the y–axis: $9x^2 + 4y^2 = 36 \Rightarrow x^2 = 4 - \frac{4}{9}y^2 \Rightarrow x = \pm\sqrt{4 - \frac{4}{9}y^2}$ , use the positive root.

$$V = 2\int_0^3 \pi\left(\sqrt{4 - \frac{4}{9}y^2}\right)^2 dy = 2\int_0^3 \pi\left(4 - \frac{4}{9}y^2\right) dy = 2\pi\left[4y - \frac{4}{27}y^3\right]_0^3 = 16\pi$$

17. $x = 2t,\ y = t^2 \Rightarrow d = \sqrt{(2t-0)^2 + (t^2-3)^2} = \sqrt{t^4 - 2t^2 + 9}.$ Minimize $d^2 = r = t^4 - 2t^2 + 9$ to minimize $d(t)$. $\frac{dr}{dt} = 4t^3 - 4t$. $\frac{dr}{dt} = 0 \Rightarrow 4t^3 - 4t = 0 \Rightarrow 4t(t^2 - 1) = 0 \Rightarrow t = 0$ or $t = \pm 1$. $\frac{d^2r}{dt^2} = 12t^2 - 4 \Rightarrow$ $\frac{d^2r}{dt^2}(0) = -4 \Rightarrow$ Relative Max. $\frac{d^2r}{dt^2}(1) = 8 \Rightarrow$ Relative Min. $r(1) = 8$. $\frac{d^2r}{dt^2}(-1) = 8 \Rightarrow$ Relative Min. $r(-1) = 8$. $\therefore$ $t = \pm 1$ both give points on the parabola that are closest to (0,3). $t = 1 \Rightarrow x = 2,\ y = 1 \Rightarrow$ (2,1) is one of the points. $t = -1 \Rightarrow x = -2,\ y = 1 \Rightarrow$ (–2,1) is the other point.

19. $x = \frac{t}{2},\ y = t + 1 \Rightarrow 2x = t \Rightarrow y = 2x + 1$

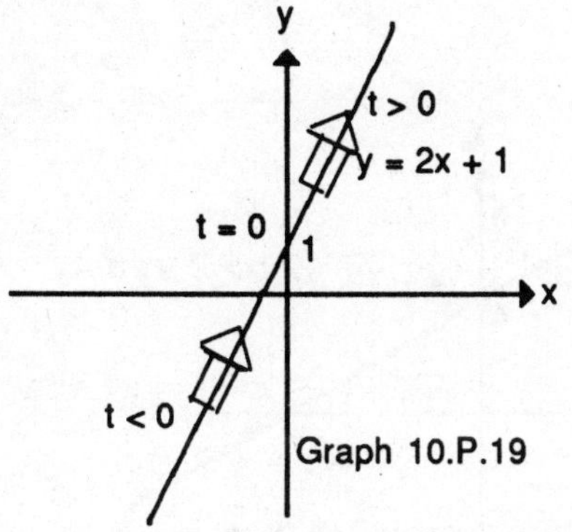

Graph 10.P.19

21. $x = \frac{1}{2}\tan t,\ y = \frac{1}{2}\sec t \Rightarrow x^2 = \frac{1}{4}\tan^2 t,$
$y^2 = \frac{1}{4}\sec^2 t \Rightarrow 4x^2 = \tan^2 t,\ 4y^2 = \sec^2 t \Rightarrow$
$4x^2 + 1 = 4y^2 \Rightarrow 1 = 4y^2 - 4x^2$

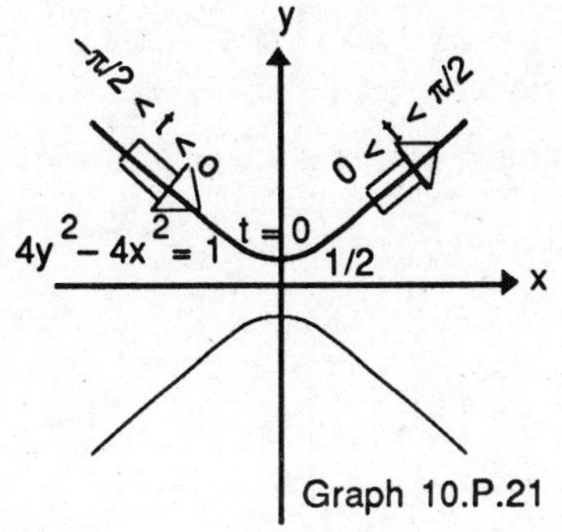

Graph 10.P.21

23. $x = -\cos t,\ y = \cos^2 t \Rightarrow y = (-x)^2 = x^2$

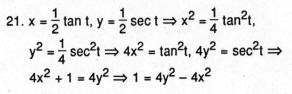

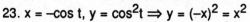

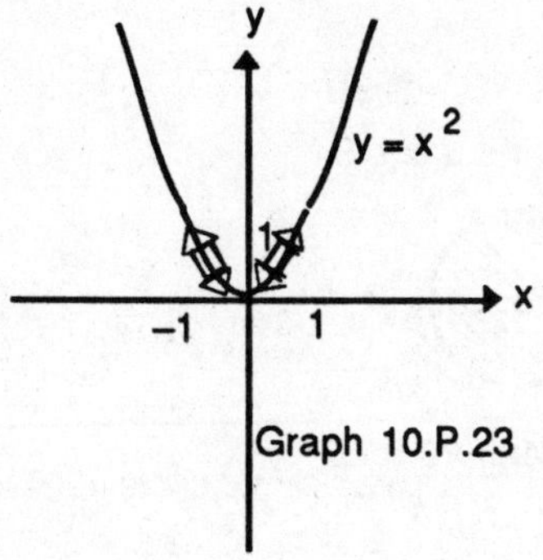

Graph 10.P.23

25. $16x^2 + 9y^2 = 144 \Rightarrow \frac{x^2}{9} + \frac{y^2}{16} = 1 \Rightarrow$
$a = 3,\ b = 4 \Rightarrow x = 3\cos t,\ y = 4\sin t,\ 0 \le t \le 2\pi$

27. $x = \frac{1}{2}\tan t,\ y = \frac{1}{2}\sec t \Rightarrow$

$$\frac{dy}{dx} = \frac{dy/dt}{dx/dt} = \frac{\frac{1}{2}\sec t\tan t}{\frac{1}{2}\sec^2 t} = \frac{\tan t}{\sec t} = \sin t$$

$\frac{dy}{dx}\left(\frac{\pi}{3}\right) = \sin\frac{\pi}{3} = \frac{\sqrt{3}}{2}$. $t = \frac{\pi}{3} \Rightarrow$

$x = \frac{1}{2}\tan\frac{\pi}{3} = \frac{\sqrt{3}}{2}$ and $y = \frac{1}{2}\sec t = 1 \Rightarrow$

$y = \frac{\sqrt{3}}{2}x + \frac{1}{4}$. $\frac{d^2y}{dx^2} = \frac{\cos t}{\frac{1}{2}\sec^2 t} = 2\cos^3 t \Rightarrow$

$$\frac{d^2y}{dx^2}\left(\frac{\pi}{3}\right) = 2\cos^3\frac{\pi}{3} = \frac{1}{4}$$

29. $x = e^{2t} - \frac{t}{8}$, $y = e^t$, $0 \le t \le \ln 2 \Rightarrow \frac{dx}{dt} = 2e^{2t} - \frac{1}{8}$, $\frac{dy}{dt} = e^t \Rightarrow$ Length $= \int_0^{\ln 2} \sqrt{\left(2e^{2t} - \frac{1}{8}\right)^2 + \left(e^t\right)^2}\, dt$

$= \int_0^{\ln 2} \sqrt{4e^{4t} + \frac{1}{2}e^{2t} + \frac{1}{16}}\, dt = \int_0^{\ln 2} \sqrt{\left(2e^{2t} + \frac{1}{8}\right)^2}\, dt = \int_0^{\ln 2} \left(2e^{2t} + \frac{1}{8}\right) dt =$

$\left[e^{2t} + \frac{t}{8}\right]_0^{\ln 2} = 3 + \frac{\ln 2}{8}$

31. $x = \frac{t^2}{2}$, $y = 2t$, $0 \le t \le \sqrt{5} \Rightarrow \frac{dx}{dt} = t$, $\frac{dy}{dt} = 2 \Rightarrow$ Area $= \int_0^{\sqrt{5}} 2\pi(2t)\sqrt{t^2 + 4}\, dt = \int_4^9 2\pi\, u^{1/2}\, du =$

$2\pi\left[\frac{2}{3}u^{3/2}\right]_4^9 = \frac{76\pi}{3}$ Let $u = t^2 + 4 \Rightarrow du = 2t\, dt$. $t = 0 \Rightarrow u = 4$, $t = \sqrt{5} \Rightarrow u = 9$

33. d 35. l 37. k 39. i

41. $r\cos\left(\theta - \frac{\pi}{3}\right) = 2\sqrt{3} \Rightarrow r\left(\cos\theta\cos\frac{\pi}{3} + \sin\theta\sin\frac{\pi}{3}\right) =$

$2\sqrt{3} \Rightarrow \frac{1}{2}r\cos\theta + \frac{\sqrt{3}}{2}r\sin\theta = 2\sqrt{3} \Rightarrow$

$r\cos\theta + \sqrt{3}\, r\sin\theta = 4\sqrt{3} \Rightarrow x + \sqrt{3}\, y = 4\sqrt{3}$

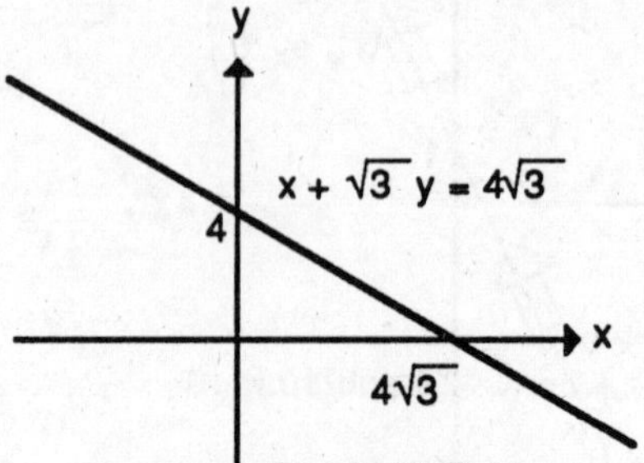

Graph 10.P.41

43. C: (0,1), r = 1

45.

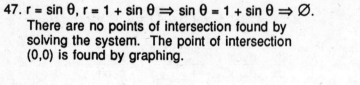

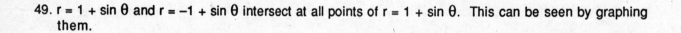

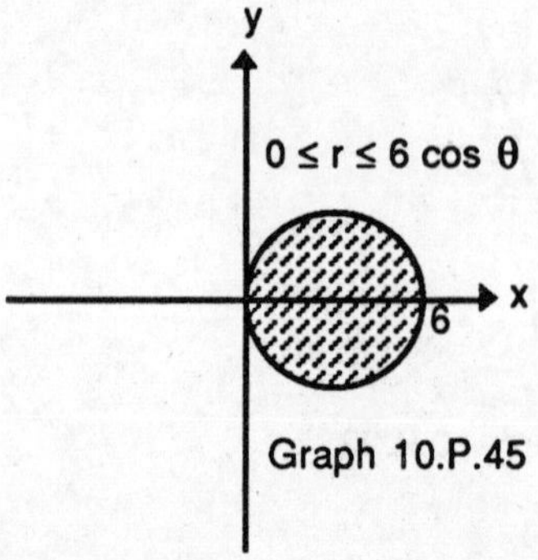

Graph 10.P.45

47. $r = \sin\theta$, $r = 1 + \sin\theta \Rightarrow \sin\theta = 1 + \sin\theta \Rightarrow \varnothing$.
There are no points of intersection found by solving the system. The point of intersection (0,0) is found by graphing.

49. $r = 1 + \sin\theta$ and $r = -1 + \sin\theta$ intersect at all points of $r = 1 + \sin\theta$. This can be seen by graphing them.

51. $r = \frac{2}{1 + \cos\theta} \Rightarrow e = 1 \Rightarrow$ Parabola

Vertex = (1,0)

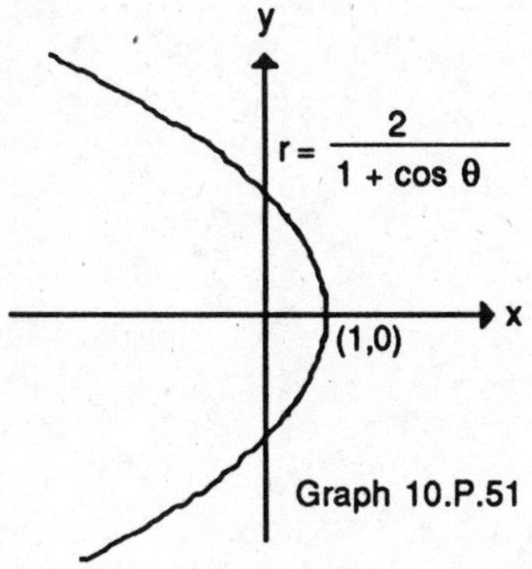


53. $r = \frac{6}{1 - 2\cos\theta} \Rightarrow e = 2 \Rightarrow$ Hyperbola

Ellipse. $ke = 4 \Rightarrow \frac{1}{2}k = 4 \Rightarrow k = 8.\ k = \frac{a}{e} - ea$

$ke = 6 \Rightarrow 2k = 6 \Rightarrow k = 3 \Rightarrow$ Vertices are $(2,\pi)$ and $(6,\pi)$.

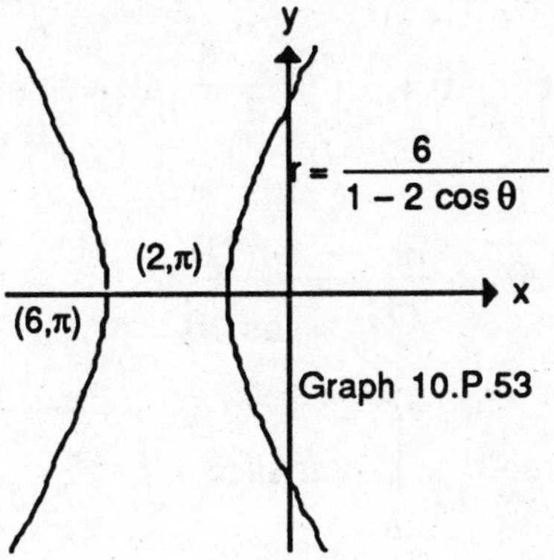


55. $e = 2,\ r\cos\theta = 2 \Rightarrow x = 2$ is directrix $\Rightarrow k = 2$ The conic is a hyperbola. $r = \frac{ke}{1 + e\cos\theta} \Rightarrow$

$r = \frac{2(2)}{1 + 2\cos\theta} \Rightarrow r = \frac{4}{1 + 2\cos\theta}$

57. $e = \frac{1}{2},\ r\sin\theta = 2 \Rightarrow y = 2$ is directrix $\Rightarrow k = 2$. The conic is an ellipse. $r = \frac{ke}{1 + e\sin\theta} \Rightarrow$

$r = \frac{2\left(\frac{1}{2}\right)}{1 + \frac{1}{2}\sin\theta} \Rightarrow r = \frac{2}{2 + \sin\theta}$

59. $r^2 = \cos 2\theta \Rightarrow r = 0$ when $\cos 2\theta = 0 \Rightarrow 2\theta = \frac{\pi}{2}, \frac{3\pi}{2} \Rightarrow \theta = \frac{\pi}{4}, \frac{3\pi}{4}$. $\theta_1 = \frac{\pi}{4} \Rightarrow m_1 = \tan\frac{\pi}{4} = 1 \Rightarrow$ $y = x$ is one tangent line. $\theta_2 = \frac{3\pi}{4} \Rightarrow m_2 = \tan\frac{3\pi}{4} = -1 \Rightarrow y = -x$ is other tangent line.

61. Tips of the petals are at $\theta = \frac{\pi}{4}, \frac{3\pi}{4}, \frac{5\pi}{4}, \frac{7\pi}{4}$, $r = 1$ at those values of $\pi$. Then for $\theta = \frac{\pi}{4}$, the line is $r\cos\left(\theta - \frac{\pi}{4}\right) = 1$; for $\theta = \frac{3\pi}{4}$, $r\cos\left(\theta - \frac{3\pi}{4}\right) = 1$; for $\theta = \frac{5\pi}{4}$, $r\cos\left(\theta - \frac{5\pi}{4}\right) = 1$; and for $\theta = \frac{7\pi}{4}$, $r\cos\left(\theta - \frac{7\pi}{4}\right) = 1$.

63. $A = 2\int_0^\pi \frac{1}{2}r^2\,d\theta = \int_0^\pi (2 - \cos\theta)^2\,d\theta = \int_0^\pi \left(4 - 2\cos\theta + \cos^2\theta\right)d\theta =$

$\int_0^\pi \left(4 - 2\cos\theta + \frac{1 + \cos 2\theta}{2}\right)d\theta = \int_0^\pi \left(\frac{9}{2} - 2\cos\theta + \frac{\cos 2\theta}{2}\right)d\theta = \left[\frac{9}{2}\theta - 2\sin\theta + \frac{\sin 2\theta}{4}\right]_0^\pi = \frac{9}{2}\pi$

65. $r = 1 + \cos 2\theta$, $r = 1 \Rightarrow 1 = 1 + \cos 2\theta \Rightarrow 0 = \cos 2\theta \Rightarrow 2\theta = \frac{\pi}{2} \Rightarrow \theta = \frac{\pi}{4}$.

$$\therefore\ A = 4\int_0^{\pi/4} \frac{1}{2}\left((1 + \cos 2\theta)^2 - 1^2\right) d\theta = 2\int_0^{\pi/4} (1 + 2\cos 2\theta + \cos^2 2\theta - 1)\, d\theta =$$

$$2\int_0^{\pi/4}\left(2\cos 2\theta + \frac{1}{2} + \frac{\cos 4\theta}{2}\right) d\theta = 2\left[\sin 2\theta + \frac{1}{2}\theta + \frac{\sin 4\theta}{8}\right]_0^{\pi/4} = 2 + \frac{\pi}{4}$$

67. $r = \sqrt{\cos 2\theta} \Rightarrow \dfrac{dr}{d\theta} = \dfrac{-\sin 2\theta}{\sqrt{\cos 2\theta}}$

$$\text{Surface Area} = \int_0^{\pi/4} 2\pi r \sin\theta \sqrt{r^2 + \left(\frac{dr}{d\theta}\right)^2}\, d\theta =$$

$$\int_0^{\pi/4} 2\pi\sqrt{\cos 2\theta}\, \sin\theta \sqrt{\left(\sqrt{\cos 2\theta}\right)^2 + \left(\frac{-\sin 2\theta}{\sqrt{\cos 2\theta}}\right)^2}\, d\theta =$$

$$\int_0^{\pi/4} 2\pi\sqrt{\cos 2\theta}\, \sin\theta \sqrt{\cos 2\theta + \frac{\sin^2 2\theta}{\cos 2\theta}}\, d\theta = \int_0^{\pi/4} 2\pi\sqrt{\cos 2\theta}\, \sin\theta \sqrt{\frac{1}{\cos 2\theta}}\, d\theta =$$

$$\int_0^{\pi/4} 2\pi \sin\theta\, d\theta = \left[2\pi(-\cos\theta)\right]_0^{\pi/4} = 2\pi\left(1 - \frac{\sqrt{2}}{2}\right)$$

69. $r = -1 + \cos\theta \Rightarrow \dfrac{dr}{d\theta} = -\sin\theta$

$$\text{Length} = \int_0^{2\pi} \sqrt{(-1 + \cos\theta) + (-\sin\theta)^2}\, d\theta = \int_0^{2\pi} \sqrt{2 - 2\cos\theta}\, d\theta =$$

$$\int_0^{2\pi} \sqrt{\frac{4(1 - \cos\theta)}{2}}\, d\theta = \int_0^{2\pi} 2\sin\frac{1}{2}\theta\, d\theta = \left[-4\cos\frac{1}{2}\theta\right]_0^{2\pi} = 8$$

71. $r = \cos^3\left(\frac{\theta}{3}\right) \Rightarrow \frac{dr}{d\theta} = -\cos^2\left(\frac{\theta}{3}\right)\sin\left(\frac{\theta}{3}\right)$

$$\text{Length} = \int_0^{\pi/4} \sqrt{\left(\cos^2\left(\frac{\theta}{3}\right)\right)^2 + \left(-\cos^2\left(\frac{\theta}{3}\right)\sin\left(\frac{\theta}{3}\right)\right)^2}\; d\theta = \int_0^{\pi/4} \sqrt{\cos^6\left(\frac{\theta}{3}\right) + \cos^4\left(\frac{\theta}{3}\right)\sin^2\left(\frac{\theta}{3}\right)}\; d\theta$$

$$= \int_0^{\pi/4} \sqrt{\cos^4\left(\frac{\theta}{3}\right)}\; d\theta = \int_0^{\pi/4} \cos^2\left(\frac{\theta}{3}\right) d\theta = \int_0^{\pi/4} \frac{1 + \cos\left(\frac{2\theta}{3}\right)}{2}\, d\theta = \left[\frac{\theta + \frac{3}{2}\sin\left(\frac{2\theta}{3}\right)}{2}\right]_0^{\pi/4} = \frac{\pi + 3}{8}$$

73. a) $r_{av} = \frac{1}{2\pi - 0}\int_0^{2\pi} a(1 - \cos\theta)\, d\theta = \frac{1}{2\pi}\left[a(\theta - \sin\theta)\right]_0^{2\pi} = a$

b) $r_{av} = \frac{1}{2\pi - 0}\int_0^{2\pi} a\, d\theta = \frac{1}{2\pi}\left[a\theta\right]_0^{2\pi} = a$

c) $r_{av} = \frac{1}{\frac{\pi}{2} - \left(-\frac{\pi}{2}\right)}\int_{-\pi/2}^{\pi/2} a\cos\theta\, d\theta = \frac{1}{\pi}\left[a\sin\theta\right]_{-\pi/2}^{\pi/2} = \frac{2a}{\pi}$

75. $r_{max} + r_{min} = 2a$. $r_{max} - r_{min} = a + ea - (a - ea) = 2ea$. $\therefore \frac{r_{max} - r_{min}}{r_{max} + r_{min}} = \frac{2ea}{2a} = e$, the eccentricity.

77. Length of $r = f(\theta) = \int_\alpha^\beta \sqrt{\left(f(\theta)\right)^2 + \left(f'(\theta)\right)^2}\; d\theta$.

Length of $r = 2f(\theta) = \int_\alpha^\beta \sqrt{\left(2f(\theta)\right)^2 + \left(2f'(\theta)\right)^2}\; d\theta = \int_\alpha^\beta 2\sqrt{\left(f(\theta)\right)^2 + \left(f'(\theta)\right)^2}\; d\theta =$

$2(\text{Length of } r = f(\theta))$

# CHAPTER 11

# VECTORS AND ANALYTIC GEOMETRY IN SPACE

## 11.1 VECTORS IN THE PLANE

1.

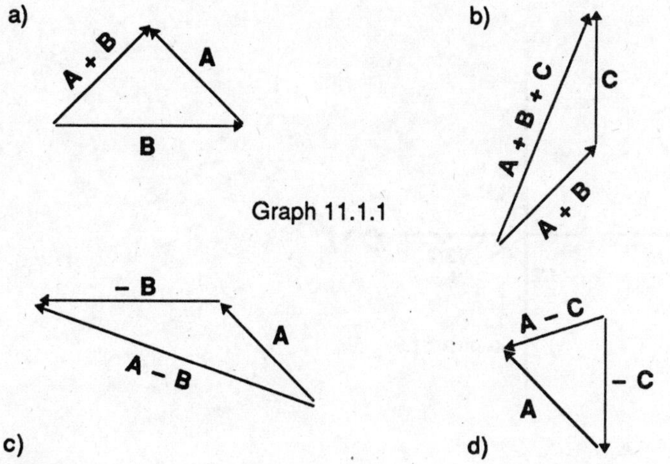


Graph 11.1.1

3. $\overrightarrow{P_1P_2} = \mathbf{i} - 4\mathbf{j}$

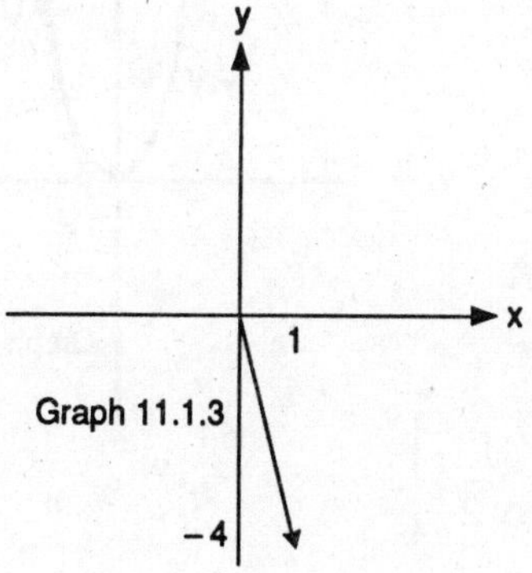


Graph 11.1.3

5. $\overrightarrow{AO} = -2\mathbf{i} - 3\mathbf{j}$

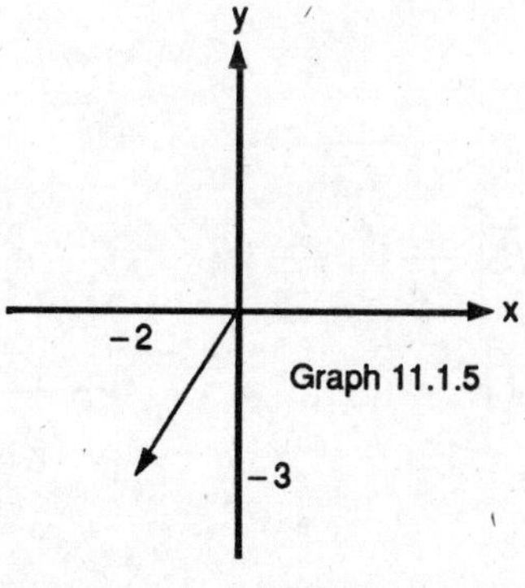


Graph 11.1.5

7. $\mathbf{U} = \frac{\sqrt{3}}{2}\mathbf{i} + \frac{1}{2}\mathbf{j}$, when $\theta = \frac{\pi}{6}$

$\mathbf{U} = -\frac{1}{2}\mathbf{i} + \frac{\sqrt{3}}{2}\mathbf{j}$, when $\theta = \frac{2\pi}{3}$

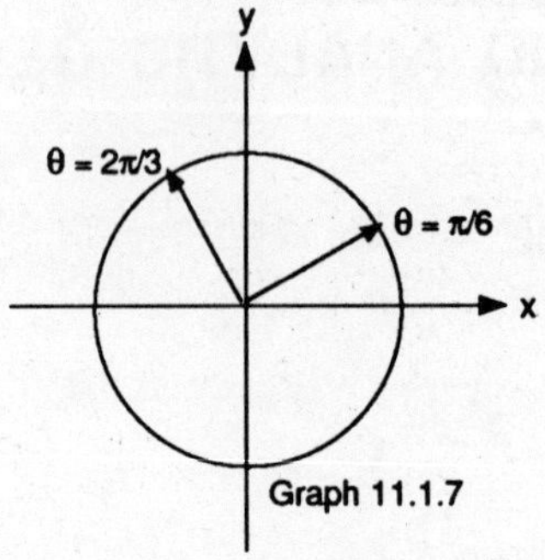

Graph 11.1.7

9. $\mathbf{U} = \frac{\sqrt{3}}{2}\mathbf{i} - \frac{1}{2}\mathbf{j}$

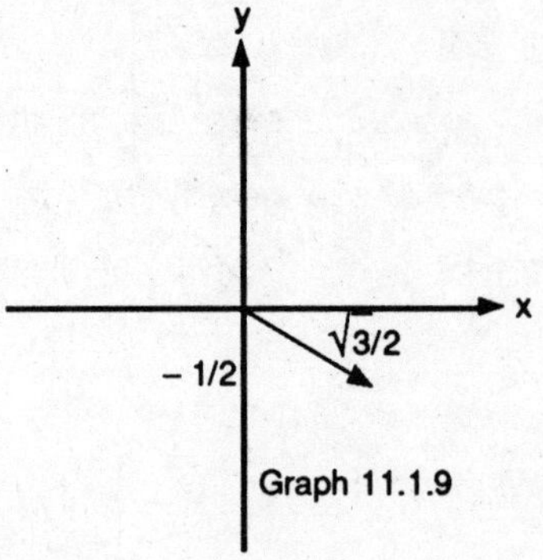

Graph 11.1.9

11. $\mathbf{U} = \frac{1}{\sqrt{17}}\mathbf{i} + \frac{4}{\sqrt{17}}\mathbf{j}, -\mathbf{U} = -\frac{1}{\sqrt{17}}\mathbf{i} - \frac{4}{\sqrt{17}}\mathbf{j},$

$\mathbf{V} = \frac{4}{\sqrt{17}}\mathbf{i} - \frac{1}{\sqrt{17}}\mathbf{j}, -\mathbf{V} = -\frac{4}{\sqrt{17}}\mathbf{i} + \frac{1}{\sqrt{17}}\mathbf{j}$

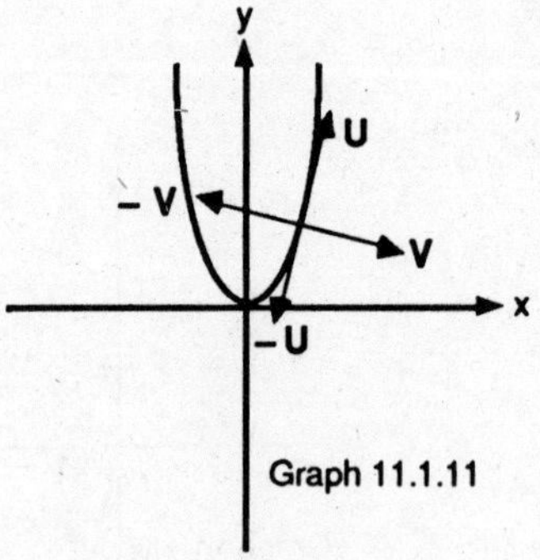

Graph 11.1.11

13. $|\mathbf{i} + \mathbf{j}| = \sqrt{1^2 + 1^2} = \sqrt{2}, \mathbf{i} + \mathbf{j} = \sqrt{2}\left[\frac{1}{\sqrt{2}}\mathbf{i} + \frac{1}{\sqrt{2}}\mathbf{j}\right]$

15. $\left|\sqrt{3}\,\mathbf{i} + \mathbf{j}\right| = \sqrt{\left(\sqrt{3}\right)^2 + 1^2} = 2, 2\left[\frac{\sqrt{3}}{2}\mathbf{i} + \frac{1}{2}\mathbf{j}\right]$

17. $|5\mathbf{i} + 12\mathbf{j}| = 13, 13\left[\frac{5}{13}\mathbf{i} + \frac{12}{13}\mathbf{j}\right]$

19. $|\mathbf{A}| = |3\mathbf{i} + 6\mathbf{j}| = \sqrt{3^2 + 6^2} = 3\sqrt{5} \Rightarrow \mathbf{A} = 3\sqrt{5}\left[\frac{1}{\sqrt{5}}\mathbf{i} + \frac{2}{\sqrt{5}}\mathbf{j}\right];$

$|\mathbf{B}| = |-\mathbf{i} - 2\mathbf{j}| = \sqrt{5} \Rightarrow \mathbf{B} = \sqrt{5}\left[-\frac{1}{\sqrt{5}}\mathbf{i} - \frac{2}{\sqrt{5}}\mathbf{j}\right]$

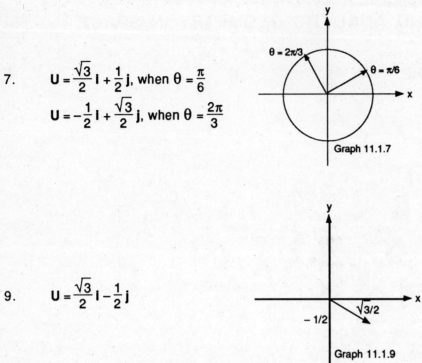

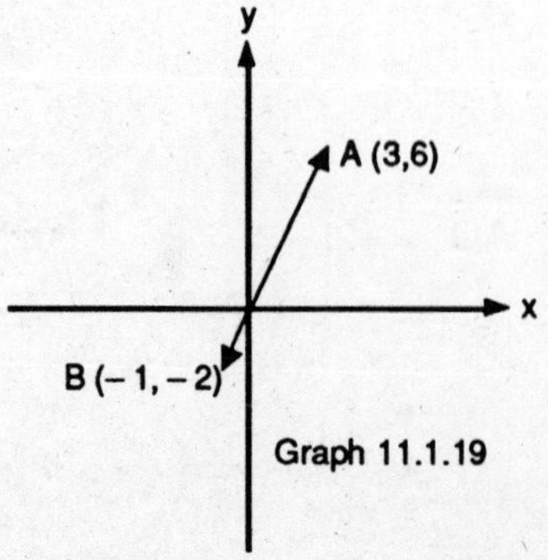

Graph 11.1.19

## 11.2 CARTESIAN (RECTANGULAR) COORDINATES AND VECTORS IN SPACE

1. a line through the point (2,3,0) parallel to the z–axis

3. the x–axis

5. the circle, $x^2 + y^2 = 4$ in the xy–plane

7. the circle, $x^2 + z^2 = 4$ in the xz–plane

9. the circle, $y^2 + z^2 = 1$ in the yz–plane

11. the circle, $x^2 + y^2 = 16$ in the xy–plane

13. a) the first quadrant of the xy–plane b) the fourth quadrant of the xy–plane

15. a) a solid sphere of radius 1 centered at the origin
    b) all points which are greater than 1 unit from the origin

17. a) the upper hemisphere of radius 1 centered at the origin
    b) the solid upper hemisphere of raduis 1 centered at the origin

19. a) $x = 3$ b) $y = -1$ c) $z = -2$

21. a) $z = 1$ b) $x = 3$ c) $y = -1$

23. a) $x^2 + (y-2)^2 = 4$ b) $(y-2)^2 + z^2 = 4$ c) $x^2 + z^2 = 4$

25. a) $y = 3, z = -1$ b) $x = 1, z = -1$ c) $x = 1, y = 3$

27. $x^2 + y^2 + z^2 = 25,\ z = 3$

29. $0 \le z \le 1$

31. $z \le 0$

33. a) $(x-1)^2 + (y-1)^2 + (z-1)^2 < 1$ b) $(x-1)^2 + (y-1)^2 + (z-1)^2 > 1$

35. length $= |2\mathbf{i} + \mathbf{j} - 2\mathbf{k}| = \sqrt{2^2 + 1^2 + (-2)^2} = 3$, $2\mathbf{i} + \mathbf{j} - 2\mathbf{k} = 3\left[\frac{2}{3}\mathbf{i} + \frac{1}{3}\mathbf{j} - \frac{2}{3}\mathbf{k}\right] \Rightarrow$
    the direction is $\frac{2}{3}\mathbf{i} + \frac{1}{3}\mathbf{j} - \frac{2}{3}\mathbf{k}$

37. length $= |\mathbf{i} + 4\mathbf{j} - 8\mathbf{k}| = \sqrt{1 + 16 + 64} = 9$, $\mathbf{i} + 4\mathbf{j} - 8\mathbf{k} = 9\left[\frac{1}{9}\mathbf{i} + \frac{4}{9}\mathbf{j} - \frac{8}{9}\mathbf{k}\right] \Rightarrow$
    the direction is $\frac{1}{9}\mathbf{i} + \frac{4}{9}\mathbf{j} - \frac{8}{9}\mathbf{k}$

39. length $= |5\mathbf{k}| = \sqrt{25} = 5$, $5\mathbf{k} = 5[\mathbf{k}] \Rightarrow$ the direction is $\mathbf{k}$

41. length $= |-4\mathbf{j}| = \sqrt{(-4)^2} = \sqrt{16} = 4$, $4[-\mathbf{j}] \Rightarrow$ the direction is $-\mathbf{j}$

43. length $= \left|-\frac{1}{3}\mathbf{j} + \frac{1}{4}\mathbf{k}\right| = \sqrt{\left(-\frac{1}{3}\right)^2 + \left(\frac{1}{4}\right)^2} = \frac{5}{12}$, $-\frac{1}{3}\mathbf{j} + \frac{1}{4}\mathbf{k} = \frac{5}{12}\left[-\frac{4}{5}\mathbf{j} + \frac{3}{5}\mathbf{k}\right] \Rightarrow$
    the direction is $-\frac{4}{5}\mathbf{j} + \frac{3}{5}\mathbf{k}$

45. length $= \left|\frac{1}{\sqrt{6}}\mathbf{i} - \frac{1}{\sqrt{6}}\mathbf{j} - \frac{1}{\sqrt{6}}\mathbf{k}\right| = \sqrt{3\left(\frac{1}{\sqrt{6}}\right)^2} = \sqrt{\frac{1}{2}}$, $\frac{1}{\sqrt{6}}\mathbf{i} - \frac{1}{\sqrt{6}}\mathbf{j} - \frac{1}{\sqrt{6}}\mathbf{k} =$
    $\sqrt{\frac{1}{2}}\left[\frac{1}{\sqrt{3}}\mathbf{i} - \frac{1}{\sqrt{3}}\mathbf{j} - \frac{1}{\sqrt{3}}\mathbf{k}\right] \Rightarrow$ the direction is $\frac{1}{\sqrt{3}}\mathbf{i} - \frac{1}{\sqrt{3}}\mathbf{j} - \frac{1}{\sqrt{3}}\mathbf{k}$

47. the length $= \left|\overrightarrow{P_1P_2}\right| = |2\mathbf{i} + 2\mathbf{j} - \mathbf{k}| = \sqrt{2^2 + 2^2 + (-1)^2} = 3$, $2\mathbf{i} + 2\mathbf{j} - \mathbf{k} = 3\left[\frac{2}{3}\mathbf{i} + \frac{2}{3}\mathbf{j} - \frac{1}{3}\mathbf{k}\right] \Rightarrow$
the direction is $\frac{2}{3}\mathbf{i} + \frac{2}{3}\mathbf{j} - \frac{1}{3}\mathbf{k}$

49. the length $= \left|\overrightarrow{P_1P_2}\right| = |(3\mathbf{i} - 6\mathbf{j} + 2\mathbf{k}) = \sqrt{9 + 36 + 4} = 7$, $3\mathbf{i} - 6\mathbf{j} + 2\mathbf{k} = 7\left[\frac{3}{7}\mathbf{i} - \frac{6}{7}\mathbf{j} + \frac{2}{7}\mathbf{k}\right] \Rightarrow$
the direction is $\frac{3}{7}\mathbf{i} - \frac{6}{7}\mathbf{j} + \frac{2}{7}\mathbf{k}$

51. the length $= \left|\overrightarrow{P_1P_2}\right| = |2\mathbf{i} - 2\mathbf{j} - 2\mathbf{k}| = \sqrt{3 \cdot 2^2} = 2\sqrt{3}$, $2\mathbf{i} - 2\mathbf{j} - 2\mathbf{k} = 2\sqrt{3}\left[\frac{1}{\sqrt{3}}\mathbf{i} - \frac{1}{\sqrt{3}}\mathbf{j} - \frac{1}{\sqrt{3}}\mathbf{k}\right] \Rightarrow$
the direction is $\frac{1}{\sqrt{3}}\mathbf{i} - \frac{1}{\sqrt{3}}\mathbf{j} - \frac{1}{\sqrt{3}}\mathbf{k}$

53. a) $2\mathbf{i}$ b) $4\mathbf{j}$ c) $\sqrt{3}\mathbf{k}$
d) $\frac{3}{10}\mathbf{j} + \frac{2}{5}\mathbf{k}$ e) $6\mathbf{i} + 2\mathbf{j} + 3\mathbf{k}$ f) $au_1\mathbf{i} + au_2\mathbf{j} + au_3\mathbf{k}$

55. $|\mathbf{A}| = |\mathbf{i} + \mathbf{j} + \mathbf{k}| = \sqrt{3}$, $\mathbf{A} = \sqrt{3}\left[\frac{1}{\sqrt{3}}\mathbf{i} + \frac{1}{\sqrt{3}}\mathbf{j} + \frac{1}{\sqrt{3}}\mathbf{k}\right] \Rightarrow$ the desired vector is $5\left[\frac{1}{\sqrt{3}}\mathbf{i} + \frac{1}{\sqrt{3}}\mathbf{j} + \frac{1}{\sqrt{3}}\mathbf{k}\right]$

57. a) center $(-2,0,2)$, radius $2\sqrt{2}$ b) center $(-1/2,-1/2,-1/2)$, $\frac{\sqrt{21}}{2}$
c) center $(\sqrt{2},\sqrt{2},-\sqrt{2})$, radius $\sqrt{2}$ d) center $(0,-1/3,1/3)$, radius $\frac{\sqrt{29}}{3}$

59. a) $\sqrt{y^2 + z^2}$ b) $\sqrt{x^2 + z^2}$ c) $\sqrt{x^2 + y^2}$

## 11.3 DOT PRODUCTS

| | $\mathbf{A \cdot B}$ | $\lvert\mathbf{A}\rvert$ | $\lvert\mathbf{B}\rvert$ | $\cos\theta$ | $\lvert\mathbf{B}\rvert\cos\theta$ | $\text{Proj}_\mathbf{A}\,\mathbf{B}$ |
|---|---|---|---|---|---|---|
| 1. | 10 | $\sqrt{13}$ | $\sqrt{26}$ | $\frac{10}{13\sqrt{2}}$ | $\frac{10}{\sqrt{13}}$ | $\frac{10}{13}[3\mathbf{i} + 2\mathbf{j}]$ |
| 3. | 4 | $\sqrt{14}$ | 2 | $\frac{2}{\sqrt{14}}$ | $\frac{4}{\sqrt{14}}$ | $\frac{2}{7}[3\mathbf{i} - 2\mathbf{j} - \mathbf{k}]$ |
| 5. | 2 | $\sqrt{34}$ | $\sqrt{3}$ | $\frac{2}{\sqrt{3}\sqrt{34}}$ | $\frac{2}{\sqrt{34}}$ | $\frac{1}{17}[5\mathbf{j} - 3\mathbf{k}]$ |
| 7. | $\sqrt{3} - \sqrt{2}$ | $\sqrt{2}$ | 3 | $\frac{\sqrt{3} - \sqrt{2}}{3\sqrt{2}}$ | $\frac{\sqrt{3} - \sqrt{2}}{\sqrt{2}}$ | $\frac{\sqrt{3} - \sqrt{2}}{2}[-\mathbf{i} + \mathbf{j}]$ |
| 9. | $-25$ | 5 | 5 | $-1$ | $-5$ | $-2\mathbf{i} + 4\mathbf{j} - \sqrt{5}\mathbf{k}$ |
| 11. | 25 | 15 | 5 | $\frac{1}{3}$ | $\frac{5}{3}$ | $\frac{1}{9}[10\mathbf{i} + 11\mathbf{j} - 2\mathbf{k}]$ |

13. $\mathbf{B} = \left(\frac{\mathbf{A \cdot B}}{\mathbf{A \cdot A}}\mathbf{A}\right) + \left(\mathbf{B} - \frac{\mathbf{A \cdot B}}{\mathbf{A \cdot A}}\mathbf{A}\right) = \frac{3}{2}[\mathbf{i} + \mathbf{j}] + \left[(3\mathbf{j} + 4\mathbf{k}) - \frac{3}{2}(\mathbf{i} + \mathbf{j})\right] = \left[\frac{3}{2}\mathbf{i} + \frac{3}{2}\mathbf{j}\right] + \left[-\frac{3}{2}\mathbf{i} + \frac{3}{2}\mathbf{j} + 4\mathbf{k}\right]$,
where $\mathbf{A \cdot B} = 3$ and $\mathbf{A \cdot A} = 2$

15. $\mathbf{B} = \left(\frac{\mathbf{A}\cdot\mathbf{B}}{\mathbf{A}\cdot\mathbf{A}}\mathbf{A}\right) + \left(\mathbf{B} - \frac{\mathbf{A}\cdot\mathbf{B}}{\mathbf{A}\cdot\mathbf{A}}\mathbf{A}\right) = \frac{14}{3}[\mathbf{i} + 2\mathbf{j} - \mathbf{k}] + \left[(8\mathbf{i} + 4\mathbf{j} - 12\mathbf{k}) - \left(\frac{14}{3}\mathbf{i} + \frac{28}{3}\mathbf{j} - \frac{14}{3}\mathbf{k}\right)\right] =$
$\left[\frac{14}{3}\mathbf{i} + \frac{28}{3}\mathbf{j} - \frac{14}{3}\mathbf{k}\right] + \left[\frac{10}{3}\mathbf{i} - \frac{16}{3}\mathbf{j} - \frac{22}{3}\mathbf{k}\right]$, where $\mathbf{A}\cdot\mathbf{B} = 28$ and $\mathbf{A}\cdot\mathbf{A} = 6$

17. $(\mathbf{i} + 2\mathbf{j}) \cdot \big((x-2)\mathbf{i} + (y-1)\mathbf{j}\big) = 0 \Rightarrow x + 2y = 4$

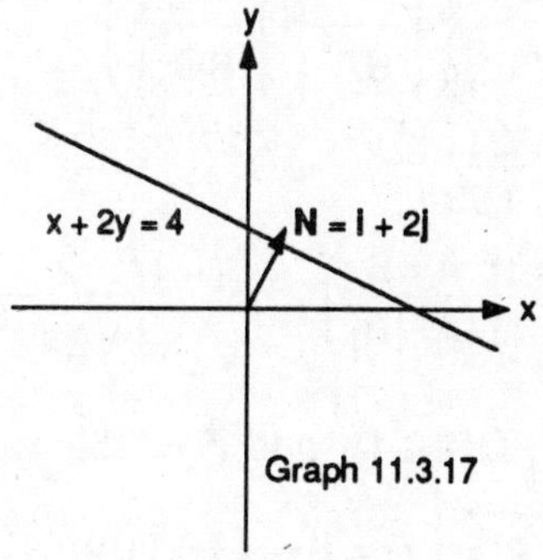

Graph 11.3.17

19. $(-2\mathbf{i} - \mathbf{j}) \cdot \big((x+1)\mathbf{i} + (y-2)\mathbf{j}\big) = 0 \Rightarrow -2x - y = 0$

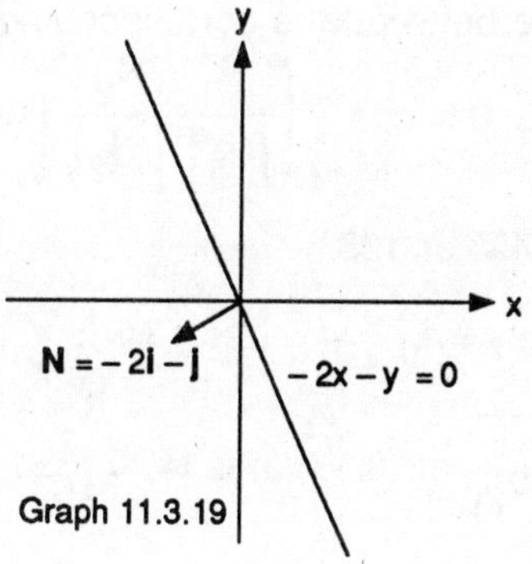

Graph 11.3.19

21. distance $= \left|\text{proj}_{\mathbf{N}}\overrightarrow{PS}\right| = \left|\frac{\mathbf{N}\cdot\overrightarrow{PS}}{|\mathbf{N}|}\right| = \left|\frac{(\mathbf{i}+3\mathbf{j})\cdot(2\mathbf{i}+6\mathbf{j})}{\sqrt{1^2+3^2}}\right| = \left|\frac{20}{\sqrt{10}}\right| = 2\sqrt{10}$, where S(2,8), P(0,2)

and $\mathbf{N} + \mathbf{i} + 3\mathbf{j}$

23. distance $= \left|\text{proj}_{\mathbf{N}}\overrightarrow{PS}\right| = \left|\frac{\mathbf{N}\cdot\overrightarrow{PS}}{|\mathbf{N}|}\right| = \left|\frac{(\mathbf{i}+\mathbf{j})\cdot(\mathbf{i}+\mathbf{j})}{\sqrt{1+1}}\right| = \sqrt{2}$, where S(2,1), P(1,0) and $\mathbf{N} = \mathbf{i} + \mathbf{j}$

25. $\mathbf{A}\cdot\mathbf{B} = \left(\frac{1}{\sqrt3}\right)(0) + \left(-\frac{1}{\sqrt3}\right)\left(\frac{1}{\sqrt2}\right) + \left(\frac{1}{\sqrt3}\right)\left(\frac{1}{\sqrt2}\right) = 0$, $\mathbf{A}\cdot\mathbf{C} = \left(\frac{1}{\sqrt3}\right)\left(-\frac{2}{\sqrt6}\right) + \left(-\frac{1}{\sqrt3}\right)\left(-\frac{1}{\sqrt6}\right) +$
$\left(\frac{1}{\sqrt3}\right)\left(\frac{1}{\sqrt6}\right) = 0$ and $\mathbf{B}\cdot\mathbf{C} = (0)\left(-\frac{2}{\sqrt6}\right) + \left(\frac{1}{\sqrt2}\right)\left(-\frac{1}{\sqrt6}\right) + \left(\frac{1}{\sqrt2}\right)\left(\frac{1}{\sqrt6}\right) = 0 \Rightarrow$ that **A**, **B** and **C** are orthogonal

27. If $\mathbf{A} = (\mathbf{i} + \mathbf{j} + \mathbf{k})$, $\mathbf{B}_1 = \mathbf{i}$, and $\mathbf{B}_2 = \mathbf{j}$, then $\mathbf{A}\cdot\mathbf{B}_1 = \mathbf{A}\cdot\mathbf{B}_2$, but $\mathbf{B}_1 \neq \mathbf{B}_2$

29. $\overrightarrow{AB} = 3\mathbf{i} + \mathbf{j} - 3\mathbf{k}$, $\overrightarrow{AC} = 2\mathbf{i} - 2\mathbf{j}$, $\overrightarrow{BA} = -3\mathbf{i} - \mathbf{j} + 3\mathbf{k}$, $\overrightarrow{CA} = -2\mathbf{i} + 2\mathbf{j}$, $\overrightarrow{CB} = \mathbf{i} + 3\mathbf{j} - 3\mathbf{k}$,

$\overrightarrow{BC} = -\mathbf{i} - 3\mathbf{j} + 3\mathbf{k} \Rightarrow \angle A = \cos^{-1}\left(\frac{\overrightarrow{AB} \cdot \overrightarrow{AC}}{\left|\overrightarrow{AB}\right|\left|\overrightarrow{AC}\right|}\right) = \cos^{-1}\left(\frac{4}{\sqrt{152}}\right) \approx 71.1°$,

$\angle B = \cos^{-1}\left(\frac{\overrightarrow{BA} \cdot \overrightarrow{BC}}{\left|\overrightarrow{BA}\right|\left|\overrightarrow{BC}\right|}\right) = \cos^{-1}\left(\frac{15}{19}\right) \approx 37.9°$, $\angle C = \cos^{-1}\left(\frac{\overrightarrow{CA} \cdot \overrightarrow{CB}}{\left|\overrightarrow{CA}\right|\left|\overrightarrow{CB}\right|}\right) =$

$\cos^{-1}\left(\frac{4}{\sqrt{152}}\right) \approx 71.1°$

31. $\theta = \cos^{-1}\left(\frac{\mathbf{A} \cdot \mathbf{B}}{|\mathbf{A}|\,|\mathbf{B}|}\right) = \cos^{-1}\left(\frac{2}{\sqrt{2}\sqrt{3}}\right) \approx 35.3°$, where $\mathbf{A} = \mathbf{i} + \mathbf{k}$ and $\mathbf{B} = \mathbf{i} + \mathbf{j} + \mathbf{k}$

33. $P(0,0,0)$, $Q(1,1,1)$ and $\mathbf{F} = -5\mathbf{k} \Rightarrow \overrightarrow{PQ} = \mathbf{i} + \mathbf{j} + \mathbf{k}$, $W = \mathbf{F} \cdot \overrightarrow{PQ} = (-5\mathbf{k}) \cdot (\mathbf{i} + \mathbf{j} + \mathbf{k}) = -5\ \mathrm{N \cdot m}$

35. $W = |\mathbf{F}|\left|\overrightarrow{PQ}\right| \cos\theta = (200)(20)\left(\cos\frac{\pi}{6}\right) = 20000\sqrt{3} \approx 3464.10\ \mathrm{N \cdot m}$

37. The angle between the corresponding normals is equal to the angle between the corresponding tangents. $\theta = \cos^{-1}\left(\frac{\mathbf{N}_1 \cdot \mathbf{N}_2}{|\mathbf{N}_1|\,|\mathbf{N}_2|}\right) = \cos^{-1}\left(\frac{1}{\sqrt{2}}\right) = 45°$, where $\mathbf{N}_1 = 3\mathbf{i} + \mathbf{j}$, $\mathbf{N}_2 = 2\mathbf{i} - \mathbf{j}$; the angle is either 45° or 135°

39. The curve $y = \sqrt{\frac{3}{4} + x}$ has $y = \frac{1}{\sqrt{3}}x + \frac{\sqrt{3}}{2}$ as its tangent line at $(0, \sqrt{3}/2)$. The normal vector for $y = \sqrt{\frac{3}{4} + x}$ at $(0, \sqrt{3}/2)$ is $\mathbf{N}_1 = \frac{1}{\sqrt{3}}\mathbf{i} - \mathbf{j}$. The curve $y = \sqrt{\frac{3}{4} - x}$ has $y = -\frac{1}{\sqrt{3}}x + \frac{\sqrt{3}}{2}$ as its tangent line at $(0, \sqrt{3}/2)$. The normal vector for $y = \sqrt{\frac{3}{4} + x}$ at $(0, \sqrt{3}/2)$ is $\mathbf{N}_2 = -\frac{1}{\sqrt{3}}\mathbf{i} - \mathbf{j}$.

$\theta = \cos^{-1}\left(\frac{\mathbf{N}_1 \cdot \mathbf{N}_2}{|\mathbf{N}_1|\,|\mathbf{N}_2|}\right) = \cos^{-1}\left(\frac{1}{2}\right) = \frac{\pi}{3}$; the angle is either $\frac{\pi}{3}$ or $\frac{2\pi}{3}$. See exercise 37 for additional details.

41. The points of intersection for the curves $y = x^2$ and $y = \sqrt[3]{x}$ are (0,0) and (1,1). At (0,0) the tangent line for $y = x^2$ is $y = 0$ and the tangent line for $y = \sqrt{x}$ is $x = 0$. Therefore, the angle of intersection at (0,0) is $\frac{\pi}{2}$. At (1,1) the tangent line for $y = x^2$ is $y = 2x - 1$ and the tangent line for $y = \sqrt[3]{x}$ is $y = \frac{1}{3}x + \frac{2}{3}$. The corresponding normal vectors are: $\mathbf{N}_1 = 2\mathbf{i} - \mathbf{j}$, $\mathbf{N}_2 = \frac{1}{3}\mathbf{i} - \mathbf{j}$.

$\theta = \cos^{-1}\left(\frac{\mathbf{N}_1 \cdot \mathbf{N}_2}{|\mathbf{N}_1|\,|\mathbf{N}_2|}\right) = \cos^{-1}\frac{1}{\sqrt{2}} = \frac{\pi}{4}$; the angle is either 45° or 135°

## 11.4 CROSS PRODUCTS

1. $\mathbf{A} \times \mathbf{B} = \begin{vmatrix} \mathbf{i} & \mathbf{j} & \mathbf{k} \\ 2 & -2 & -1 \\ 1 & 0 & -1 \end{vmatrix} = 3\left[\frac{2}{3}\mathbf{i} + \frac{1}{3}\mathbf{j} + \frac{2}{3}\mathbf{k}\right] \Rightarrow$ length = 3 and the direction is $\frac{2}{3}\mathbf{i} + \frac{1}{3}\mathbf{j} + \frac{2}{3}\mathbf{k}$

$\mathbf{B} \times \mathbf{A} = \begin{vmatrix} \mathbf{i} & \mathbf{j} & \mathbf{k} \\ 1 & 0 & -1 \\ 2 & -2 & -1 \end{vmatrix} = -3\left[\frac{2}{3}\mathbf{i} + \frac{1}{3}\mathbf{j} + \frac{2}{3}\mathbf{k}\right] \Rightarrow$ length = 3 and the direction is $-\frac{2}{3}\mathbf{i} - \frac{1}{3}\mathbf{j} - \frac{2}{3}\mathbf{k}$

3. $\mathbf{A} \times \mathbf{B} = \begin{vmatrix} \mathbf{i} & \mathbf{j} & \mathbf{k} \\ 2 & -2 & 4 \\ -1 & 1 & -2 \end{vmatrix} = \mathbf{0} \Rightarrow$ length = 0 and has no direction

$\mathbf{B} \times \mathbf{A} = \begin{vmatrix} \mathbf{i} & \mathbf{j} & \mathbf{k} \\ -1 & 1 & -2 \\ 2 & -2 & 4 \end{vmatrix} = \mathbf{0} \Rightarrow$ length = 0 and has no direction

5. $\mathbf{A} \times \mathbf{B} = \begin{vmatrix} \mathbf{i} & \mathbf{j} & \mathbf{k} \\ 2 & 0 & 0 \\ 0 & -3 & 0 \end{vmatrix} = -6\,[\mathbf{k}] \Rightarrow$ length = 6 and the direction is $-\mathbf{k}$

$\mathbf{B} \times \mathbf{A} = \begin{vmatrix} \mathbf{i} & \mathbf{j} & \mathbf{k} \\ 0 & -3 & 0 \\ 2 & 0 & 0 \end{vmatrix} = 6\,[\mathbf{k}] \Rightarrow$ length = 6 and the direction is $\mathbf{k}$

7. $\mathbf{A} \times \mathbf{B} = \begin{vmatrix} \mathbf{i} & \mathbf{j} & \mathbf{k} \\ -8 & -2 & -4 \\ 2 & 2 & 1 \end{vmatrix} = [6\mathbf{i} - 12\mathbf{k}] \Rightarrow$ length = $6\sqrt{5}$ and the direction is $\frac{1}{\sqrt{5}}\mathbf{i} - \frac{2}{\sqrt{5}}\mathbf{k}$

$\mathbf{B} \times \mathbf{A} = \begin{vmatrix} \mathbf{i} & \mathbf{j} & \mathbf{k} \\ 2 & 2 & 1 \\ -8 & -2 & -4 \end{vmatrix} = -[6\mathbf{i} - 12\mathbf{k}] \Rightarrow$ length = $6\sqrt{5}$ and the direction is $-\frac{1}{\sqrt{5}}\mathbf{i} + \frac{2}{\sqrt{5}}\mathbf{k}$

9. $\mathbf{A} \times \mathbf{B} = \begin{vmatrix} \mathbf{i} & \mathbf{j} & \mathbf{k} \\ 1 & 0 & 0 \\ 0 & 1 & 0 \end{vmatrix} = \mathbf{k}$

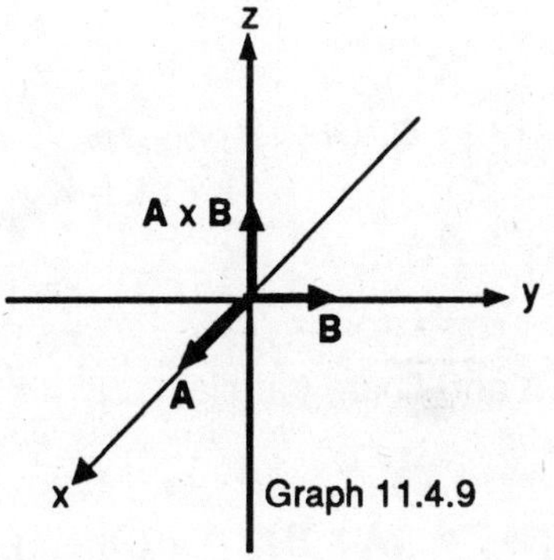


Graph 11.4.9

11. $\mathbf{A} \times \mathbf{B} = \begin{vmatrix} \mathbf{i} & \mathbf{j} & \mathbf{k} \\ 1 & 0 & -1 \\ 0 & 1 & 1 \end{vmatrix} = \mathbf{i} - \mathbf{j} + \mathbf{k}$

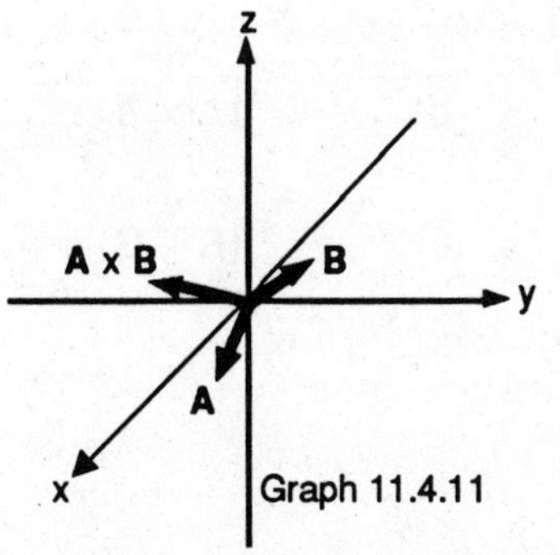


Graph 11.4.11

13. $\mathbf{A} \times \mathbf{B} = \begin{vmatrix} \mathbf{i} & \mathbf{j} & \mathbf{k} \\ 1 & 3 & 2 \\ 0 & 0 & 1 \end{vmatrix} = 3\mathbf{i} - \mathbf{j}$

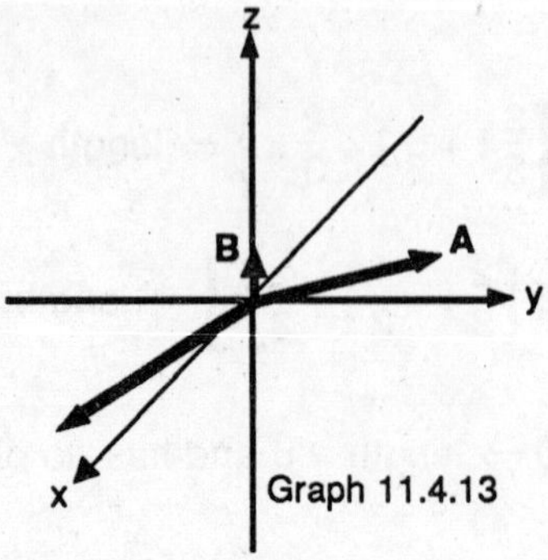

Graph 11.4.13

15. a) $\overrightarrow{PQ} = \mathbf{i} + \mathbf{j} - 3\mathbf{k}$, $\overrightarrow{PR} = -\mathbf{i} + 3\mathbf{j} - \mathbf{k} \Rightarrow \pm\left(\overrightarrow{PQ} \times \overrightarrow{PR}\right) = \pm\begin{vmatrix} \mathbf{i} & \mathbf{j} & \mathbf{k} \\ 1 & 1 & -3 \\ -1 & 3 & -1 \end{vmatrix} = \pm(8\mathbf{i} + 4\mathbf{j} + 4\mathbf{k})$

b) $\dfrac{\left|\overrightarrow{PQ} \times \overrightarrow{PR}\right|}{2} = \dfrac{\sqrt{64 + 16 + 16}}{2} = 2\sqrt{6}$

c) $\pm\left[\dfrac{2}{\sqrt{6}}\mathbf{i} + \dfrac{1}{\sqrt{6}}\mathbf{j} + \dfrac{1}{\sqrt{6}}\mathbf{k}\right]$

17. a) $\overrightarrow{PQ} = \mathbf{i} + \mathbf{j} + \mathbf{k}$, $\overrightarrow{PR} = \mathbf{i} + \mathbf{j} \Rightarrow \pm\left(\overrightarrow{PQ} \times \overrightarrow{PR}\right) = \pm\begin{vmatrix} \mathbf{i} & \mathbf{j} & \mathbf{k} \\ 1 & 1 & 1 \\ 1 & 1 & 0 \end{vmatrix} = \pm(-\mathbf{i} + \mathbf{j})$

b) $\dfrac{\left|\overrightarrow{PQ} \times \overrightarrow{PR}\right|}{2} = \dfrac{\sqrt{2}}{2}$

c) $\pm\left[\dfrac{-1}{\sqrt{2}}\mathbf{i} + \dfrac{1}{\sqrt{2}}\mathbf{j}\right]$

19. a) $\mathbf{A} \cdot \mathbf{B} = -6$, $\mathbf{A} \cdot \mathbf{C} = -81$, $\mathbf{B} \cdot \mathbf{C} = 18 \Rightarrow$ none

b) $\mathbf{A} \times \mathbf{B} = \begin{vmatrix} \mathbf{i} & \mathbf{j} & \mathbf{k} \\ 5 & -1 & 1 \\ 0 & 1 & -5 \end{vmatrix} \neq \mathbf{0}$, $\mathbf{A} \times \mathbf{C} = \begin{vmatrix} \mathbf{i} & \mathbf{j} & \mathbf{k} \\ 5 & -1 & 1 \\ -15 & 3 & -3 \end{vmatrix} = \mathbf{0}$,

$\mathbf{B} \times \mathbf{C} = \begin{vmatrix} \mathbf{i} & \mathbf{j} & \mathbf{k} \\ 0 & 1 & -5 \\ -15 & 3 & -3 \end{vmatrix} \neq \mathbf{0} \Rightarrow$ **A** and **C** are parallel

21. $\mathbf{A} \times \mathbf{B} = \begin{vmatrix} \mathbf{i} & \mathbf{j} & \mathbf{k} \\ 2 & -1 & 0 \\ 1 & 3 & -2 \end{vmatrix} = 2\mathbf{i} + 4\mathbf{j} + 7\mathbf{k}$, $(\mathbf{A} \times \mathbf{B}) \cdot \mathbf{A} = (2\mathbf{i} + 4\mathbf{j} + 7\mathbf{k}) \cdot (2\mathbf{i} - \mathbf{j}) = 0$,

$(\mathbf{A} \times \mathbf{B}) \cdot \mathbf{B} = (2\mathbf{i} + 4\mathbf{j} + 7\mathbf{k}) \cdot (\mathbf{i} + 3\mathbf{j} - 2\mathbf{k}) = 0$

23. a) $\text{proj}_{\mathbf{B}}\,\mathbf{A} = \dfrac{\mathbf{A} \cdot \mathbf{B}}{\mathbf{B} \cdot \mathbf{B}}\mathbf{B}$ b) $(\pm)(\mathbf{A} \times \mathbf{B})$ c) $\sqrt{\mathbf{A} \cdot \mathbf{A}}\,\dfrac{\mathbf{B}}{\sqrt{\mathbf{B} \cdot \mathbf{B}}}$

d) $(\pm)(\mathbf{A} \times \mathbf{B}) \times \mathbf{C}$ e) $(\pm)(\mathbf{B} \times \mathbf{C}) \times \mathbf{A}$

25. $\left|\overrightarrow{PQ} \times \mathbf{F}\right| = \left|\overrightarrow{PQ}\right| |\mathbf{F}| \sin(60°) = 10\sqrt{3}$ ft · lb

27. If $\mathbf{A} = a_1\mathbf{i} + a_2\mathbf{j} + a_3\mathbf{k}$, $\mathbf{B} = b_1\mathbf{i} + b_2\mathbf{j} + b_3\mathbf{k}$, and $\mathbf{C} = c_1\mathbf{i} + c_2\mathbf{j} + c_3\mathbf{k}$, then $\mathbf{A} \cdot (\mathbf{B} \times \mathbf{C}) = \begin{vmatrix} a_1 & a_2 & a_3 \\ b_1 & b_2 & b_3 \\ c_1 & c_2 & c_3 \end{vmatrix}$,

$\mathbf{B} \cdot (\mathbf{C} \times \mathbf{A}) = \begin{vmatrix} b_1 & b_2 & b_3 \\ c_1 & c_2 & c_3 \\ a_1 & a_2 & a_3 \end{vmatrix}$ and $\mathbf{C} \cdot (\mathbf{A} \times \mathbf{B}) = \begin{vmatrix} c_1 & c_2 & c_3 \\ a_1 & a_2 & a_3 \\ b_1 & b_2 & b_3 \end{vmatrix}$ which all have the same value, since the interchanging of two rows, in a determinant, does not change its value. The volume is $|(\mathbf{A} \times \mathbf{B}) \cdot \mathbf{C}| = \begin{vmatrix} 2 & 0 & 0 \\ 0 & 2 & 0 \\ 0 & 0 & 2 \end{vmatrix} = 8$.

29. $|(\mathbf{A} \times \mathbf{B}) \cdot \mathbf{C}| = \begin{vmatrix} 2 & 1 & 0 \\ 2 & -1 & 1 \\ 1 & 0 & 2 \end{vmatrix} = 7$. For details about verification, see exercise 27.

## 11.5 LINES AND PLANES IN SPACE

1. the direction $\mathbf{i} + \mathbf{j} + \mathbf{k}$ and $P(3,-4,-1) \Rightarrow x = 3 + t,\ y = -4 + t,\ z = -1 + t$

3. the direction $\overrightarrow{PQ} = 5\mathbf{i} + 5\mathbf{j} - 5\mathbf{k}$ and$P(-2,0,3) \Rightarrow x = -2 + 5t,\ y = 5t,\ z = 3 - 5t$

5. the direction $2\mathbf{j} + \mathbf{k}$ and $P(0,0,0) \Rightarrow x = 0,\ y = 2t,\ z = t$

7. the direction $\mathbf{k}$ and $P(1,1,1) \Rightarrow x = 1,\ y = 1,\ z = 1 + t$

9. the direction $\mathbf{i} + 2\mathbf{j} + 2\mathbf{k}$ and $(0,-7,0) \Rightarrow x = t,\ y = -7 + 2t,\ z = 2t$

11. the direction $\mathbf{i}$ and $P(0,0,0) \Rightarrow x = t,\ y = 0,\ z = 0$

13. the direction $\overrightarrow{PQ} = \mathbf{i} + \mathbf{j} + \mathbf{k}$ and $P(0,0,0) \Rightarrow$
$x = t,\ y = t,\ z = t$, where $0 \le t \le 1$

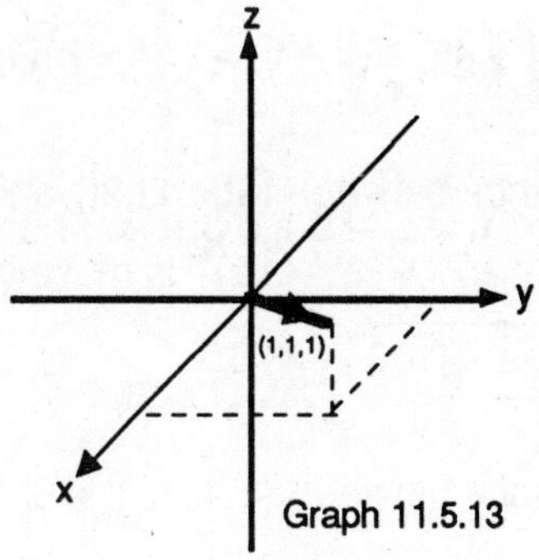

Graph 11.5.13

15. the direction $\overrightarrow{PQ} = \mathbf{j}$ and $P(1,0,0) \Rightarrow$
$x = 1,\ y = 1 + t,\ z = 0$, where $-1 \le t \le 0$

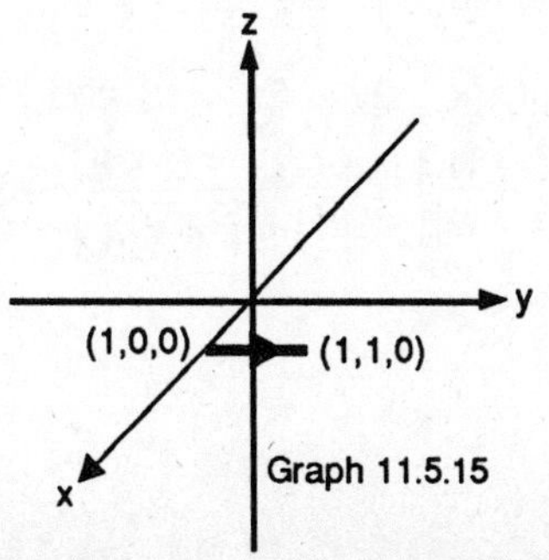

Graph 11.5.15

17. the direction $\overrightarrow{PQ} = 2\mathbf{j}$ and $P(0,-1,1) \Rightarrow$
$x = 0,\ y = -1 + 2t,\ z = 1$, where $0 \le t \le 1$

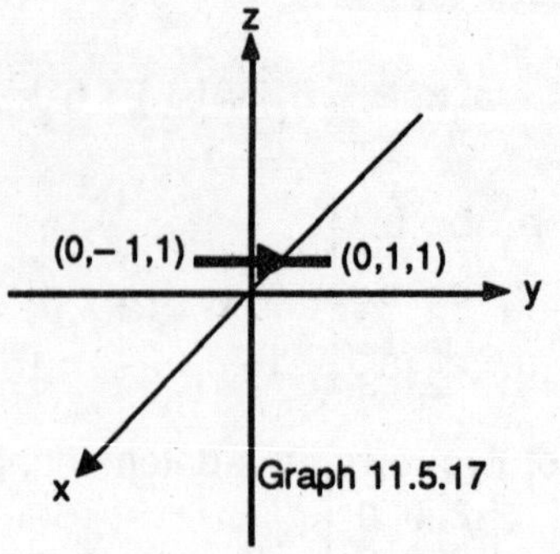

Graph 11.5.17

19. the direction $\overrightarrow{PQ} = -\mathbf{i} - 2\mathbf{k}$ and $P(2,2,0) \Rightarrow$
$x = 2 - t,\ y = 2,\ z = -2t$, where $0 \le t \le 1$

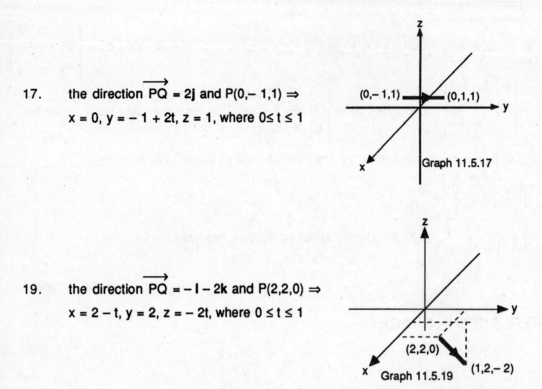

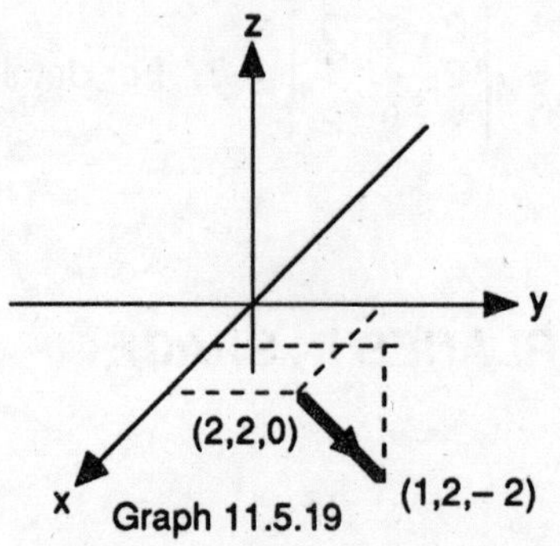

Graph 11.5.19

21. $3(x) + (-2)(y - 2) + (-1)(z + 1) = 0 \Rightarrow 3x - 2y - z = -3$

23. $\overrightarrow{PQ} = \mathbf{i} - \mathbf{j} + 3\mathbf{k},\ \overrightarrow{PS} = -\mathbf{i} - 3\mathbf{j} + 2\mathbf{k} \Rightarrow \overrightarrow{PQ} \times \overrightarrow{PS} = \begin{vmatrix} \mathbf{i} & \mathbf{j} & \mathbf{k} \\ 1 & -1 & 3 \\ -1 & -3 & 2 \end{vmatrix} = 7\mathbf{i} - 5\mathbf{j} - 4\mathbf{k}$, the normal;

$(x - 2)(7) + (y - 0)(-5) + (z - 2)(-4) = 0 \Rightarrow 7x - 5y - 4z = 6$

25. $\mathbf{N} = \mathbf{i} + 3\mathbf{j} + 4\mathbf{k},\ P(2,4,5) \Rightarrow (x - 2)(1) + (y - 4)(3) + (z - 5)(4) = 0 \Rightarrow x + 3y + 4z = 34$

27. The distance between $(4t, -2t, 2t)$ and $(0,0,12)$ is $d = \sqrt{(4t)^2 + (-2t)^2 + (2t - 12)^2}$. If $f(t) = (4t)^2 + (-2t)^2 + (2t - 12)^2$ is minimized, then d is minimized. $f'(t) = 0 \Rightarrow t = 1 \Rightarrow$
$d = \sqrt{16 + 4 + 100} = 2\sqrt{30}$

29. The distance between $(2 + 2t, 1 + 6t, 3)$ and $(2,1,3)$ is $d = \sqrt{(2t)^2 + (6t)^2}$. If $f(t) = (2t)^2 + (6t)^2$ is minimized, then d is minimized. $f'(t) = 0 \Rightarrow t = 0 \Rightarrow d = 0$

31. $S(2,-3,4)$, $x + 2y + 2z = 13$ and $P(13,0,0)$ is on the plane $\Rightarrow \overrightarrow{PS} = -11\mathbf{i} - 3\mathbf{j} + 4\mathbf{k},\ \mathbf{N} = \mathbf{i} + 2\mathbf{j} + 2\mathbf{k}$;
$d = \left| \overrightarrow{PS} \cdot \frac{\mathbf{N}}{|\mathbf{N}|} \right| = \left| \frac{-11 - 6 + 8}{\sqrt{1 + 4 + 4}} \right| = 3$

33. S(0,1,1), $4y + 3z = -12$ and P(0,– 3,0) is on the plane $\Rightarrow \overrightarrow{PS} = 4\mathbf{j} + \mathbf{k}$, $\mathbf{N} = 4\mathbf{j} + 3\mathbf{k}$;

$$d = \left|\overrightarrow{PS} \cdot \frac{\mathbf{N}}{|\mathbf{N}|}\right| = \left|\frac{16+3}{\sqrt{16+9}}\right| = \frac{19}{5}$$

35. S(0,– 1,0), $2x + y + 2z = 4$ and P(2,0,0) is on the plane $\Rightarrow \overrightarrow{PS} = -2\mathbf{i} - \mathbf{j}$, $\mathbf{N} = 2\mathbf{i} + \mathbf{j} + 2\mathbf{k}$;

$$d = \left|\overrightarrow{PS} \cdot \frac{\mathbf{N}}{|\mathbf{N}|}\right| = \left|\frac{-4-1+0}{\sqrt{4+1+4}}\right| = \frac{5}{3}$$

37. $2(1-t) - (3t) + 3(1+t) = 6 \Rightarrow t = -\ 1/2 \Rightarrow (3/2, -3/2, 1/2)$

39. $1(1+2t) + 1(1+5t) + 1(3t) = 2 \Rightarrow t = 0 \Rightarrow (1,1,0)$

41. $\mathbf{N}_1 = \mathbf{i} + \mathbf{j}$, $\mathbf{N}_2 = 2\mathbf{i} + \mathbf{j} - 2\mathbf{k} \Rightarrow \theta = \cos^{-1}\left(\frac{\mathbf{N}_1 \cdot \mathbf{N}_2}{|\mathbf{N}_1|\,|\mathbf{N}_2|}\right) = \cos^{-1}\left(\frac{2+1}{\sqrt{2}\sqrt{9}}\right) = \cos^{-1}\frac{1}{\sqrt{2}} = \frac{\pi}{4}$

43. $\mathbf{N}_1 = 2\mathbf{i} + 2\mathbf{j} + 2\mathbf{k}$, $\mathbf{N}_2 = 2\mathbf{i} - 2\mathbf{j} - \mathbf{k} \Rightarrow \theta = \cos^{-1}\left(\frac{\mathbf{N}_1 \cdot \mathbf{N}_2}{|\mathbf{N}_1|\,|\mathbf{N}_2|}\right) = \cos^{-1}\left(\frac{4-4-2}{\sqrt{12}\sqrt{9}}\right) =$

$\cos^{-1}\left(\frac{-1}{3\sqrt{3}}\right) \approx 101.1°$

45. $\mathbf{N}_1 = 2\mathbf{i} + 2\mathbf{j} - \mathbf{k}$, $\mathbf{N}_2 = \mathbf{i} + 2\mathbf{j} + \mathbf{k} \Rightarrow \theta = \cos^{-1}\left(\frac{\mathbf{N}_1 \cdot \mathbf{N}_2}{|\mathbf{N}_1|\,|\mathbf{N}_2|}\right) = \cos^{-1}\left(\frac{2+4-1}{\sqrt{9}\sqrt{6}}\right) =$

$\cos^{-1}\left(\frac{5}{3\sqrt{6}}\right) \approx 47.1°$

47. $\mathbf{N}_1 = \mathbf{i} + \mathbf{j} + \mathbf{k}$, $\mathbf{N}_2 = \mathbf{i} + \mathbf{j} \Rightarrow \mathbf{N}_1 \times \mathbf{N}_2 = \begin{vmatrix} \mathbf{i} & \mathbf{j} & \mathbf{k} \\ 1 & 1 & 1 \\ 1 & 1 & 0 \end{vmatrix} = -\mathbf{i} + \mathbf{j}$, the direction of the desired line; (1,1,– 1) is on both planes; the desired line is $x = 1 - t$, $y = 1 + t$, $z = -1$

49. $\mathbf{N}_1 = \mathbf{i} - 2\mathbf{j} + 4\mathbf{k}$, $\mathbf{N}_2 = \mathbf{i} + \mathbf{j} - 2\mathbf{k} \Rightarrow \mathbf{N}_1 \times \mathbf{N}_2 = \begin{vmatrix} \mathbf{i} & \mathbf{j} & \mathbf{k} \\ 1 & -2 & 4 \\ 1 & 1 & -2 \end{vmatrix} = 6\mathbf{j} + 3\mathbf{k}$, the direction of the desired line; (4,3,1) is on both planes; the desired line is $x = 4$, $y = 3 + 6t$, $z = 1 + 3t$

## 11.6 SURFACES IN SPACE

1. $x^2 + y^2 = 4$

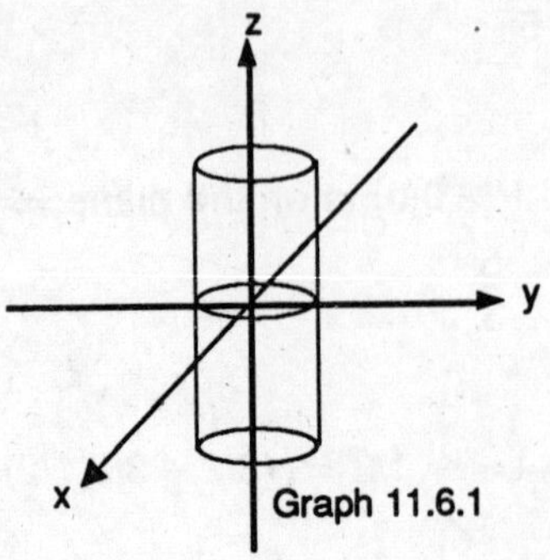

Graph 11.6.1

3. $y^2 + z^2 = 1$

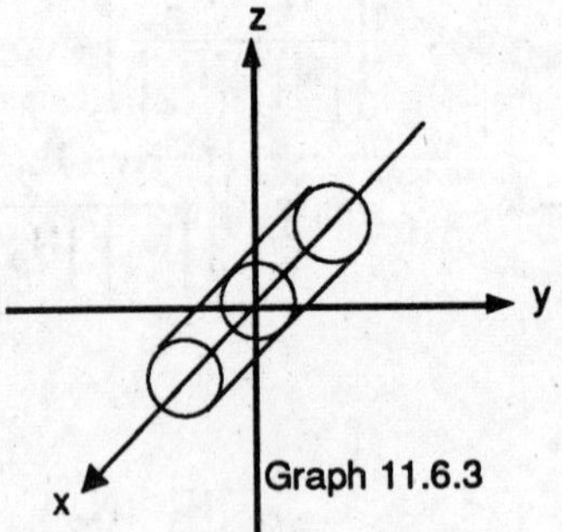

Graph 11.6.3

5. $z = y^2 - 1$

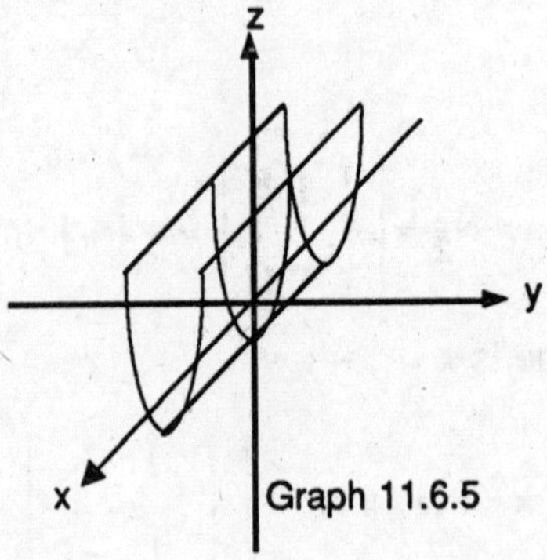

Graph 11.6.5

7. $z = 4 - x^2$

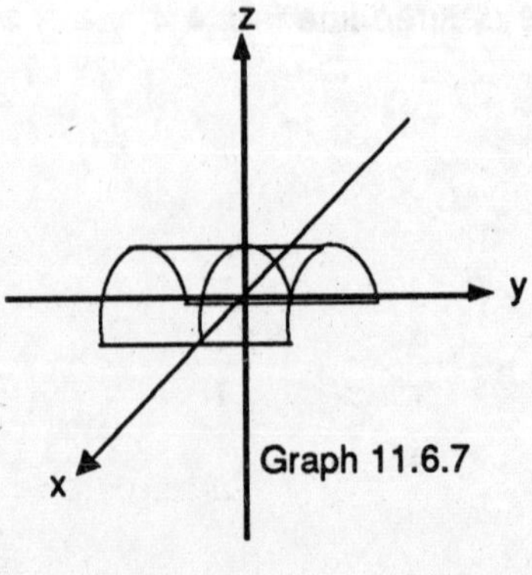

Graph 11.6.7

9. $y = x^2$

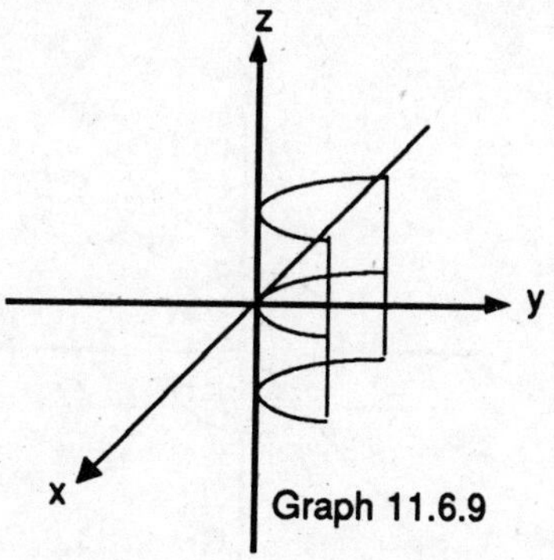


Graph 11.6.9

11. $y^2 + 4z^2 = 16$

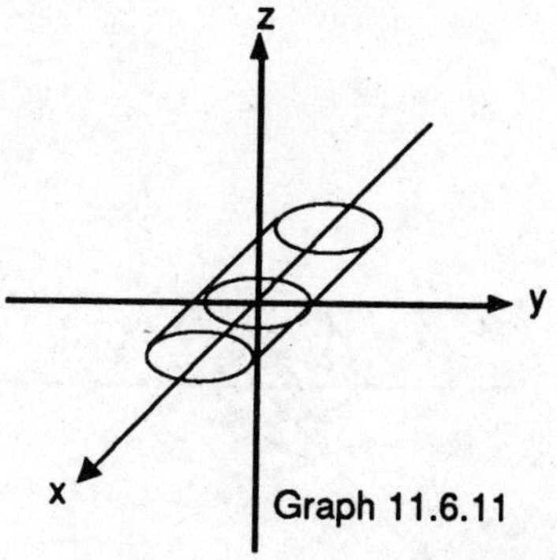


Graph 11.6.11

13. $z^2 + 4y^2 = 9$

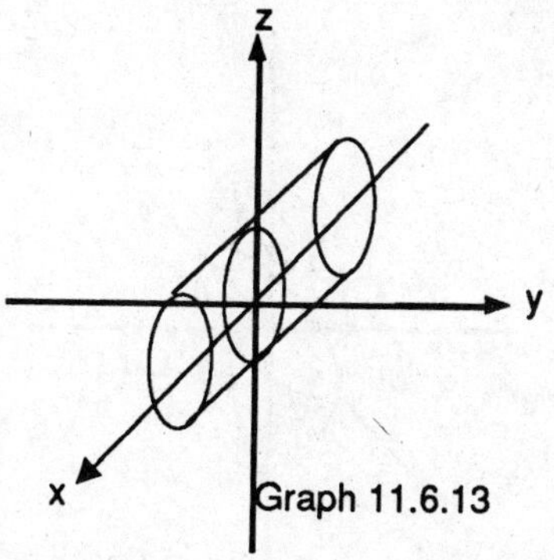


Graph 11.6.13

15. $z^2 - y^2 = 1$

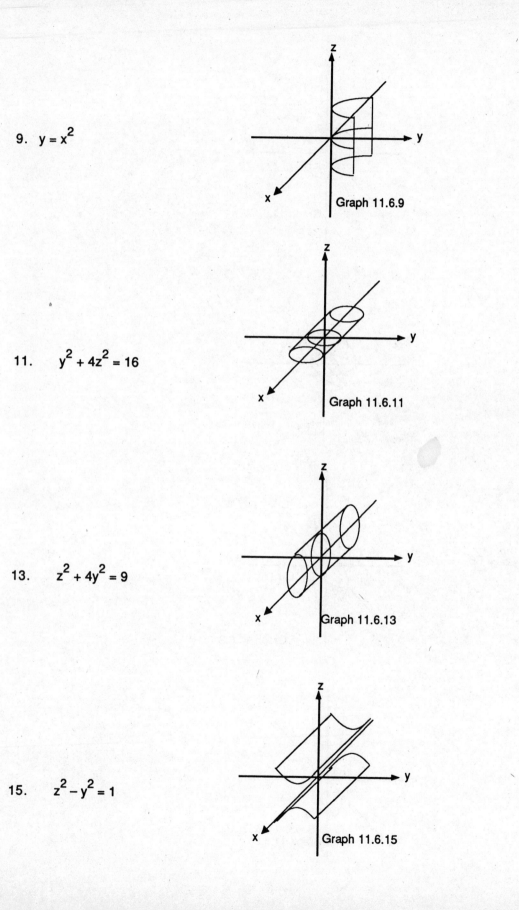

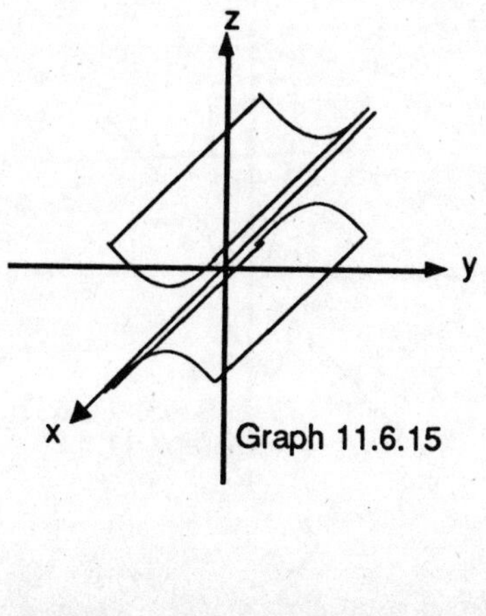


Graph 11.6.15

17. $9x^2 + y^2 + z^2 = 9$

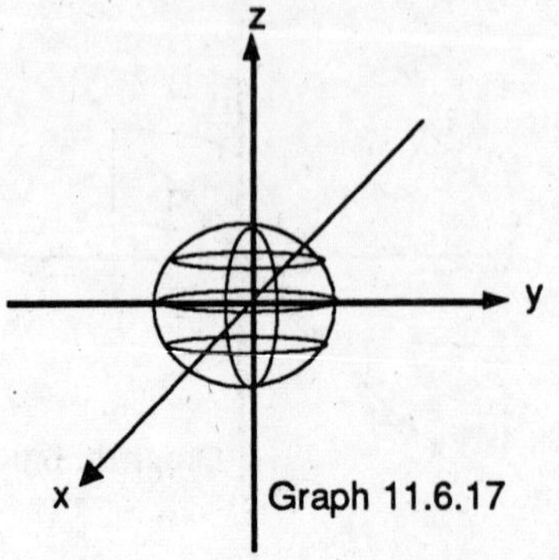

Graph 11.6.17

19. $x^2 + y^2 + z^2 = 4$

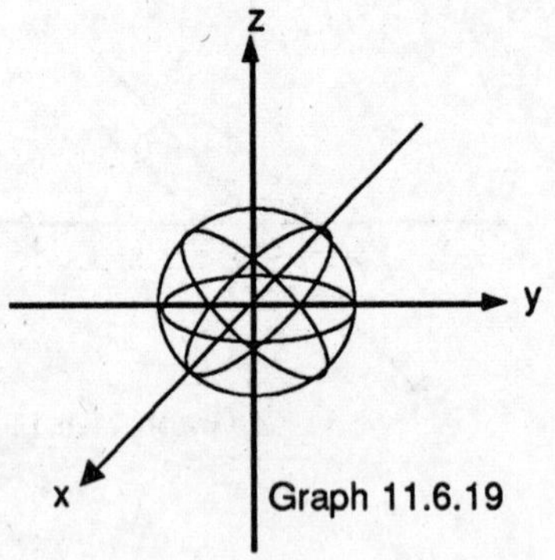

Graph 11.6.19

21. $4x^2 + 9y^2 + 4z^2 = 36$

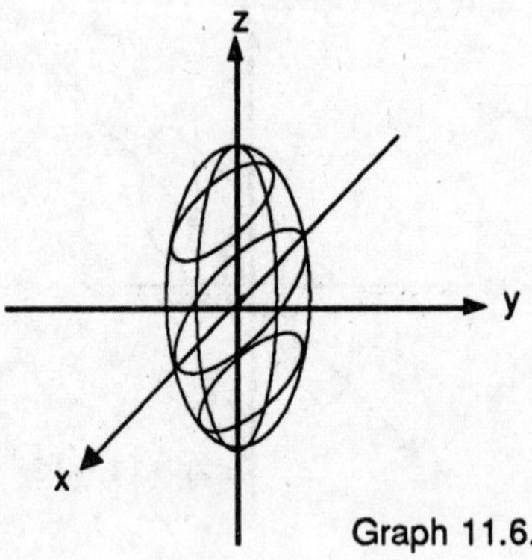

Graph 11.6.21

23. $x^2 + y^2 = z$

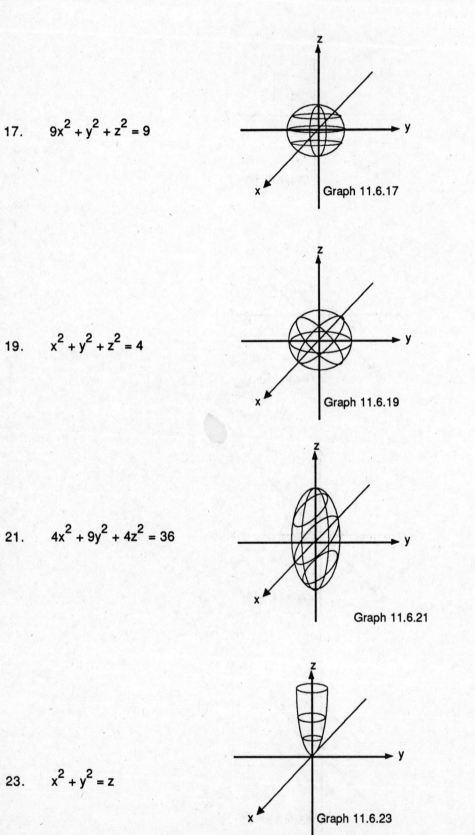

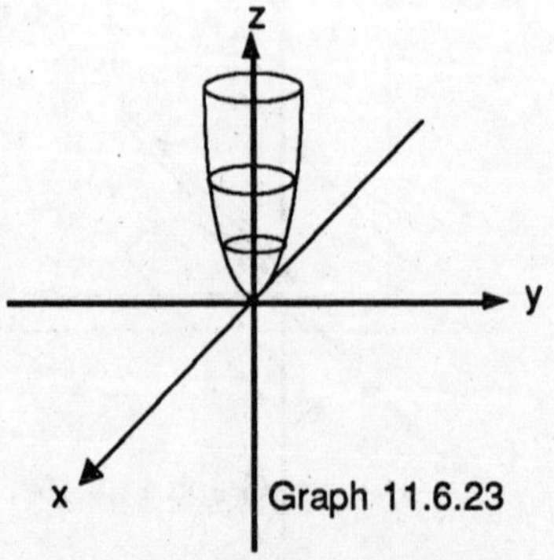

Graph 11.6.23

25. $x^2 + 4y^2 = z$

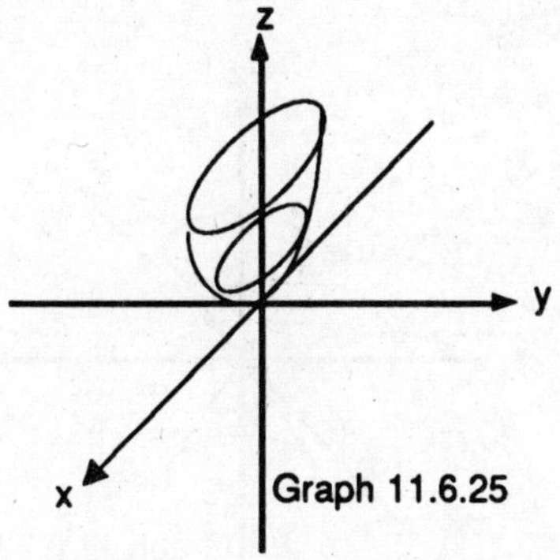

Graph 11.6.25

27. $z = 8 - x^2 - y^2$

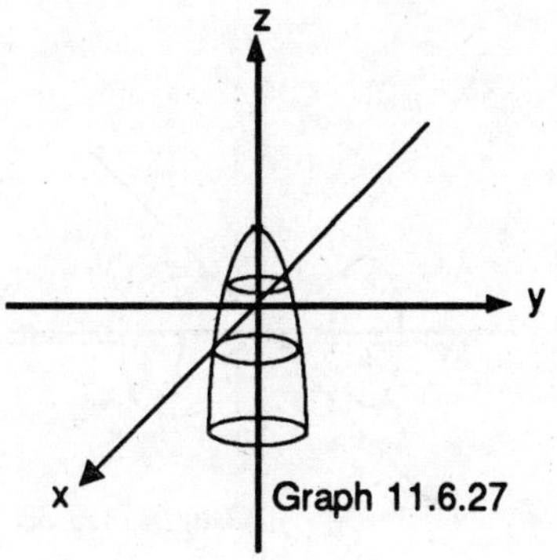

Graph 11.6.27

29. $x = 4 - 4y^2 - z^2$

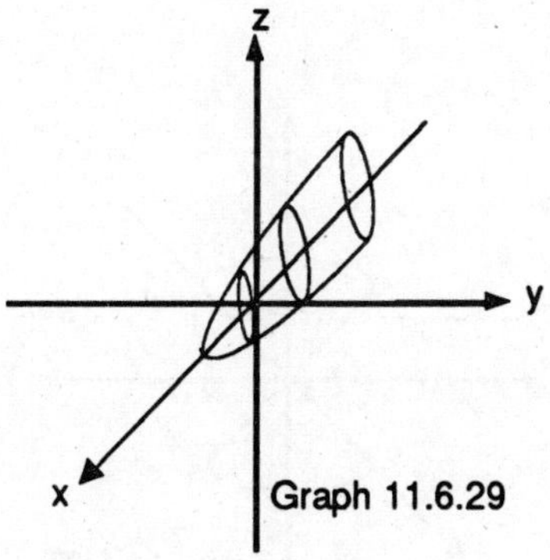

Graph 11.6.29

31. $z = x^2 + y^2 + 1$

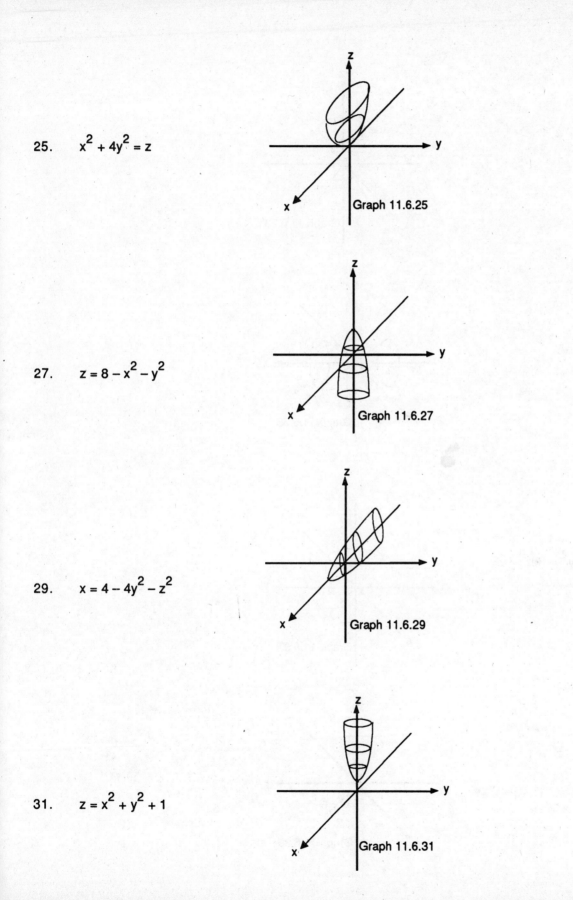

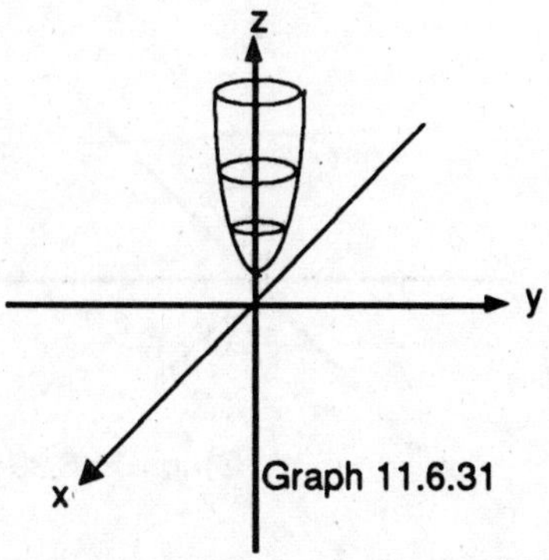

Graph 11.6.31

33. $x^2 + y^2 = z^2$

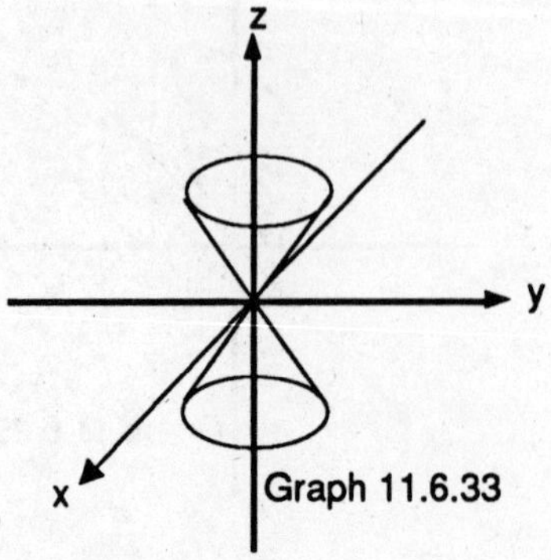

Graph 11.6.33

35. $x^2 + z^2 = y^2$

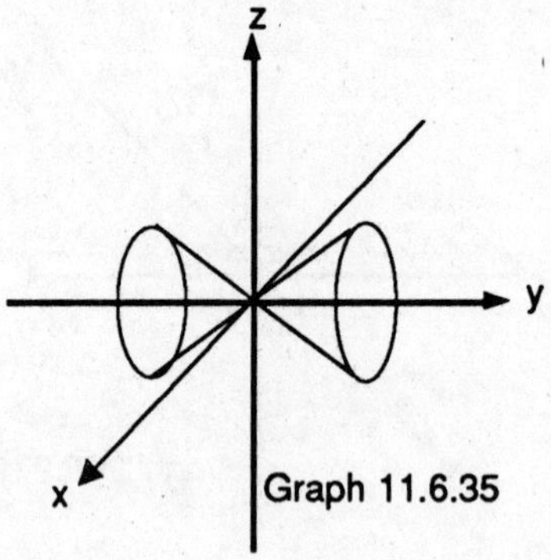

Graph 11.6.35

37. $9x^2 + 4y^2 = 36z^2$

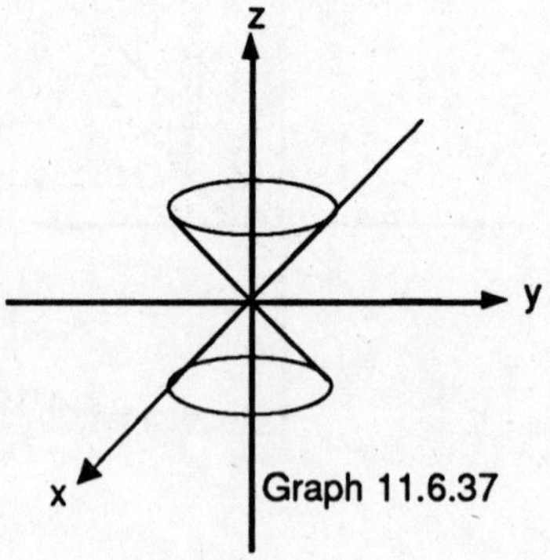

Graph 11.6.37

39. $x^2 + y^2 - z^2 = 1$

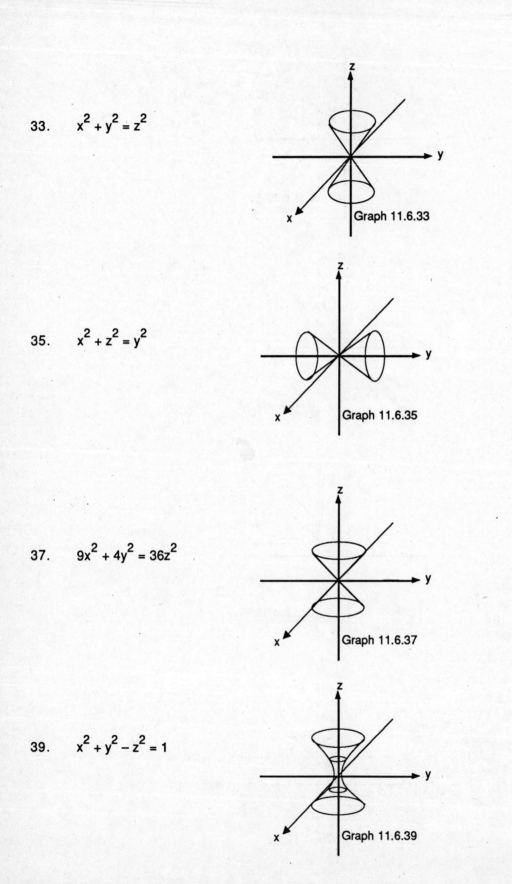

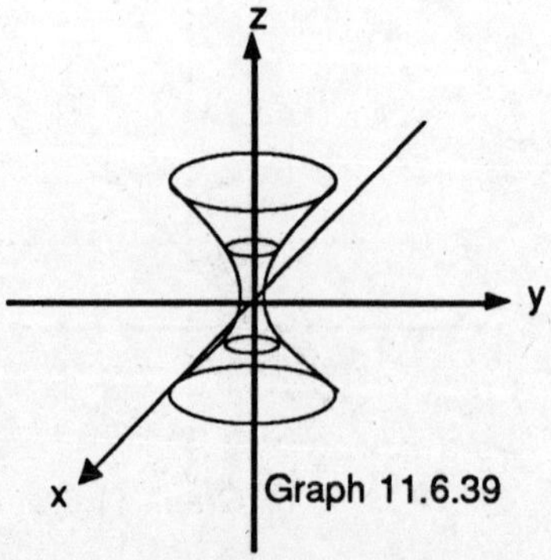

Graph 11.6.39

41. $\left(y^2/4\right) + \left(z^2/9\right) - \left(x^2/4\right) = 1$

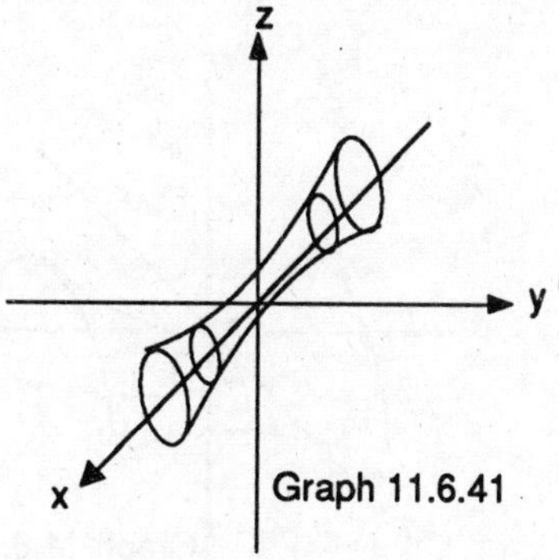

Graph 11.6.41

43. $\left(x^2/4\right) + y^2 - z^2 = 1$

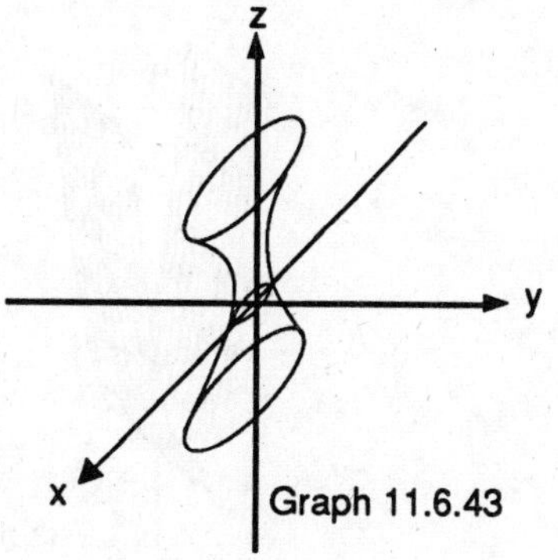

Graph 11.6.43

45. $z^2 - \left(x^2/4\right) - y^2 = 1$

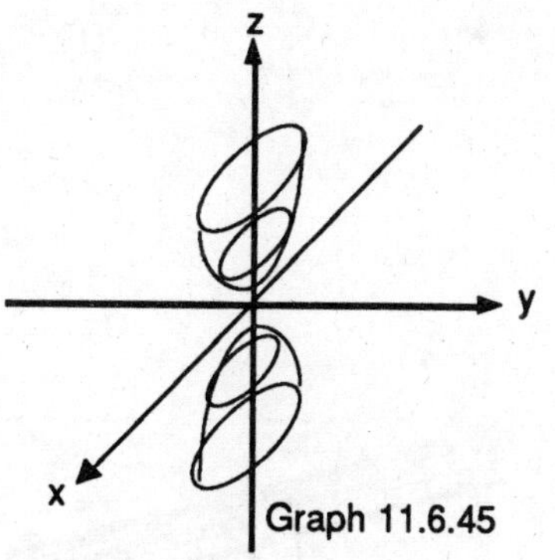

Graph 11.6.45

47. $x^2 - y^2 - \left(z^2/4\right) = 1$

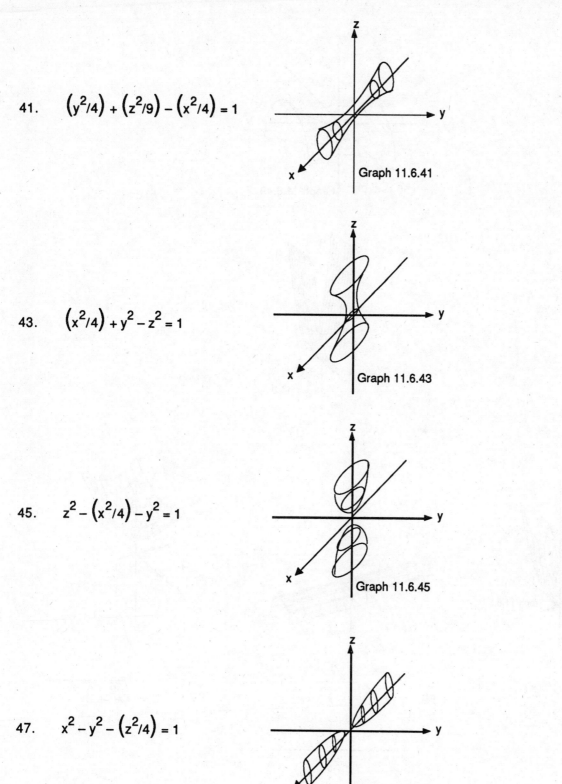

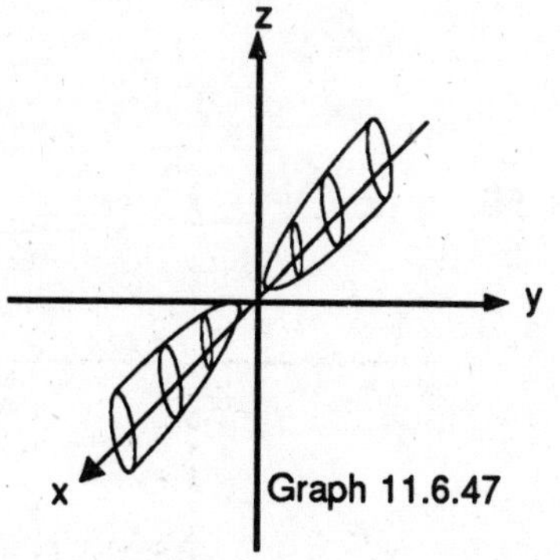

Graph 11.6.47

49. $y^2 - x^2 = z$

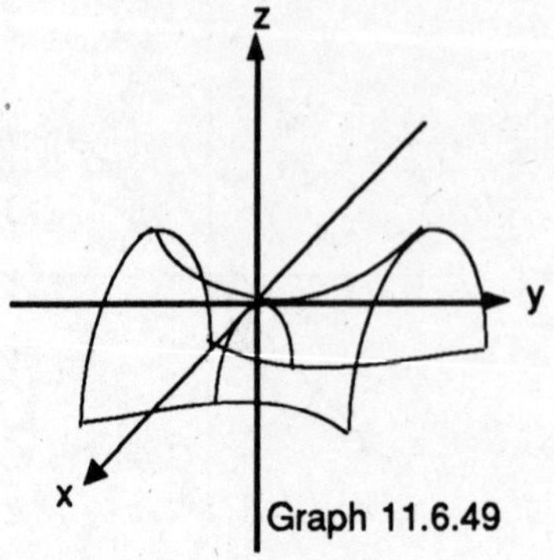

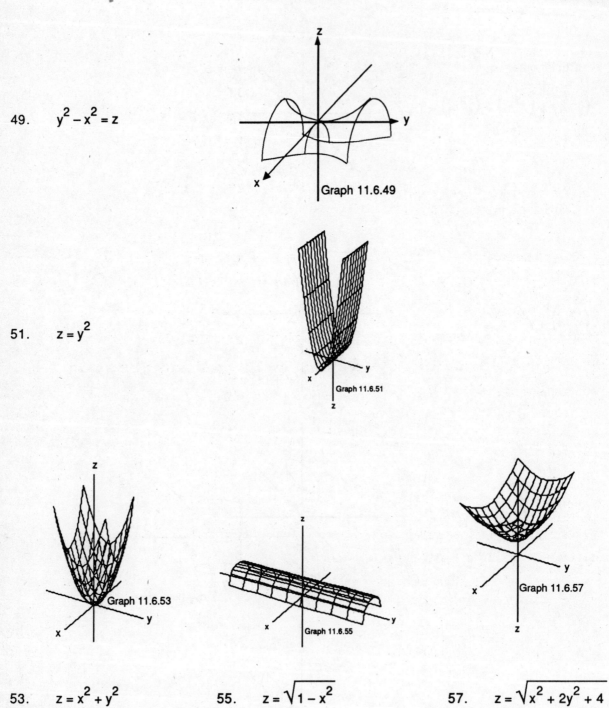

Graph 11.6.49

51. $z = y^2$

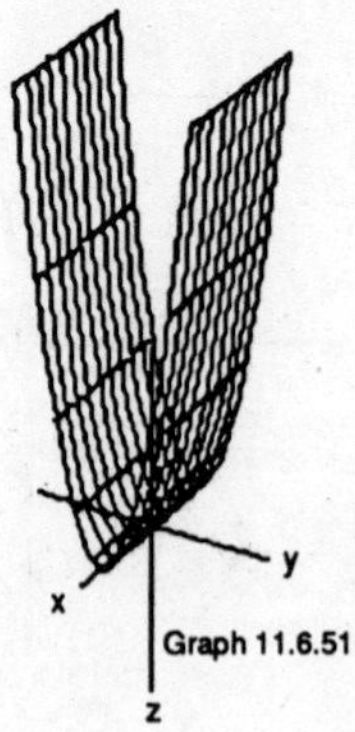

Graph 11.6.51

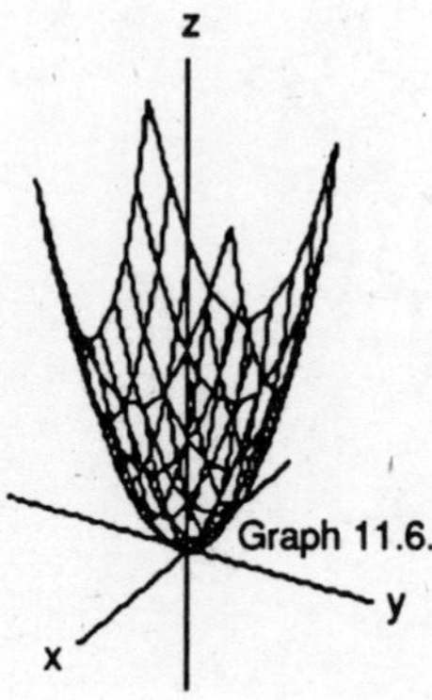

Graph 11.6.53

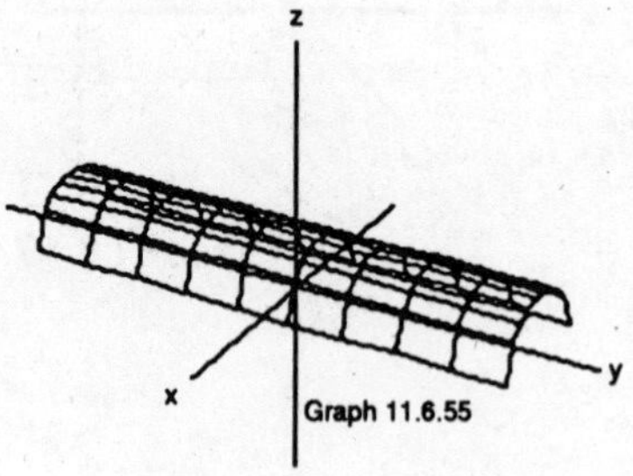

Graph 11.6.55

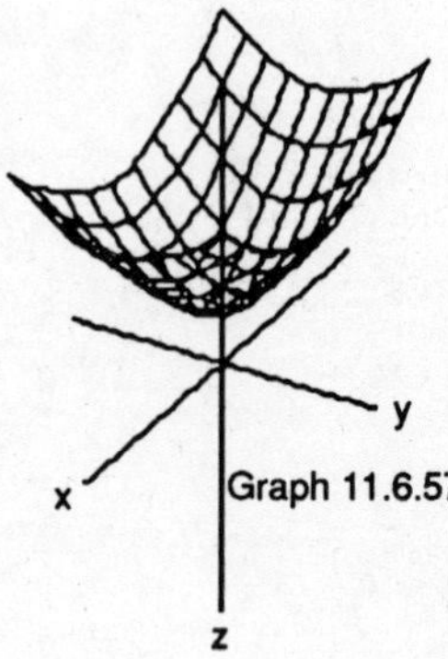

Graph 11.6.57

53. $z = x^2 + y^2$

55. $z = \sqrt{1 - x^2}$

57. $z = \sqrt{x^2 + 2y^2 + 4}$

## 11.7 CYLINDRICAL AND SPHERICAL COORDINATES

| | Rectangular | Cylindrical | Spherical |
|---|---|---|---|
| 1. | (0,0,0) | (0,0,0) | (0,0,0) |
| 3. | (0,1,0) | $(1,\pi/2,0)$ | $(1,\pi/2,\pi/2)$ |
| 5. | (1,0,0) | (1,0,0) | $(1,\pi/2,0)$ |
| 7. | (0,1,1) | $(1,\pi/2,1)$ | $(\sqrt{2},\pi/4,\pi/2)$ |
| 9. | $(0,-2\sqrt{2},0)$ | $(2\sqrt{2},3\pi/2,0)$ | $(2\sqrt{2},\pi/2,3\pi/2)$ |

* any angle may be used

11. $r = 0 \Rightarrow$ rectangular, $x^2 + y^2 = 0$; spherical, $\phi = 0$ and $\phi = \pi$; the z–axis
13. $z = 0 \Rightarrow$ cylindrical, $z = 0$; spherical, $\phi = \frac{\pi}{2}$; the xy–plane
15. $\rho \cos \phi = 3 \Rightarrow$ rectangular, $z = 3$; cylindrical, $z = 3$; the plane $z = 3$
17. $\rho \sin \phi \cos \phi = 0 \Rightarrow$ rectangular, $x = 0$; cylindrical $\theta = \frac{\pi}{2}$; the yz–plane
19. $x^2 + y^2 + z^2 = 4 \Rightarrow$ cylindrical, $r^2 + z^2 = 4$; spherical, $\rho = 2$; a sphere centered at the origin with a radius of 2

21. a right circular cylinder whose generating curve is a circle of radius 4 in the rθ–plane

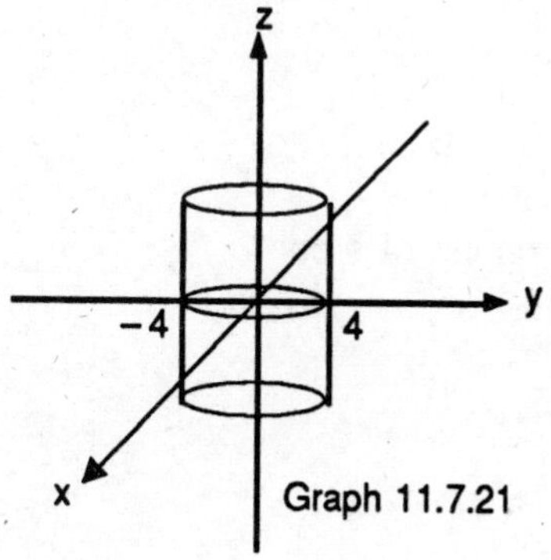

Graph 11.7.21

23. a cylinder whose generation curve is a cardioid in the rθ–plane

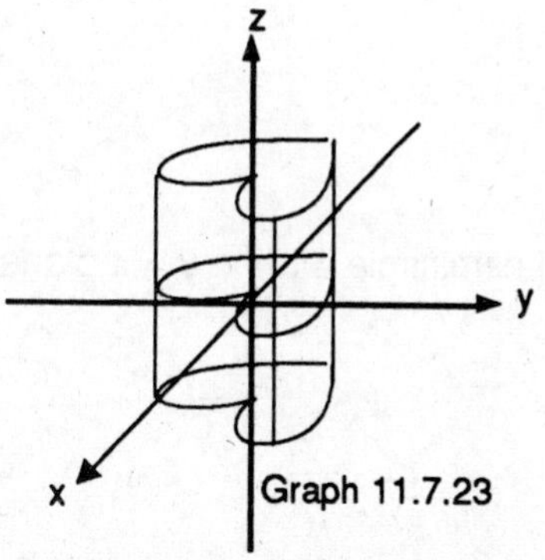

Graph 11.7.23

25. a circle contained in the plane z = 3 having a radius of 2 and center at (0,0,3)

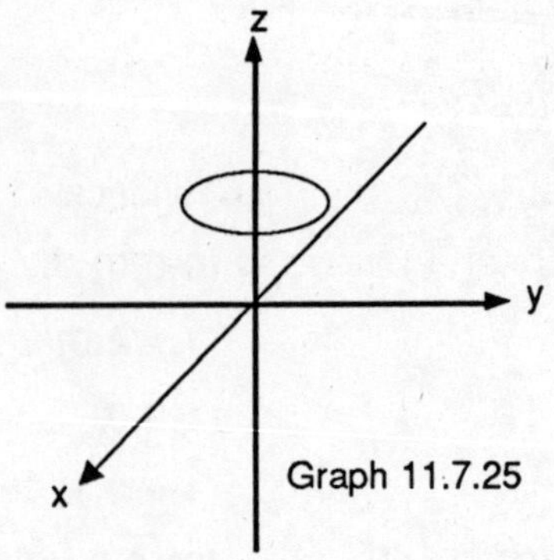


Graph 11.7.25

27. a space curve called a helix

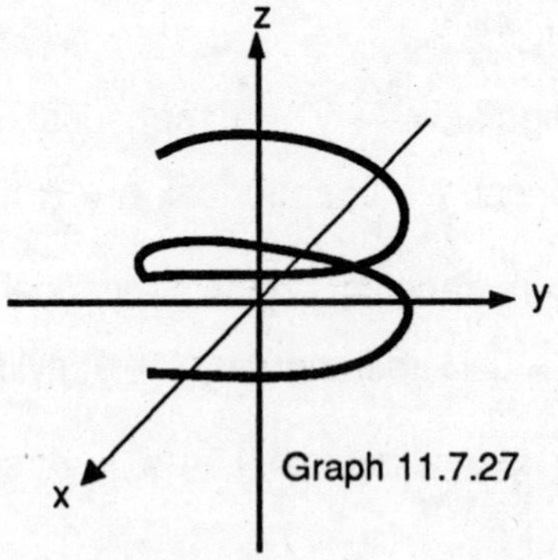


Graph 11.7.27

29. the upper nappe of a cone

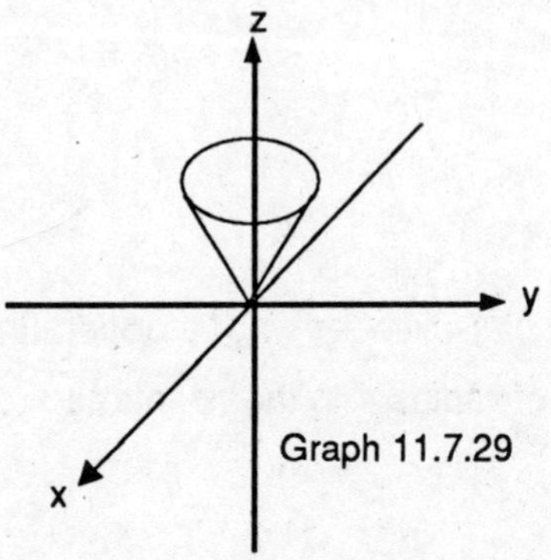


Graph 11.7.29

31. a " vertical semicircle " in the y = x plane

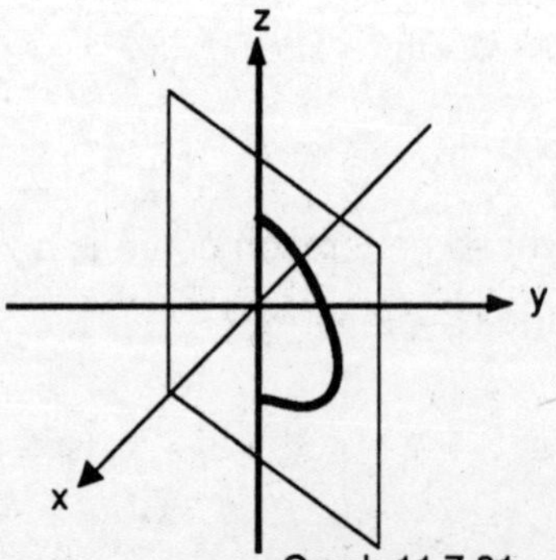


Graph 11.7.31

33. the intersection of a torus and the yz–plane when $y \geq 0$

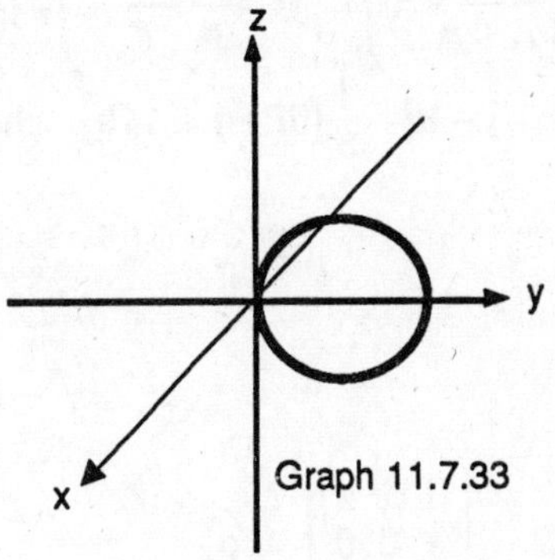

Graph 11.7.33

## PRACTICE EXERCISES

1. $\theta = 0 \Rightarrow \mathbf{u} = \mathbf{i}$; $\theta = \frac{\pi}{2} \Rightarrow \mathbf{u} = \mathbf{j}$; $\theta = \frac{2\pi}{3} \Rightarrow \mathbf{u} = -\frac{1}{2}\mathbf{i} + \frac{\sqrt{3}}{2}\mathbf{j}$;
$\theta = \frac{5\pi}{4} \Rightarrow \mathbf{u} = -\frac{1}{\sqrt{2}}\mathbf{i} - \frac{1}{\sqrt{2}}\mathbf{j}$; $\theta = \frac{5\pi}{3} \Rightarrow \mathbf{u} = \frac{1}{2}\mathbf{i} - \frac{\sqrt{3}}{2}\mathbf{j}$

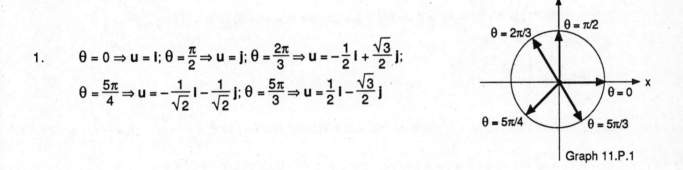

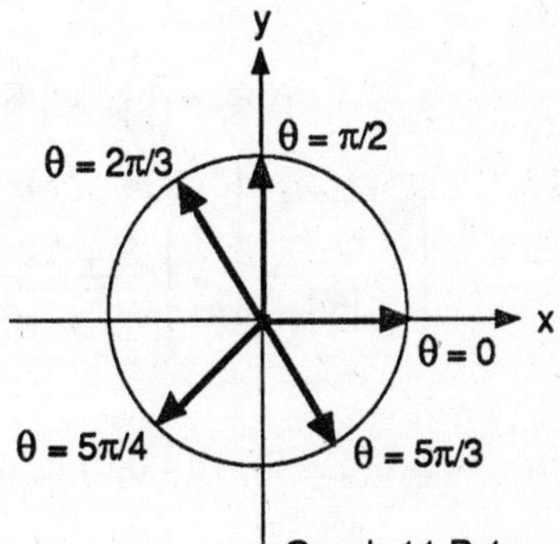

Graph 11.P.1

3. $y = \tan x \Rightarrow [y']_{\pi/4} = \left[\sec^2 x\right]_{\pi/4} = 2 = \frac{2}{1} \Rightarrow \mathbf{T} = \mathbf{i} + 2\mathbf{j} \Rightarrow$ the unit tangents are $\pm\left(\frac{1}{\sqrt{5}}\mathbf{i} + \frac{2}{\sqrt{5}}\mathbf{j}\right)$ and the unit normals are $\pm\left(-\frac{2}{\sqrt{5}}\mathbf{i} + \frac{1}{\sqrt{5}}\mathbf{j}\right)$

5. length $= \left|\sqrt{2}\mathbf{i} + \sqrt{2}\mathbf{j}\right| = \sqrt{2+2} = 2$, $\sqrt{2}\mathbf{i} + \sqrt{2}\mathbf{j} = 2\left[\frac{1}{\sqrt{2}}\mathbf{i} + \frac{1}{\sqrt{2}}\mathbf{j}\right] \Rightarrow$
the direction is $\frac{1}{\sqrt{2}}\mathbf{i} + \frac{1}{\sqrt{2}}\mathbf{j}$

7. length $= |2\mathbf{i} - 3\mathbf{j} + 6\mathbf{k}| = \sqrt{4+9+36} = 7$, $2\mathbf{i} - 3\mathbf{j} + 6\mathbf{k} = 7\left[\frac{2}{7}\mathbf{i} - \frac{3}{7}\mathbf{j} + \frac{6}{7}\mathbf{k}\right] \Rightarrow$
the direction is $= \frac{2}{7}\mathbf{i} - \frac{3}{7}\mathbf{j} + \frac{6}{7}\mathbf{k}$

9. $\frac{2\,\mathbf{A}}{|\mathbf{A}|} = \frac{2}{\sqrt{33}}[4\mathbf{i} - \mathbf{j} + 4\mathbf{k}]$

11. $|\mathbf{A}| = \sqrt{2}$, $|\mathbf{B}| = 3$, $\mathbf{A}\cdot\mathbf{B} = 3$, $\mathbf{B}\cdot\mathbf{A} = 3$, $\mathbf{A}\times\mathbf{B} = \begin{vmatrix} \mathbf{i} & \mathbf{j} & \mathbf{k} \\ 1 & 1 & 0 \\ 2 & 1 & -2 \end{vmatrix} = -2\mathbf{i} + 2\mathbf{j} - \mathbf{k}$,

$\mathbf{B}\times\mathbf{A} = \begin{vmatrix} \mathbf{i} & \mathbf{j} & \mathbf{k} \\ 2 & 1 & -2 \\ 1 & 1 & 0 \end{vmatrix} = 2\mathbf{i} - 2\mathbf{j} + \mathbf{k}$, $|\mathbf{A}\times\mathbf{B}| = \sqrt{4+4+1} = 3$, $\theta = \cos^{-1}\left(\frac{\mathbf{A}\cdot\mathbf{B}}{|\mathbf{A}|\,|\mathbf{B}|}\right) = \cos^{-1}\left(\frac{1}{\sqrt{2}}\right) = \frac{\pi}{4}$,

$\text{comp}_{\mathbf{A}}\,\mathbf{B} = \frac{\mathbf{A}\cdot\mathbf{B}}{|\mathbf{A}|} = \frac{3}{\sqrt{2}}$, $\text{proj}_{\mathbf{A}}\,\mathbf{B} = \frac{\mathbf{A}\cdot\mathbf{B}}{\mathbf{A}\cdot\mathbf{A}}\,\mathbf{A} = \frac{3}{2}[\mathbf{i} + \mathbf{j}]$

13. $\mathbf{B} = \left(\frac{\mathbf{A}\cdot\mathbf{B}}{\mathbf{A}\cdot\mathbf{A}}\mathbf{A}\right) + \left(\mathbf{B} - \frac{\mathbf{A}\cdot\mathbf{B}}{\mathbf{A}\cdot\mathbf{A}}\mathbf{A}\right) = \frac{4}{3}[2\mathbf{i} + \mathbf{j} - \mathbf{k}] + \left[(\mathbf{i} + \mathbf{j} - 5\mathbf{k}) - \frac{4}{3}(2\mathbf{i} + \mathbf{j} - \mathbf{k})\right] =$

$\frac{4}{3}[2\mathbf{i} + \mathbf{j} - \mathbf{k}] - \frac{1}{3}[5\mathbf{i} + \mathbf{j} + 11\mathbf{k}]$, where $\mathbf{A}\cdot\mathbf{B} = 8$ and $\mathbf{A}\cdot\mathbf{A} = 6$

15. $\mathbf{A} \times \mathbf{B} = \begin{vmatrix} \mathbf{i} & \mathbf{j} & \mathbf{k} \\ 1 & 0 & 0 \\ 1 & 1 & 0 \end{vmatrix} = \mathbf{k}$

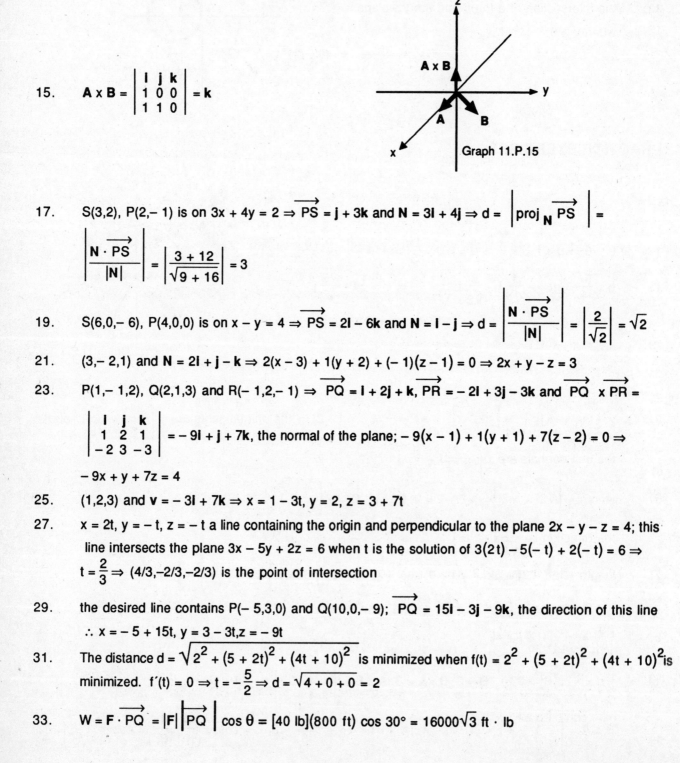

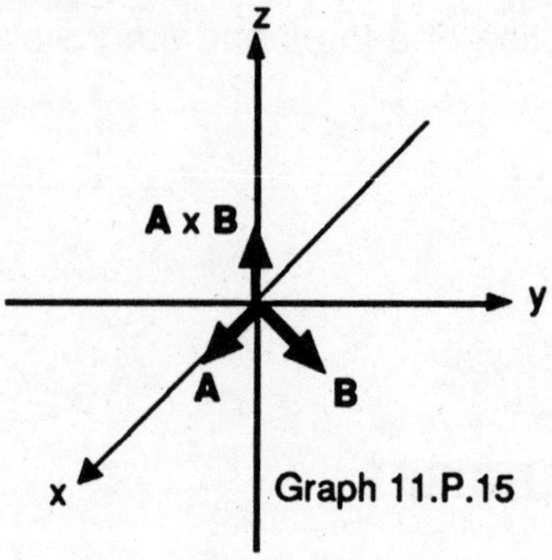


Graph 11.P.15

17. S(3,2), P(2,– 1) is on $3x + 4y = 2 \Rightarrow \overrightarrow{PS} = \mathbf{j} + 3\mathbf{k}$ and $\mathbf{N} = 3\mathbf{i} + 4\mathbf{j} \Rightarrow d = \left|\text{proj}_{\mathbf{N}}\overrightarrow{PS}\right| =$

$\left|\frac{\mathbf{N}\cdot\overrightarrow{PS}}{|\mathbf{N}|}\right| = \left|\frac{3 + 12}{\sqrt{9 + 16}}\right| = 3$

19. S(6,0,– 6), P(4,0,0) is on $x - y = 4 \Rightarrow \overrightarrow{PS} = 2\mathbf{i} - 6\mathbf{k}$ and $\mathbf{N} = \mathbf{i} - \mathbf{j} \Rightarrow d = \left|\frac{\mathbf{N}\cdot\overrightarrow{PS}}{|\mathbf{N}|}\right| = \left|\frac{2}{\sqrt{2}}\right| = \sqrt{2}$

21. (3,– 2,1) and $\mathbf{N} = 2\mathbf{i} + \mathbf{j} - \mathbf{k} \Rightarrow 2(x - 3) + 1(y + 2) + (-1)(z - 1) = 0 \Rightarrow 2x + y - z = 3$

23. P(1,– 1,2), Q(2,1,3) and R(– 1,2,– 1) $\Rightarrow \overrightarrow{PQ} = \mathbf{i} + 2\mathbf{j} + \mathbf{k}$, $\overrightarrow{PR} = -2\mathbf{i} + 3\mathbf{j} - 3\mathbf{k}$ and $\overrightarrow{PQ} \times \overrightarrow{PR} =$

$\begin{vmatrix} \mathbf{i} & \mathbf{j} & \mathbf{k} \\ 1 & 2 & 1 \\ -2 & 3 & -3 \end{vmatrix} = -9\mathbf{i} + \mathbf{j} + 7\mathbf{k}$, the normal of the plane; $-9(x - 1) + 1(y + 1) + 7(z - 2) = 0 \Rightarrow$

$-9x + y + 7z = 4$

25. (1,2,3) and $\mathbf{v} = -3\mathbf{i} + 7\mathbf{k} \Rightarrow x = 1 - 3t,\ y = 2,\ z = 3 + 7t$

27. $x = 2t,\ y = -t,\ z = -t$ a line containing the origin and perpendicular to the plane $2x - y - z = 4$; this line intersects the plane $3x - 5y + 2z = 6$ when t is the solution of $3(2t) - 5(-t) + 2(-t) = 6 \Rightarrow$ $t = \frac{2}{3} \Rightarrow (4/3, -2/3, -2/3)$ is the point of intersection

29. the desired line contains P(– 5,3,0) and Q(10,0,– 9); $\overrightarrow{PQ} = 15\mathbf{i} - 3\mathbf{j} - 9\mathbf{k}$, the direction of this line $\therefore x = -5 + 15t,\ y = 3 - 3t, z = -9t$

31. The distance $d = \sqrt{2^2 + (5 + 2t)^2 + (4t + 10)^2}$ is minimized when $f(t) = 2^2 + (5 + 2t)^2 + (4t + 10)^2$ is minimized. $f'(t) = 0 \Rightarrow t = -\frac{5}{2} \Rightarrow d = \sqrt{4 + 0 + 0} = 2$

33. $W = \mathbf{F}\cdot\overrightarrow{PQ} = |\mathbf{F}|\left|\overrightarrow{PQ}\right|\cos\theta = [40\text{ lb}](800\text{ ft})\cos 30° = 16000\sqrt{3}\text{ ft}\cdot\text{lb}$

35. a) area = $|\mathbf{A} \times \mathbf{B}| = \left|\left|\begin{matrix} \mathbf{i} & \mathbf{j} & \mathbf{k} \\ 1 & 1 & -1 \\ 2 & 1 & 1 \end{matrix}\right|\right| = |2\mathbf{i} - 3\mathbf{j} - \mathbf{k}| = \sqrt{4+9+1} = \sqrt{14}$

b) volume = $\mathbf{A} \cdot (\mathbf{B} \times \mathbf{C}) = \left|\begin{matrix} 1 & 1 & -1 \\ 2 & 1 & 1 \\ -1 & -2 & 3 \end{matrix}\right| = 1$

37. a) true b) false, $(2\mathbf{i}) \cdot (2\mathbf{i}) = 4$ while $|2\mathbf{i}| = \sqrt{2^2} = 2$ c) true

d) true e) false, they have opposite directions f) true

g) true h) true

39. the y–axis in the xy–plane and the yz–plane in three dimensional space

41. a circle centered at (0,0) with a radius of 2 in the xy–plane; a cylinder parallel with the z–axis in three dimensional space with the circle as its generating curve

43. a horizontal parabola opening to the right with its vertex at (0,0) in the xy–plane; a cylinder parallel with the z–axis in three dimensional space with the parabola as the generating curve

45. a horizontal cardioid in the $r\theta$–plane; a cylinder parallel with the z–axis in three dimensional space with the cardioid as the generating curve

47. a horizontal lemniscate of length $2\sqrt{2}$ in the $r\theta$–plane; a cylinder parallel with the z–axis in three dimensional space with the lemniscate as the generating curve

49. a sphere with a radius of 2 centered at the origin

51. the upper nappe of a cone whose surface makes a $\frac{\pi}{6}$ angle with the z–axis

53. the upper hemisphere of a sphere with a radius 1 centered at the origin

| | Rectangular | Cylindrical | Spherical |
|---|---|---|---|
| 55. | (1,0,0) | (1,0,0) | $(1,\pi/2,0)$ |
| 57. | (0,1,1) | $(1,\pi/2,1)$ | $(\sqrt{2},\pi/4,\pi/2)$ |
| 59. | (1,1,1) | $(\sqrt{2},\pi/4,1)$ | $(\sqrt{3},\tan^{-1}\sqrt{2},\pi/4)$ |

61. $z = 2 \Rightarrow$ cylindrical, $z = 2$; spherical, $\rho \cos \phi = 2$; a plane parallel with the xy–plane

63. $z = r^2 \Rightarrow$ rectangular, $z = x^2 + y^2$; spherical, $\rho = 0$ or $\rho = \dfrac{\cos \phi}{\sin^2 \phi}$ when $0 < \phi < \frac{\pi}{2}$; a paraboloid symmetric to the z–axis opening upward

65. $\rho = 4 \Rightarrow$ rectangular, $x^2 + y^2 + z^2 = 16$; cylindrical, $r^2 + z^2 = 16$; a sphere with a radius of 4 centered at the origin

67. $x^2 + y^2 + z^2 = 4$

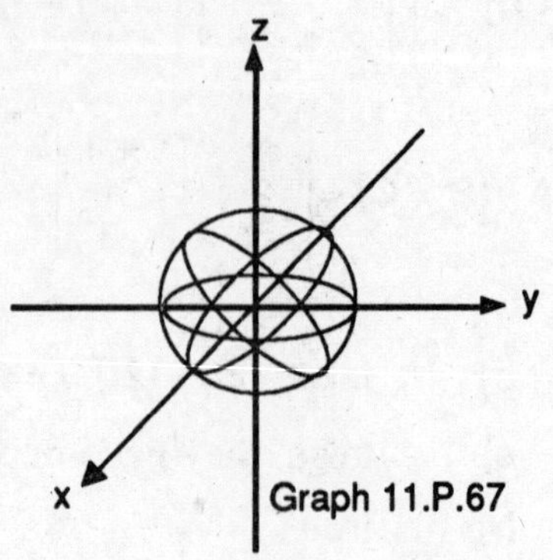

Graph 11.P.67

69. $4x^2 + 4y^2 + z^2 = 4$

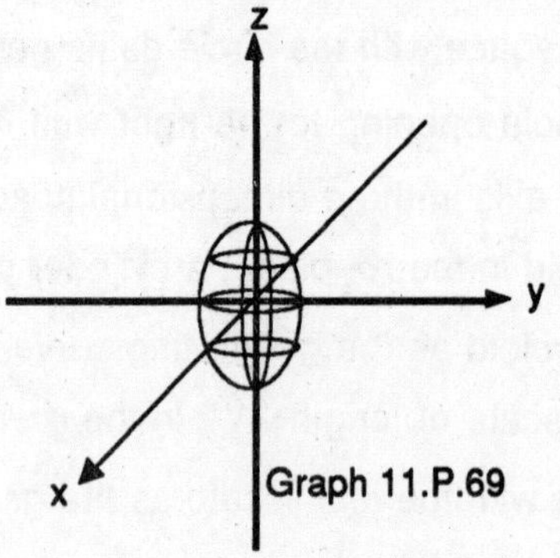

Graph 11.P.69

71. $z = -\left(x^2 + y^2\right)$

Graph 11.P.71
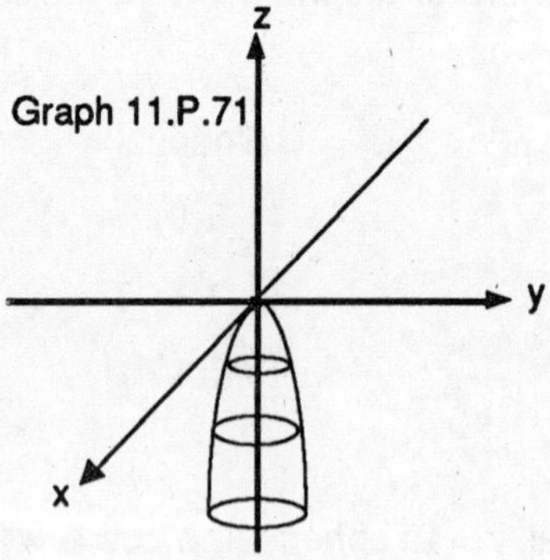


73. $x^2 + y^2 = z^2$

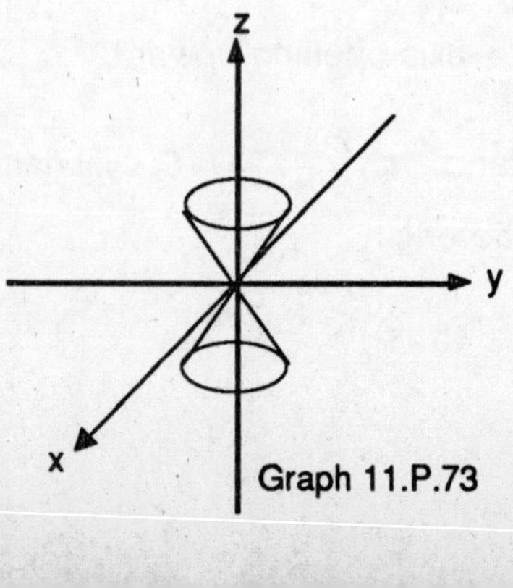

Graph 11.P.73

75. $x^2 + y^2 - z^2 = 4$

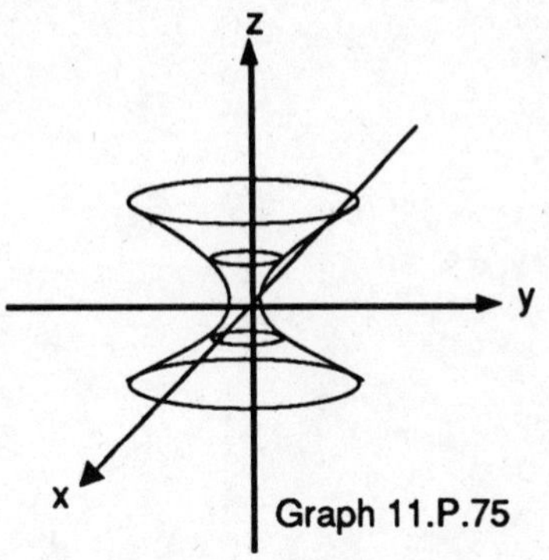

Graph 11.P.75

77. $y^2 - x^2 - z^2 = 1$

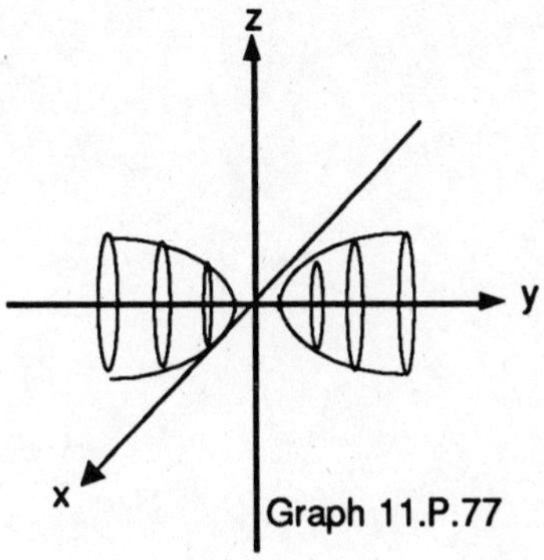

Graph 11.P.77

# CHAPTER 12

# VECTOR-VALUED FUNCTIONS AND MOTION IN SPACE

## 12.1 VECTOR-VALUED FUNCTIONS AND CURVES IN SPACE. DERIVATIVES AND INTEGRALS

1. $\mathbf{r} = (2\cos t)\,\mathbf{i} + (3\sin t)\,\mathbf{j} + 4t\,\mathbf{k} \Rightarrow \mathbf{v} = \frac{d\mathbf{r}}{dt} = (-2\sin t)\,\mathbf{i} + (3\cos t)\,\mathbf{j} + 4\,\mathbf{k}$

$\mathbf{a} = \frac{d^2\mathbf{r}}{dt^2} = (-2\cos t)\,\mathbf{i} - (3\sin t)\,\mathbf{j}$. Speed: $\left|\mathbf{v}\left(\frac{\pi}{2}\right)\right| = \sqrt{\left(-2\sin\frac{\pi}{2}\right)^2 + \left(3\cos\frac{\pi}{2}\right)^2 + 4^2} = 2\sqrt{5}$

Direction: $\frac{\mathbf{v}\left(\frac{\pi}{2}\right)}{\left|\mathbf{v}\left(\frac{\pi}{2}\right)\right|} = \left(-\frac{2}{2\sqrt{5}}\sin\frac{\pi}{2}\right)\mathbf{i} + \left(\frac{3}{2\sqrt{5}}\cos\frac{\pi}{2}\right)\mathbf{j} + \frac{4}{2\sqrt{5}}\mathbf{k} = -\frac{1}{\sqrt{5}}\mathbf{i} + \frac{2}{\sqrt{5}}\mathbf{k}$

$\mathbf{v}\left(\frac{\pi}{2}\right) = 2\sqrt{5}\left[-\frac{1}{\sqrt{5}}\mathbf{i} + \frac{2}{\sqrt{5}}\mathbf{k}\right]$

3. $\mathbf{r} = (\cos 2t)\,\mathbf{j} + (2\sin t)\,\mathbf{k} \Rightarrow \mathbf{v} = (-2\sin 2t)\,\mathbf{j} + (2\cos t)\,\mathbf{k}$, $\mathbf{a} = (-4\cos 2t)\,\mathbf{j} - (2\sin t)\,\mathbf{k}$. Speed: $|\mathbf{v}(0)| = \sqrt{(-2\sin 2(0))^2 + (2\cos 0)^2} = 2$ Direction: $\frac{\mathbf{v}(0)}{|\mathbf{v}(0)|} = \frac{-2\sin 2(0)\,\mathbf{j} + 2\cos 0\,\mathbf{k}}{2} = \mathbf{k}$. $\mathbf{v}(0) = 2\,\mathbf{k}$

5. $\mathbf{r} = (\sec t)\,\mathbf{i} + (\tan t)\,\mathbf{j} + \frac{4}{3}t\,\mathbf{k} \Rightarrow \mathbf{v} = \frac{d\mathbf{r}}{dt} = (\sec t\tan t)\,\mathbf{i} + (\sec^2 t)\,\mathbf{j} + \frac{4}{3}\mathbf{k}$, $\mathbf{a} = \frac{d^2\mathbf{r}}{dt^2} = (\sec t\tan^2 t + \sec^3 t)\,\mathbf{i} +$ $(2\sec^2 t\tan t)\,\mathbf{j}$. Speed: $\left|\mathbf{v}\left(\frac{\pi}{6}\right)\right| = \sqrt{\left(\sec\frac{\pi}{6}\tan\frac{\pi}{6}\right)^2 + \left(\sec^2\frac{\pi}{6}\right)^2 + \left(\frac{4}{3}\right)^2} = 2$. Direction: $\frac{\mathbf{v}\left(\frac{\pi}{6}\right)}{\left|\mathbf{v}\left(\frac{\pi}{6}\right)\right|}$ $= \frac{\sec\frac{\pi}{6}\tan\frac{\pi}{6}\,\mathbf{i} + \sec^2\frac{\pi}{6}\,\mathbf{j} + \frac{4}{3}\mathbf{k}}{2} = \frac{1}{3}\mathbf{i} + \frac{2}{3}\mathbf{j} + \frac{2}{3}\mathbf{k}$. $\mathbf{v}\left(\frac{\pi}{6}\right) = 2\left(\frac{1}{3}\mathbf{i} + \frac{2}{3}\mathbf{j} + \frac{2}{3}\mathbf{k}\right)$

7. $\mathbf{r} = \left(e^{-t}\right)\mathbf{i} + (2\cos 3t)\,\mathbf{j} + (2\sin 3t)\,\mathbf{k} \Rightarrow \mathbf{v} = \frac{d\mathbf{r}}{dt} = \left(-e^{-t}\right)\mathbf{i} - (6\sin 3t)\,\mathbf{j} + (6\cos 3t)\,\mathbf{k}$, $\mathbf{a} = \frac{d^2\mathbf{r}}{dt^2} =$ $\left(e^{-t}\right)\mathbf{i} - (18\cos 3t)\,\mathbf{j} - (18\sin 3t)\,\mathbf{k}$. Speed: $|\mathbf{v}(0)| = \sqrt{\left(-e^0\right)^2 + (-6\sin 3(0))^2 + (6\cos 3(0))^2} = \sqrt{37}$

Direction: $\frac{\mathbf{v}(0)}{|\mathbf{v}(0)|} = \frac{\left(-e^0\right)\mathbf{i} - 6\sin 3(0)\,\mathbf{j} + 6\cos 3(0)\,\mathbf{k}}{\sqrt{37}} = -\frac{1}{\sqrt{37}}\mathbf{i} + \frac{6}{\sqrt{37}}\mathbf{k}$.

$\mathbf{v}(0) = \sqrt{37}\left[-\frac{1}{\sqrt{37}}\mathbf{i} + \frac{6}{\sqrt{37}}\mathbf{k}.\right]$

9. $\mathbf{v} = 3\,\mathbf{i} + \sqrt{3}\,\mathbf{j} + 2t\,\mathbf{k}$, $\mathbf{a} = 2\,\mathbf{k}$. $\mathbf{v}(0) = 3\,\mathbf{i} + \sqrt{3}\,\mathbf{j}$, $\mathbf{a}(0) = 2\,\mathbf{k} \Rightarrow |\mathbf{v}(0)| = \sqrt{3^2 + \left(\sqrt{3}\right)^2 + 0^2} = \sqrt{12}$, $|\mathbf{a}(0)| = \sqrt{2^2} = 2$. $\mathbf{v}(0)\cdot\mathbf{a}(0) = 0 \Rightarrow \cos\theta = \frac{0}{2\sqrt{12}} = 0 \Rightarrow \theta = \frac{\pi}{2}$

11. $\mathbf{v} = \frac{2t}{t^2+1}\mathbf{i} + \frac{1}{t^2+1}\mathbf{j} + t\left(t^2+1\right)^{-1/2}\mathbf{k}$, $\mathbf{a} = \frac{-2t^2+2}{\left(t^2+1\right)^2}\mathbf{i} - \frac{2t}{\left(t^2+1\right)^2}\mathbf{j} + \frac{1}{\left(t^2+1\right)^{3/2}}\mathbf{k}$. $\mathbf{v}(0) = \mathbf{j}$, $\mathbf{a}(0) = 2\,\mathbf{i} + \mathbf{k} \Rightarrow |\mathbf{v}(0)| = 1$, $|\mathbf{a}(0)| = \sqrt{2^2+1^2} = \sqrt{5}$. $\mathbf{v}(0)\cdot\mathbf{a}(0) = 0 \Rightarrow \cos\theta = \frac{0}{1\sqrt{5}} = 0 \Rightarrow \theta = \frac{\pi}{2}$

13. $\mathbf{v} = (1-\cos t)\,\mathbf{i} + (\sin t)\,\mathbf{j}$, $\mathbf{a} = (\sin t)\,\mathbf{i} + (\cos t)\,\mathbf{j} \Rightarrow \mathbf{v}\cdot\mathbf{a} = \sin t(1-\cos t) + \sin t(\cos t) = \sin t$.
$\mathbf{v}\cdot\mathbf{a} = 0 \Rightarrow \sin t = 0 \Rightarrow t = 0, \pi, 2\pi$

15. $$\int_0^1 \left(t^3\,\mathbf{i} + 7\,\mathbf{j} + (t+1)\,\mathbf{k}\right)dt = \left[\frac{t^4}{4}\right]_0^1\mathbf{i} + [7t]_0^1\,\mathbf{j} + \left[\frac{t^2}{2}+t\right]_0^1\mathbf{k} = \frac{1}{4}\,\mathbf{i} + 7\,\mathbf{j} + \frac{3}{2}\,\mathbf{k}$$

17. $$\int_{-\pi/4}^{\pi/4} \left((\sin t)\,\mathbf{i} + (1+\cos t)\,\mathbf{j} + (\sec^2 t)\,\mathbf{k}\right)dt = [-\cos t]_{-\pi/4}^{\pi/4}\,\mathbf{i} + [t+\sin t]_{-\pi/4}^{\pi/4}\,\mathbf{j} + [\tan t]_{-\pi/4}^{\pi/4}\,\mathbf{k} = \left(\frac{\pi+2\sqrt{2}}{2}\right)\mathbf{j} + 2\,\mathbf{k}$$

19. $$\int_1^4 \left(\frac{1}{t}\,\mathbf{i} + \frac{1}{5-t}\,\mathbf{j} + \frac{1}{2t}\,\mathbf{k}\right)dt = [\ln t]_1^4\,\mathbf{i} + [-\ln(5-t)]_1^4\,\mathbf{j} + \left[\frac{1}{2}\ln 2t\right]_1^4\mathbf{k} = (\ln 4)\,\mathbf{i} + (\ln 4)\,\mathbf{j} + (\ln 2)\,\mathbf{k}$$

21. $\mathbf{v} = (\cos t)\,\mathbf{i} - (\sin t)\,\mathbf{j}$, $\mathbf{a} = -(\sin t)\,\mathbf{i} - (\cos t)\,\mathbf{j} \Rightarrow$ For $t = \frac{\pi}{4}$, $\mathbf{v}\left(\frac{\pi}{4}\right) = \frac{\sqrt{2}}{2}\,\mathbf{i} - \frac{\sqrt{2}}{2}\,\mathbf{j}$, $\mathbf{a}\left(\frac{\pi}{4}\right) = -\frac{\sqrt{2}}{2}\,\mathbf{i} - \frac{\sqrt{2}}{2}\,\mathbf{j}$;
For $t = \frac{\pi}{2}$, $\mathbf{v}\left(\frac{\pi}{2}\right) = -\,\mathbf{j}$, $\mathbf{a}\left(\frac{\pi}{2}\right) = -\,\mathbf{i}$

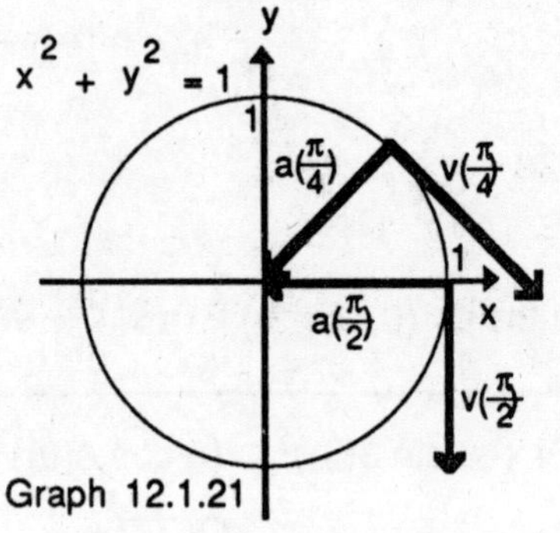


Graph 12.1.21

23. $\mathbf{v} = (1 - \cos t)\,\mathbf{i} + (\sin t)\,\mathbf{j}$, $\mathbf{a} = (\sin t)\,\mathbf{i} + (\cos t)\,\mathbf{j} \Rightarrow$ For $t = \pi$, $\mathbf{v}(\pi) = 2\,\mathbf{i}$, $\mathbf{a}(\pi) = -\mathbf{j}$; For $t = \frac{3\pi}{2}$,

$\mathbf{v}\left(\frac{3\pi}{2}\right) = \mathbf{i} - \mathbf{j}$, $\mathbf{a}\left(\frac{3\pi}{2}\right) = -\mathbf{i}$

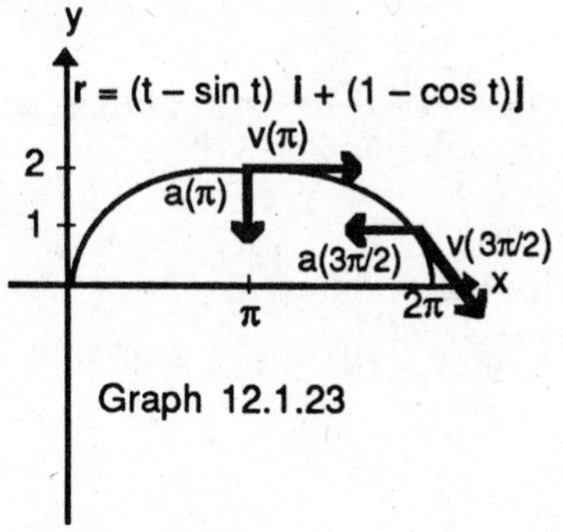


Graph 12.1.23

25. $\mathbf{r} = \int (-t\,\mathbf{i} - t\,\mathbf{j} - t\,\mathbf{k})\,dt = -\frac{t^2}{2}\mathbf{i} - \frac{t^2}{2}\mathbf{j} - \frac{t^2}{2}\mathbf{k} + \mathbf{C}$. $\mathbf{r}(0) = 0\,\mathbf{i} - 0\,\mathbf{j} - 0\,\mathbf{k} + \mathbf{C} = \mathbf{i} + 2\,\mathbf{j} + 3\,\mathbf{k} \Rightarrow$

$\mathbf{C} = \mathbf{i} + 2\,\mathbf{j} + 3\,\mathbf{k}$. $\therefore\ \mathbf{r} = \left(-\frac{t^2}{2} + 1\right)\mathbf{i} + \left(-\frac{t^2}{2} + 2\right)\mathbf{j} + \left(-\frac{t^2}{2} + 3\right)\mathbf{k}$

27. $\mathbf{r} = \int \left(\frac{3}{2}(t+1)^{1/2}\,\mathbf{i} + e^{-t}\,\mathbf{j} + \frac{1}{t+1}\mathbf{k}\right) dt = (t+1)^{3/2}\,\mathbf{i} - e^{-t}\,\mathbf{j} + \ln(t+1)\,\mathbf{k} + \mathbf{C}$. $\mathbf{r}(0) = \mathbf{k} \Rightarrow$

$(0+1)^{3/2}\,\mathbf{i} - e^{-0}\,\mathbf{j} + \ln(0+1)\,\mathbf{k} + \mathbf{C} = \mathbf{k} \Rightarrow \mathbf{C} = -\mathbf{i} + \mathbf{j} + \mathbf{k}$.

$\therefore\ \mathbf{r} = \left((t+1)^{3/2} - 1\right)\mathbf{i} + \left(1 - e^{-t}\right)\mathbf{j} + (1 + \ln(t+1))\,\mathbf{k}$

29. $\frac{d\mathbf{r}}{dt} = \int (-32\,\mathbf{k})\,dt = -32t\,\mathbf{k} + \mathbf{C}_1$. $\frac{d\mathbf{r}}{dt}(0) = 8\,\mathbf{i} + 8\,\mathbf{j} \Rightarrow -32(0)\,\mathbf{k} + \mathbf{C}_1 = 8\,\mathbf{i} + 8\,\mathbf{j} \Rightarrow \mathbf{C}_1 = 8\,\mathbf{i} + 8\,\mathbf{j}$

$\therefore\ \frac{d\mathbf{r}}{dt} = 8\,\mathbf{i} + 8\,\mathbf{j} - 32t\,\mathbf{k}$. $\mathbf{r} = \int (8\,\mathbf{i} + 8\,\mathbf{j} - 32t\,\mathbf{k})\,dt = 8t\,\mathbf{i} + 8t\,\mathbf{j} - 16t^2\,\mathbf{k} + \mathbf{C}_2$. $\mathbf{r}(0) = 100\,\mathbf{k} \Rightarrow$

$8(0)\,\mathbf{i} + 8(0)\,\mathbf{j} - 16(0)^2\,\mathbf{k} + \mathbf{C}_2 = 100\,\mathbf{k} \Rightarrow \mathbf{C}_2 = 100\,\mathbf{k}$. $\therefore\ \mathbf{r} = 8t\,\mathbf{i} + 8t\,\mathbf{j} + (100 - 16t^2)\,\mathbf{k}$

31. $\mathbf{v} = (1 - \cos t)\,\mathbf{i} + (\sin t)\,\mathbf{j}$, $\mathbf{a} = (\sin t)\,\mathbf{i} + (\cos t)\,\mathbf{j}$. $|\mathbf{v}|^2 = (1 - \cos t)^2 + \sin^2 t = 2 - 2\cos t$. $|\mathbf{v}|^2$ is at a max when $\cos t = -1 \Rightarrow t = \pi, 3\pi, 5\pi$, etc. At these values of t, $|\mathbf{v}|^2 = 4 \Rightarrow \max|\mathbf{v}| = \sqrt{4} = 2$. $|\mathbf{v}|^2$ is at a min when $\cos t = 1 \Rightarrow t = 0, 2\pi, 4\pi$, etc. At these values of t, $|\mathbf{v}|^2 = 0 \Rightarrow \min|\mathbf{v}| = 0$.
$|\mathbf{a}|^2 = \sin^2 t + \cos^2 t = 1$ for every $t \Rightarrow \max|\mathbf{a}| = \min|\mathbf{a}| = \sqrt{1} = 1$.

33. Let $\mathbf{f} = \mathbf{C}$, a constant vector. Then $\mathbf{f} = c_1\,\mathbf{i} + c_2\,\mathbf{j} + c_3\,\mathbf{k}$ where $c_1, c_2, c_3$ are Real Numbers.

$\frac{d\mathbf{f}}{dt} = 0\,\mathbf{i} + 0\,\mathbf{j} + 0\,\mathbf{k} = \mathbf{0}$

35. Let $\mathbf{u} = f_1(t)\,\mathbf{i} + f_2(t)\,\mathbf{j} + f_3(t)\,\mathbf{k}$, $\mathbf{v} = g_1(t)\,\mathbf{i} + g_2(t)\,\mathbf{j} + g_3(t)\,\mathbf{k}$. Then $\mathbf{u} + \mathbf{v} = (f_1(t) + g_1(t))\,\mathbf{i} + (f_2(t) + g_2(t))\,\mathbf{j} +$

$(f_3(t) + g_3(t))\,\mathbf{k} \Rightarrow \frac{d}{dt}(\mathbf{u} + \mathbf{v}) = (f'_1(t) + g'_1(t))\,\mathbf{i} + (f'_2(t) + g'_2(t))\,\mathbf{j} + (f'_3(t) + g'_3(t))\,\mathbf{k} =$

$(f'_1(t)\,\mathbf{i} + f'_2(t)\,\mathbf{j} + f'_3(t)\,\mathbf{k}) + (g'_1(t)\,\mathbf{i} + g'_2(t)\,\mathbf{j} + g'_3(t)\,\mathbf{k} = \frac{d\mathbf{u}}{dt} + \frac{d\mathbf{v}}{dt}$.

$\mathbf{u} - \mathbf{v} = (f_1(t) - g_1(t))\,\mathbf{i} + (f_2(t) - g_2(t))\,\mathbf{j} + (f_3(t) - g_3(t))\,\mathbf{k} \Rightarrow \frac{d}{dt}(\mathbf{u} - \mathbf{v}) = (f'_1(t) - g'_1(t))\,\mathbf{i} + (f'_2(t) - g'_2(t))\,\mathbf{j} +$

$(f'_3(t) - g'_3(t))\,\mathbf{k} = (f'_1(t)\,\mathbf{i} + f'_2(t)\,\mathbf{j} + f'_3(t)\,\mathbf{k}) - (g'_1(t)\,\mathbf{i} + g'_2(t)\,\mathbf{j} + g'_3(t)\,\mathbf{k} = \frac{d\mathbf{u}}{dt} - \frac{d\mathbf{v}}{dt}$.

## SECTION 12.2 MODELING PROJECTILE MOTION

1. $x = \left(v_0 \cos\alpha\right)t \Rightarrow (21\text{ km})\left(\frac{1000\text{ m}}{1\text{ km}}\right) = 840\text{ m/s}(\cos 60°)t \Rightarrow t = \frac{21\,000\text{ m}}{(840\text{ km/s})(\cos 60°)} = 50\text{ seconds}$

3. a) $t = \frac{2v_0 \sin\alpha}{g} = \frac{2(500\text{ m/s})\sin 45°}{9.8\text{ m/s}^2} = 72.2\text{ seconds.}$ $R = \frac{v_0^2}{g}\sin 2\alpha = \frac{(500\text{ m/s})^2}{9.8\text{ m/s}^2}(\sin 2(45°)) =$ $25\,510.2\text{ m}$

   b) $x = (v_0 \cos\alpha)t \Rightarrow 5000\text{ m} = (500\text{ m/s})(\cos 45°)t \Rightarrow t = \frac{5000\text{ m}}{(500\text{ m/s})\cos 45°} = 14.14\text{ s}$

   $y = (v_0 \sin\alpha)t - \frac{1}{2}gt^2 \Rightarrow y = (500\text{ m/s})(\sin 45°)(14.14\text{ s}) - \frac{1}{2}\left(9.8\text{ m/s}^2\right)(14.14\text{ s})^2 = 4020.3\text{ m}$

   c) $y_{max} = \frac{\left(v_0 \sin\alpha\right)^2}{2g} = \frac{\left((500\text{ m/s})\sin 45°\right)^2}{2\left(9.8\text{ m/s}^2\right)} = 6377.6\text{ m}$

5. $R = \frac{v_0^2}{g}\sin 2\alpha = \frac{v_0^2}{g}\left(2\sin\alpha\cos\alpha\right) = \frac{v_0^2}{g}\left(2\cos(90° - \alpha)\sin(90° - \alpha)\right) = \frac{v_0^2}{g}\left(\sin 2(90° - \alpha)\right)$

7. $R = \frac{v_0^2}{g}\sin 2\alpha \Rightarrow 10\text{ m} = \frac{v_0^2}{9.8\text{ m/s}^2}\sin 2(45°) \Rightarrow v_0^2 = 98\text{ m}^2/\text{s}^2 \Rightarrow v_0 = 9.9\text{ m/s.}$

   $6\text{ m} = \frac{(9.9\text{ m/s})^2}{9.8\text{ m/s}^2}\sin 2\alpha \Rightarrow \sin 2\alpha = 0.59999 \Rightarrow 2\alpha = 36.87°\text{ or }143.12° \Rightarrow \alpha = 18.44°\text{ or }71.56°$

9. $R = \frac{v_0^2}{g}\sin 2\alpha \Rightarrow (746.4\text{ ft}) = \frac{v_0^2}{32\text{ ft/sec}^2}\sin 2(9°) \Rightarrow v_0^2 = 77\,292.84\text{ ft}^2/\text{sec}^2 \Rightarrow v_0 = 278.01\text{ ft/sec} =$ 189.6 mph

11. $y_{max} = \frac{\left(v_0 \sin\alpha\right)^2}{2g} \Rightarrow \frac{3}{4}y_{max} = \frac{3\left(v_0 \sin\alpha\right)^2}{8g}$. $y = \left(v_0 \sin\alpha\right)t - \frac{1}{2}gt^2 \Rightarrow \frac{3\left(v_0 \sin\alpha\right)^2}{8g} =$ $\left(v_0 \sin\alpha\right)t - \frac{1}{2}gt^2 \Rightarrow 3\left(v_0 \sin\alpha\right)^2 = \left(8gv_0 \sin\alpha\right)t - 4g^2t^2 \Rightarrow$

    $4g^2t^2 - \left(8gv_0 \sin\alpha\right)t + 3\left(v_0 \sin\alpha\right)^2 = 0 \Rightarrow 2gt - 3v_0 \sin\alpha = 0\text{ or }2gt - v_0 \sin\alpha = 0 \Rightarrow t = \frac{3v_0 \sin\alpha}{2g}$ or $t = \frac{v_0 \sin\alpha}{2g}$. Since the time it takes to reach $y_{max}$ is $t_{max} = \frac{v_0 \sin\alpha}{g}$, then the time it takes the projectile to reach $\frac{3}{4}$ of $y_{max}$ is $t = \frac{v_0 \sin\alpha}{2g}$ or $\frac{1}{2}t_{max}$.

13. $x = \left(v_0 \cos\alpha\right)t \Rightarrow 135\text{ ft} = (90\text{ ft/sec})(\cos 30°)t \Rightarrow t = 1.732\text{ sec.}$ $y = \left(v_0 \sin\alpha\right)t - \frac{1}{2}gt^2 \Rightarrow$ $y = (90\text{ ft/sec})(\sin 30°)(1.732\text{ sec}) - \frac{1}{2}\left(32\text{ ft/sec}^2\right)(1.732\text{ sec})^2 \Rightarrow y = 29.94\text{ ft.}$ The golf ball will clip the leaves at the top.

15. $x = \left(v_0 \cos\alpha\right)t \Rightarrow 315\text{ ft} = (v_0 \cos 20°)t \Rightarrow v_0 = \frac{315}{t\cos 20°}$. $y = \left(v_0 \sin\alpha\right)t - \frac{1}{2}gt^2 \Rightarrow$

    $34\text{ ft} = \frac{315}{t\cos 20°}(t\sin 20°) - \frac{1}{2}(32)t^2 \Rightarrow 34 = 315\tan 20° - 16t^2 \Rightarrow t^2 = 5.04\text{ sec}^2 \Rightarrow t = 2.25\text{ sec}$

    $t = 2.25\text{ sec} \Rightarrow v_0 = \frac{315}{(2.25)\cos 20°} = 148.98\text{ ft/sec}$

17. Height of the Marble A, R units downrange: $x = (v_0 \cos\alpha)\, t$ and $x = R \Rightarrow R = (v_0 \cos\alpha)\, t \Rightarrow$

$t = \frac{R}{v_0 \cos\alpha}$. $y = (v_0 \sin\alpha)t - \frac{1}{2}gt^2 \Rightarrow y = (v_0 \sin\alpha)\left(\frac{R}{v_0 \cos\alpha}\right) - \frac{1}{2}g\left(\frac{R}{v_0 \cos\alpha}\right)^2 \Rightarrow$

$y = R\tan\alpha - \frac{1}{2}g\left(\frac{R^2}{v_0^2 \cos^2\alpha}\right)$ is the height of Marble A after $t = \frac{R}{v_0 \cos\alpha}$ seconds.

Height of the Marble B, at $t = \frac{R}{v_0 \cos\alpha}$ seconds: $y = R\tan\alpha - \frac{1}{2}gt^2 = R\tan\alpha - \frac{1}{2}g\left(\frac{R}{v_0 \cos\alpha}\right)^2 =$

$R\tan\alpha - \frac{1}{2}g\left(\frac{R^2}{v_0^2 \cos^2\alpha}\right)$ which is the height of Marble A. ∴ They collide regardless of the initial velocity.

19. $\frac{d\mathbf{r}}{dt} = \int (-g\,\mathbf{j})\,dt = -gt\,\mathbf{j} + \mathbf{C}_1$. $\frac{d\mathbf{r}}{dt}(0) = (v_0 \cos\alpha)\,\mathbf{i} + (v_0 \sin\alpha)\,\mathbf{j} \Rightarrow -g(0)\,\mathbf{j} + \mathbf{C}_1 = (v_0 \cos\alpha)\,\mathbf{i} +$

$(v_0 \sin\alpha)\,\mathbf{j} \Rightarrow \mathbf{C}_1 = (v_0 \cos\alpha)\,\mathbf{i} + (v_0 \sin\alpha)\,\mathbf{j} \Rightarrow \frac{d\mathbf{r}}{dt} = (v_0 \cos\alpha)\,\mathbf{i} + (v_0 \sin\alpha - gt)\,\mathbf{j}$.

$\mathbf{r} = \int \left((v_0 \cos\alpha)\,\mathbf{i} + (v_0 \sin\alpha - gt)\,\mathbf{j}.\right) dt = (v_0 t \cos\alpha)\,\mathbf{i} + (v_0 t \sin\alpha - \frac{1}{2}gt^2)\,\mathbf{j} + \mathbf{C}_2$.

$\mathbf{r}(0) = x_0\,\mathbf{i} + y_0\,\mathbf{j} \Rightarrow (v_0(0)\cos\alpha)\,\mathbf{i} + (v_0(0)\sin\alpha - \frac{1}{2}g(0)^2)\,\mathbf{j} + \mathbf{C}_2 = x_0\,\mathbf{i} + y_0\,\mathbf{j} \Rightarrow$

$\mathbf{C}_2 = x_0\,\mathbf{i} + y_0\,\mathbf{j}$. ∴ $\mathbf{r} = (x_0 + v_0 t\cos\alpha)\,\mathbf{i} + (y_0 + v_0 t \sin\alpha - \frac{1}{2}gt^2)\,\mathbf{j} \Rightarrow x = x_0 + v_0 t\cos\alpha$,

$y = y_0 + v_0 t \sin\alpha - \frac{1}{2}gt^2$

## SECTION 12.3 DIRECTED DISTANCE AND THE UNIT TANGENT VECTOR T

1. $\mathbf{r} = (2\cos t)\,\mathbf{i} + (2\sin t)\,\mathbf{j} + \sqrt{5}\,t\,\mathbf{k} \Rightarrow \mathbf{v} = (-2\sin t)\,\mathbf{i} + (2\cos t)\,\mathbf{j} + \sqrt{5}\,\mathbf{k} \Rightarrow$

$|\mathbf{v}| = \sqrt{(-2\sin t)^2 + (2\cos t)^2 + (\sqrt{5})^2} = \sqrt{4\sin^2 t + 4\cos^2 t + 5} = 3$. $\mathbf{T} = \frac{\mathbf{v}}{|\mathbf{v}|} = \left(-\frac{2}{3}\sin t\right)\mathbf{i} +$

$\left(\frac{2}{3}\cos t\right)\mathbf{j} + \frac{\sqrt{5}}{3}\mathbf{k}$. Length $= \int_0^{\pi} |\mathbf{v}|\,dt = \int_0^{\pi} 3\,dt = [3t]_0^{\pi} = 3\pi$

3. $\mathbf{r} = t\,\mathbf{i} + \frac{2}{3}t^{3/2}\,\mathbf{k} \Rightarrow \mathbf{v} = \mathbf{i} + t^{1/2}\,\mathbf{k} \Rightarrow |\mathbf{v}| = \sqrt{1^2 + \left(t^{1/2}\right)^2} = \sqrt{1+t}$. $\mathbf{T} = \frac{\mathbf{v}}{|\mathbf{v}|} = \frac{1}{\sqrt{1+t}}\mathbf{i} + \frac{\sqrt{t}}{\sqrt{1+t}}\mathbf{k}$

Length $= \int_0^{8} \sqrt{1+t}\,dt = \left[\frac{2}{3}(1+t)^{3/2}\right]_0^{8} = \frac{52}{3}$

5. $\mathbf{r} = (2+t)\,\mathbf{i} - (t+1)\,\mathbf{j} + t\,\mathbf{k} \Rightarrow \mathbf{v} = \mathbf{i} - \mathbf{j} + \mathbf{k} \Rightarrow |\mathbf{v}| = \sqrt{1^2 + (-1)^2 + 1^2} = \sqrt{3}$. $\mathbf{T} = \frac{\mathbf{v}}{|\mathbf{v}|} = \frac{1}{\sqrt{3}}\mathbf{i} - \frac{1}{\sqrt{3}}\mathbf{j} + \frac{1}{\sqrt{3}}\mathbf{k}$

Length $= \int_0^{3} \sqrt{3}\,dt = \left[\sqrt{3}\,t\right]_0^{3} = 3\sqrt{3}$

7. $\mathbf{r} = (t\cos t)\,\mathbf{i} + (t\sin t)\,\mathbf{j} + \frac{2\sqrt{2}}{3}t^{3/2}\,\mathbf{k} \Rightarrow \mathbf{v} = (\cos t - t\sin t)\,\mathbf{i} + (\sin t + t\cos t)\,\mathbf{j} + (\sqrt{2}\;t^{1/2})\,\mathbf{k} \Rightarrow$
$|\mathbf{v}| = \sqrt{(\cos t - t\sin t)^2 + (\sin t + t\cos t)^2 + (\sqrt{2}\;t^{1/2})^2} = \sqrt{1 + t^2 + 2t} = \sqrt{(t+1)^2} = |t+1| = t + 1$ since
$t \ge 0$. $\mathbf{T} = \frac{\mathbf{v}}{|\mathbf{v}|} = \left(\frac{\cos t - t\sin t}{t+1}\right)\mathbf{i} + \left(\frac{\sin t + t\cos t}{t+1}\right)\mathbf{j} + \left(\frac{\sqrt{2}\;t^{1/2}}{t+1}\right)\mathbf{k}$. Length $= \int_0^{\pi} (t+1)\,dt =$
$\left[\frac{t^2}{2} + t\right]_0^{\pi} = \frac{\pi^2}{2} + \pi$

9. $\mathbf{r} = (4\cos t)\,\mathbf{i} + (4\sin t)\,\mathbf{j} + 3t\,\mathbf{k} \Rightarrow \mathbf{v} = (-4\sin t)\,\mathbf{i} + (4\cos t)\,\mathbf{j} + 3\,\mathbf{k} \Rightarrow |\mathbf{v}| = \sqrt{(-4\sin t)^2 + (4\cos t)^2 + 3^2}$
$= \sqrt{25} = 5$. $s(t) = \int_0^t 5\,d\tau = 5t$ Length $= s\left(\frac{\pi}{2}\right) = \frac{5\pi}{2}$

11. $\mathbf{r} = (e^t\cos t)\,\mathbf{i} + (e^t\sin t)\,\mathbf{j} + e^t\,\mathbf{k} \Rightarrow \mathbf{v} = (e^t\cos t - e^t\sin t)\,\mathbf{i} + (e^t\sin t + e^t\cos t)\,\mathbf{j} + e^t\,\mathbf{k} \Rightarrow$
$|\mathbf{v}| = \sqrt{(e^t\cos t - e^t\sin t)^2 + (e^t\sin t + e^t\cos t)^2 + (e^t)^2} = \sqrt{3e^{2t}} = \sqrt{3}\,e^t$
$s(t) = \int_0^t \sqrt{3}\,e^{\tau}\,d\tau = \sqrt{3}\,e^t - \sqrt{3}$. Length $= s(\ln 4) = \sqrt{3}\,e^{\ln 4} - \sqrt{3} = 3\sqrt{3}$

13. $\mathbf{r} = (\sqrt{2}\,t)\,\mathbf{i} + (\sqrt{2}\,t)\,\mathbf{j} + (1 - t^2)\,\mathbf{k} \Rightarrow \mathbf{v} = \sqrt{2}\,\mathbf{i} + \sqrt{2}\,\mathbf{j} - 2t\,\mathbf{k} \Rightarrow |\mathbf{v}| = \sqrt{(\sqrt{2})^2 + (\sqrt{2})^2 + (-2t)^2} = \sqrt{4 + 4t^2}$
$= 2\sqrt{1+t^2}$ Length $= \int_0^1 2\sqrt{1+t^2}\,dt = \left[2\left(\frac{t}{2}\sqrt{1+t^2} + \frac{1}{2}\ln\left(t + \sqrt{1+t^2}\right)\right)\right]_0^1 = \sqrt{2} + \ln\left(1 + \sqrt{2}\right)$

15. a) $\mathbf{r} = (\cos 4t)\,\mathbf{i} + (\sin 4t)\,\mathbf{j} + 4t\,\mathbf{k} \Rightarrow \mathbf{v} = (-4\sin 4t)\,\mathbf{i} + (4\cos 4t)\,\mathbf{j} + 4\,\mathbf{k} \Rightarrow$
$|\mathbf{v}| = \sqrt{(-4\sin 4t)^2 + (4\cos 4t)^2 + 4^2} = \sqrt{32} = 4\sqrt{2}$. Length $= \int_0^{\pi/2} 4\sqrt{2}\,dt = \left[4\sqrt{2}\,t\right]_0^{\pi/2} = 2\pi\sqrt{2}$

b) $\mathbf{r} = \left(\cos\frac{t}{2}\right)\mathbf{i} + \left(\sin\frac{t}{2}\right)\mathbf{j} + \frac{t}{2}\mathbf{k} \Rightarrow \mathbf{v} = \left(-\frac{1}{2}\sin\frac{t}{2}\right)\mathbf{i} + \left(\frac{1}{2}\cos\frac{t}{2}\right)\mathbf{j} + \frac{1}{2}\mathbf{k} \Rightarrow$
$|\mathbf{v}| = \sqrt{\left(-\frac{1}{2}\sin\frac{t}{2}\right)^2 + \left(\frac{1}{2}\cos\frac{t}{2}\right)^2 + \left(\frac{1}{2}\right)^2} = \sqrt{\frac{1}{4} + \frac{1}{4}} = \frac{\sqrt{2}}{2}$. Length $= \int_0^{4\pi} \frac{\sqrt{2}}{2}\,dt = \left[\frac{\sqrt{2}}{2}\,t\right]_0^{4\pi} =$
$2\pi\sqrt{2}$

c) $\mathbf{r} = (\cos t)\,\mathbf{i} - (\sin t)\,\mathbf{j} - t\,\mathbf{k} \Rightarrow \mathbf{v} = (-\sin t)\,\mathbf{i} - (\cos t)\,\mathbf{j} - \mathbf{k} \Rightarrow |\mathbf{v}| = \sqrt{(-\sin t)^2 + (-\cos t)^2 + (-1)^2} =$
$\sqrt{1+1} = \sqrt{2}$. Length $= \int_{-2\pi}^{0} \sqrt{2}\,dt = \left[\sqrt{2}\,t\right]_{-2\pi}^{0} = 2\pi\sqrt{2}$

## SECTION 12.4 CURVATURE, TORSION AND THE TNB FRAME

1. $\mathbf{r} = t\,\mathbf{i} + \ln(\cos t)\,\mathbf{j} \Rightarrow \mathbf{v} = \mathbf{i} + \frac{-\sin t}{\cos t}\mathbf{j} = \mathbf{i} - \tan t\,\mathbf{j} \Rightarrow |\mathbf{v}| = \sqrt{1^2 + (-\tan t)^2} = \sqrt{\sec^2 t} = |\sec t| = \sec t$ since $-\frac{\pi}{2} < t < \frac{\pi}{2}$. $\mathbf{T} = \frac{\mathbf{v}}{|\mathbf{v}|} = \frac{1}{\sec t}\mathbf{i} - \frac{\tan t}{\sec t}\mathbf{j} = \cos t\,\mathbf{i} - \sin t\,\mathbf{j}$. $\frac{d\mathbf{T}}{dt} = -\sin t\,\mathbf{i} - \cos t\,\mathbf{j} \Rightarrow$ $\left|\frac{d\mathbf{T}}{dt}\right| = \sqrt{(-\sin t)^2 + (-\cos t)^2} = 1$. $\mathbf{N} = \frac{d\mathbf{T}/dt}{|d\mathbf{T}/dt|} = (-\sin t)\,\mathbf{i} - (\cos t)\,\mathbf{j}$. $\mathbf{a} = (-\sec^2 t)\,\mathbf{j} \Rightarrow$

$$\mathbf{v} \times \mathbf{a} = \begin{vmatrix} \mathbf{i} & \mathbf{j} & \mathbf{k} \\ 1 & -\tan t & 0 \\ 0 & -\sec^2 t & 0 \end{vmatrix} = (-\sec^2 t)\,\mathbf{k}.\ |\mathbf{v} \times \mathbf{a}| = \sqrt{(-\sec^2 t)^2} = \sec^2 t \Rightarrow \kappa = \frac{|\mathbf{v} \times \mathbf{a}|}{|\mathbf{v}|^3} = \frac{\sec^2 t}{\sec^3 t} = \cos t$$

3. $\mathbf{r} = (2t + 3)\,\mathbf{i} + (5 - t^2)\,\mathbf{j} \Rightarrow \mathbf{v} = 2\,\mathbf{i} - 2t\,\mathbf{j} \Rightarrow |\mathbf{v}| = \sqrt{2^2 + (-2t)^2} = 2\sqrt{1 + t^2}$. $\mathbf{T} = \frac{\mathbf{v}}{|\mathbf{v}|} = \frac{2}{2\sqrt{1+t^2}}\mathbf{i} + \frac{-2t}{2\sqrt{1+t^2}}\mathbf{j} = \frac{1}{\sqrt{1+t^2}}\mathbf{i} - \frac{t}{\sqrt{1+t^2}}\mathbf{j}$. $\frac{d\mathbf{T}}{dt} = = \frac{-t}{\left(\sqrt{1+t^2}\right)^3}\mathbf{i} - \frac{1}{\left(\sqrt{1+t^2}\right)^3}\mathbf{j} \Rightarrow$

$$\left|\frac{d\mathbf{T}}{dt}\right| = \sqrt{\left(\frac{-t}{\left(\sqrt{1+t^2}\right)^3}\right)^2 + \left(-\frac{1}{\left(\sqrt{1+t^2}\right)^3}\right)^2} = \sqrt{\frac{1}{(1+t^2)^2}} = \frac{1}{1+t^2}.\ \mathbf{N} = \frac{d\mathbf{T}/dt}{|d\mathbf{T}/dt|} = \frac{-t}{\sqrt{1+t^2}}\mathbf{i}$$

$$-\frac{1}{\sqrt{1+t^2}}\mathbf{j}.\ \mathbf{a} = -2\,\mathbf{j} \Rightarrow \mathbf{v} \times \mathbf{a} = \begin{vmatrix} \mathbf{i} & \mathbf{j} & \mathbf{k} \\ 2 & -2t & 0 \\ 0 & -2 & 0 \end{vmatrix} = -4\,\mathbf{k} \Rightarrow |\mathbf{v} \times \mathbf{a}| = \sqrt{(-4)^2} = 4.\ \kappa = \frac{|\mathbf{v} \times \mathbf{a}|}{|\mathbf{v}|^3} =$$

$$\frac{4}{\left(2\sqrt{1+t^2}\right)^3} = \frac{1}{2\left(\sqrt{1+t^2}\right)^3}$$

5. $\mathbf{r} = (3\sin t)\,\mathbf{i} + (3\cos t)\,\mathbf{j} + 4t\,\mathbf{k} \Rightarrow \mathbf{v} = (3\cos t)\,\mathbf{i} + (-3\sin t)\,\mathbf{j} + 4\,\mathbf{k} \Rightarrow |\mathbf{v}| = \sqrt{(3\cos t)^2 + (-3\sin t)^2 + 4^2}$ $= \sqrt{25} = 5$. $\mathbf{T} = \frac{\mathbf{v}}{|\mathbf{v}|} = \frac{3\cos t}{5}\mathbf{i} - \frac{3\sin t}{5}\mathbf{j} + \frac{4}{5}\mathbf{k} \Rightarrow \frac{d\mathbf{T}}{dt} = \left(-\frac{3}{5}\sin t\right)\mathbf{i} - \left(\frac{3}{5}\cos t\right)\mathbf{j}$

$$\left|\frac{d\mathbf{T}}{dt}\right| = \sqrt{\left(-\frac{3}{5}\sin t\right)^2 + \left(-\frac{3}{5}\cos t\right)^2} = \frac{3}{5}.\ \mathbf{N} = \frac{d\mathbf{T}/dt}{|d\mathbf{T}/dt|} = (-\sin t)\,\mathbf{i} - (\cos t)\,\mathbf{j}$$

$$\mathbf{a} = (-3\sin t)\,\mathbf{i} + (-3\cos t)\,\mathbf{j} \Rightarrow \mathbf{v} \times \mathbf{a} = \begin{vmatrix} \mathbf{i} & \mathbf{j} & \mathbf{k} \\ 3\cos t & -3\sin t & 4 \\ -3\sin t & -3\cos t & 0 \end{vmatrix} = (12\cos t)\,\mathbf{i} - (12\sin t)\,\mathbf{j} - 9\,\mathbf{k} \Rightarrow$$

$$|\mathbf{v} \times \mathbf{a}| = \sqrt{(12\cos t)^2 + (-12\sin t)^2 + (-9)^2} = \sqrt{225} = 15.\ \kappa = \frac{|\mathbf{v} \times \mathbf{a}|}{|\mathbf{v}|^3} = \frac{15}{5^3} = \frac{3}{25}$$

$$\mathbf{B} = \mathbf{T} \times \mathbf{N} = \begin{vmatrix} \mathbf{i} & \mathbf{j} & \mathbf{k} \\ \frac{3}{5}\cos t & -\frac{3}{5}\sin t & \frac{4}{5} \\ -\sin t & -\cos t & 0 \end{vmatrix} = \left(\frac{4}{5}\cos t\right)\mathbf{i} - \left(\frac{4}{5}\sin t\right)\mathbf{j} + \left(-\frac{3}{5}\cos^2 t - \frac{3}{5}\sin^2 t\right)\mathbf{k} =$$

$$\left(\frac{4}{5}\cos t\right)\mathbf{i} - \left(\frac{4}{5}\sin t\right)\mathbf{j} - \frac{3}{5}\mathbf{k}.\ \dot{\mathbf{a}} = (-3\cos t)\,\mathbf{i} + (3\sin t)\,\mathbf{j} \Rightarrow \tau = \frac{\begin{vmatrix} 3\cos t & -3\sin t & 4 \\ -3\sin t & -3\cos t & 0 \\ -3\cos t & 3\sin t & 0 \end{vmatrix}}{|\mathbf{v} \times \mathbf{a}|^2} =$$

5. (Continued)

$\frac{|-36 \sin^2 t - 36 \cos^2 t|}{15^2} = \frac{4}{25}$

7. $\mathbf{r} = (e^t \cos t)\,\mathbf{i} + (e^t \sin t)\,\mathbf{j} + 2\,\mathbf{k} \Rightarrow \mathbf{v} = (e^t \cos t - e^t \sin t)\,\mathbf{i} + (e^t \sin t + e^t \cos t)\,\mathbf{j} \Rightarrow$

$|\mathbf{v}| = \sqrt{(e^t \cos t - e^t \sin t)^2 + (e^t \sin t + e^t \cos t)^2} = \sqrt{2e^{2t}} = e^t\sqrt{2}.$

$\mathbf{a} = \left(e^t(\cos t - \sin t) + e^t(-\sin t - \cos t)\right)\mathbf{i} + \left(e^t(\sin t + \cos t) + e^t(\cos t - \sin t)\right)\mathbf{j} =$

$(-2e^t \sin t)\,\mathbf{i} + (2e^t \cos t)\,\mathbf{j} \Rightarrow \mathbf{v} \times \mathbf{a} = \begin{vmatrix} \mathbf{i} & \mathbf{j} & \mathbf{k} \\ e^t \cos t - e^t \sin t & e^t \sin t + e^t \cos t & 0 \\ -2e^t \sin t & 2e^t \cos t & 0 \end{vmatrix} = (2e^{2t})\,\mathbf{k} \Rightarrow$

$|\mathbf{v} \times \mathbf{a}| = \sqrt{(2e^{2t})^2} = 2e^{2t}.\ \kappa = \frac{|\mathbf{v} \times \mathbf{a}|}{|\mathbf{v}|^3} = \frac{2e^{2t}}{\left(e^t\sqrt{2}\right)^3} = \frac{1}{e^t\sqrt{2}}.$

$\mathbf{T} = \frac{\mathbf{v}}{|\mathbf{v}|} = \left(\frac{e^t \cos t - e^t \sin t}{e^t\sqrt{2}}\right)\mathbf{i} + \left(\frac{e^t \cos t + e^t \sin t}{e^t\sqrt{2}}\right)\mathbf{j} \Rightarrow \frac{d\mathbf{T}}{dt} = \left(\frac{-\sin t - \cos t}{\sqrt{2}}\right)\mathbf{i} + \left(\frac{\cos t - \sin t}{\sqrt{2}}\right)\mathbf{j} \Rightarrow$

$\left|\frac{d\mathbf{T}}{dt}\right| = \sqrt{\left(\frac{-\sin t - \cos t}{\sqrt{2}}\right)^2 + \left(\frac{\cos t - \sin t}{\sqrt{2}}\right)^2} = 1.$

$\mathbf{N} = \frac{d\mathbf{T}/dt}{|d\mathbf{T}/dt|} = \left(\frac{-\cos t - \sin t}{\sqrt{2}}\right)\mathbf{i} + \left(\frac{-\sin t + \cos t}{\sqrt{2}}\right)\mathbf{j}.\ \dot{\mathbf{a}} = (-2e^t \sin t - 2e^t \cos t)\,\mathbf{i} + (2e^t \cos t - 2e^t \sin t)\,\mathbf{j} \Rightarrow$

$$\tau = \frac{\begin{vmatrix} e^t \cos t - e^t \sin t & e^t \sin t + e^t \cos t & 0 \\ -2e^t \sin t & 2e^t \cos t & 0 \\ -2e^t \sin t - 2e^t \cos t & 2e^t \cos t - 2e^t \sin t & 0 \end{vmatrix}}{|\mathbf{v} \times \mathbf{a}|^2} = 0.$$

$$\mathbf{B} = \mathbf{T} \times \mathbf{N} = \begin{vmatrix} \mathbf{i} & \mathbf{j} & \mathbf{k} \\ \frac{\cos t - \sin t}{\sqrt{2}} & \frac{\sin t + \cos t}{\sqrt{2}} & 0 \\ \frac{-\cos t - \sin t}{\sqrt{2}} & \frac{-\sin t + \cos t}{\sqrt{2}} & 0 \end{vmatrix} = \mathbf{k}$$

9. $\mathbf{r} = (2t + 3)\,\mathbf{i} + (t^2 - 1)\,\mathbf{j} \Rightarrow \mathbf{v} = 2\,\mathbf{i} + 2t\,\mathbf{j} \Rightarrow |\mathbf{v}| = \sqrt{2^2 + (2t)^2} = 2\sqrt{1 + t^2}$

$a_{\mathbf{T}} = 2\left(\frac{1}{2}\right)(1 + t^2)^{-1/2}(2t) = \frac{2t}{\sqrt{1 + t^2}};\ \mathbf{a} = 2\,\mathbf{j} \Rightarrow |\mathbf{a}| = 2 \Rightarrow a_{\mathbf{N}} = \sqrt{|\mathbf{a}|^2 - a_{\mathbf{T}}^2} = \sqrt{2^2 - \left(\frac{2t}{\sqrt{1 + t^2}}\right)^2}$

$= \frac{2}{\sqrt{1 + t^2}}.\ \therefore \mathbf{a} = \frac{2t}{\sqrt{1 + t^2}}\mathbf{T} + \frac{2}{\sqrt{1 + t^2}}\mathbf{N}.$

11. $\mathbf{r} = (a \cos t)\,\mathbf{i} + (a \sin t)\,\mathbf{j} + bt\,\mathbf{k} \Rightarrow \mathbf{v} = (-a \sin t)\,\mathbf{i} + (a \cos t)\,\mathbf{j} + b\,\mathbf{k} \Rightarrow |\mathbf{v}| = \sqrt{(-a \sin t)^2 + (a \cos t)^2 + b^2}$

$= \sqrt{a^2 + b^2}.\ a_{\mathbf{T}} = 0.\ \mathbf{a} = (-a \cos t)\,\mathbf{i} + (-a \sin t)\,\mathbf{j} \Rightarrow |\mathbf{a}| = \sqrt{(-a \cos t)^2 + (-a \sin t)^2} = \sqrt{a^2} = |a|.$

$a_{\mathbf{N}} = \sqrt{|\mathbf{a}|^2 - a_{\mathbf{T}}^2} = \sqrt{|\mathbf{a}|^2 - 0^2} = |\mathbf{a}| = |a|.\ \therefore \mathbf{a} = (0)\,\mathbf{T} + |a|\,\mathbf{N} = |a|\,\mathbf{N}$

13. $\mathbf{r} = (t+1)\,\mathbf{i} + 2t\,\mathbf{j} + t^2\,\mathbf{k} \Rightarrow \mathbf{v} = \mathbf{i} + 2\,\mathbf{j} + 2t\,\mathbf{k} \Rightarrow \mathbf{v}(1) = \mathbf{i} + 2\,\mathbf{j} + 2\,\mathbf{k} \Rightarrow |\mathbf{v}| = \sqrt{1^2 + 2^2 + (2t)^2} = \sqrt{5 + 4t^2}$. $a_{\mathbf{T}} = \frac{1}{2}\left(5 + 4t^2\right)^{-1/2}(8t) = 4t\left(5 + 4t^2\right)^{-1/2} \Rightarrow a_{\mathbf{T}}(1) = \frac{4}{\sqrt{9}} = \frac{4}{3}$. $\mathbf{a} = 2\,\mathbf{k} \Rightarrow \mathbf{a}(1) = 2\,\mathbf{k} \Rightarrow$

$|\mathbf{a}(1)| = 2$. $a_{\mathbf{N}} = \sqrt{|\mathbf{a}|^2 - a_{\mathbf{T}}^2} = \sqrt{2^2 - \left(\frac{4}{3}\right)^2} = \frac{2\sqrt{5}}{3}$. $\therefore\ \mathbf{a}(1) = \frac{4}{3}\mathbf{T} + \frac{2\sqrt{5}}{3}\mathbf{N}$

15. $\mathbf{r} = t^2\,\mathbf{i} + (t + \frac{1}{3}t^3)\,\mathbf{j} + (t - \frac{1}{3}t^3)\,\mathbf{k} \Rightarrow \mathbf{v} = 2t\,\mathbf{i} + (1 + t^2)\,\mathbf{j} + (1 - t^2)\,\mathbf{k} \Rightarrow |\mathbf{v}| = \sqrt{(2t)^2 + (1 + t^2)^2 + (1 - t^2)^2}$
$= \sqrt{2}\left(t^2 + 1\right)$. $a_{\mathbf{T}} = 2t\sqrt{2} \Rightarrow a_{\mathbf{T}}(0) = 0$. $\mathbf{a} = 2\,\mathbf{i} + 2t\,\mathbf{j} - 2t\,\mathbf{k} \Rightarrow$

$\mathbf{a}(0) = 2\,\mathbf{i} \Rightarrow |\mathbf{a}(0)| = 2$. $a_{\mathbf{N}} = \sqrt{|\mathbf{a}|^2 - a_{\mathbf{T}}^2} = \sqrt{2^2 - 0^2} = 2$. $\therefore\ \mathbf{a}(0) = (0)\,\mathbf{T} + 2\,\mathbf{N} = 2\,\mathbf{N}$

17. $\mathbf{r} = (\cos t)\,\mathbf{i} + (\sin t)\,\mathbf{j} - \mathbf{k} \Rightarrow \mathbf{v} = (-\sin t)\,\mathbf{i} + (\cos t)\,\mathbf{j} \Rightarrow |\mathbf{v}| = \sqrt{(-\sin t)^2 + (\cos t)^2} = 1$. $\mathbf{T} = \frac{\mathbf{v}}{|\mathbf{v}|} =$
$(-\sin t)\,\mathbf{i} + (\cos t)\,\mathbf{j} \Rightarrow \mathbf{T}\left(\frac{\pi}{4}\right) = -\frac{\sqrt{2}}{2}\mathbf{i} + \frac{\sqrt{2}}{2}\mathbf{j}$. $\frac{d\mathbf{T}}{dt} = (-\cos t)\,\mathbf{i} - (\sin t)\,\mathbf{j} \Rightarrow \left|\frac{d\mathbf{T}}{dt}\right| = \sqrt{(-\cos t)^2 + (-\sin t)^2}$
$= 1$. $\mathbf{N} = \frac{d\mathbf{T}/dt}{|d\mathbf{T}/dt|} = (-\cos t)\,\mathbf{i} - (\sin t)\,\mathbf{j} \Rightarrow \mathbf{N}\left(\frac{\pi}{4}\right) = -\frac{\sqrt{2}}{2}\mathbf{i} - \frac{\sqrt{2}}{2}\mathbf{j}$. $\mathbf{r}\left(\frac{\pi}{4}\right) = \frac{\sqrt{2}}{2}\mathbf{i} + \frac{\sqrt{2}}{2}\mathbf{j} - \mathbf{k}$

$\mathbf{B} = \mathbf{T} \times \mathbf{N} = \begin{vmatrix} \mathbf{i} & \mathbf{j} & \mathbf{k} \\ -\sin t & \cos t & 0 \\ -\cos t & -\sin t & 0 \end{vmatrix} = \mathbf{k} \Rightarrow \mathbf{B}\left(\frac{\pi}{4}\right) = \mathbf{k}$. $P = \left(\frac{\sqrt{2}}{2}, \frac{\sqrt{2}}{2}, -1\right)\left(\text{see } \mathbf{r}\left(\frac{\pi}{4}\right)\right)$, the osculating

plane is $z = -1$ since $\mathbf{B}$ is the normal vector and $(0)x + (0)y + (1)z = (0)\left(\frac{\sqrt{2}}{2}\right) + (0)\left(\frac{\sqrt{2}}{2}\right) + (1)(-1)$.

The normal plane is $-x + y = 0$ since $\mathbf{T}$ is the normal vector and $-\frac{\sqrt{2}}{2}x + \frac{\sqrt{2}}{2}y + (0)z = \left(-\frac{\sqrt{2}}{2}\right)\left(\frac{\sqrt{2}}{2}\right) + \left(\frac{\sqrt{2}}{2}\right)\left(\frac{\sqrt{2}}{2}\right) + (-1)(0) \Rightarrow -\frac{\sqrt{2}}{2}x + \frac{\sqrt{2}}{2}y = 0$. The rectifying plane is $x + y = \sqrt{2}$ since $\mathbf{N}$ is the normal vector and $-\frac{\sqrt{2}}{2}x - \frac{\sqrt{2}}{2}y + (0)z = \left(-\frac{\sqrt{2}}{2}\right)\left(\frac{\sqrt{2}}{2}\right) - \left(\frac{\sqrt{2}}{2}\right)\left(\frac{\sqrt{2}}{2}\right) + (-1)(0) \Rightarrow -\frac{\sqrt{2}}{2}x - \frac{\sqrt{2}}{2}y = -1$.

$= \frac{1}{\sqrt{2}}$. $\mathbf{N} = \frac{d\mathbf{T}/dt}{|d\mathbf{T}/dt|} = \frac{(-\cos t/\sqrt{2})\,\mathbf{i} - (\sin t/\sqrt{2})\,\mathbf{j}}{1/\sqrt{2}} = (-\cos t)\,\mathbf{i} - (\sin t)\,\mathbf{j} \Rightarrow \mathbf{N}(0) = -\mathbf{i}$

19. If $|\mathbf{v}|$ is constant, then $a_{\mathbf{T}} = 0 \Rightarrow \mathbf{a} = a_{\mathbf{N}}\,\mathbf{N} \Rightarrow \mathbf{a}$ is normal to $\mathbf{T} \Rightarrow$ acceleration is normal to the path.

21. $\mathbf{r} = t\,\mathbf{i} + (\sin t)\,\mathbf{j} \Rightarrow \mathbf{v} = \mathbf{i} + (\cos t)\,\mathbf{j} \Rightarrow |\mathbf{v}| = \sqrt{1^2 + (\cos t)^2} = \sqrt{1 + \cos^2 t} \Rightarrow \left|\mathbf{v}\left(\frac{\pi}{2}\right)\right| = \sqrt{1 + \cos^2\left(\frac{\pi}{2}\right)}$
$= 1$. $\mathbf{a} = (-\sin t)\,\mathbf{j} \Rightarrow \mathbf{v} \times \mathbf{a} = \begin{vmatrix} \mathbf{i} & \mathbf{j} & \mathbf{k} \\ 1 & \cos t & 0 \\ 0 & -\sin t & 0 \end{vmatrix} = (-\sin t)\,\mathbf{k} \Rightarrow |\mathbf{v} \times \mathbf{a}| = \sqrt{(-\sin t)^2} = |\sin t| \Rightarrow$

$|\mathbf{v} \times \mathbf{a}|\left(\frac{\pi}{2}\right) = \left|\sin\left(\frac{\pi}{2}\right)\right| = 1 \Rightarrow \kappa = \frac{|\mathbf{v} \times \mathbf{a}|}{|\mathbf{v}|^3} = \frac{1}{1^3} = 1$. $\therefore\ \rho = \frac{1}{1} = 1 \Rightarrow$ center is $\left(\frac{\pi}{2}, 0\right)$, $r = 1 \Rightarrow$
$\left(x - \frac{\pi}{2}\right)^2 + y^2 = 1$

23. $\mathbf{v} = \mathbf{T}(ds/dt)$ and $\mathbf{a}\ (d^2s/dt^2)\mathbf{T} + \kappa(ds/dt)^2\mathbf{N} \Rightarrow |\mathbf{v} \times \mathbf{a}| = \left|(ds/dt)\mathbf{T} \times \left((d^2s/dt^2)\mathbf{T} + \kappa(ds/dt)^2\mathbf{N}\right)\right| =$
$\left|\kappa(ds/dt)^3(\mathbf{T} \times \mathbf{N})\right| = \kappa\left|(ds/dt)^3\mathbf{B}\right| = \kappa\left||\mathbf{v}|^3\mathbf{B}\right| = \kappa\,|\mathbf{v}|^3|\mathbf{B}| = \kappa\,|\mathbf{v}|^3$. $\therefore\ |\mathbf{v} \times \mathbf{a}| = \kappa\,|\mathbf{v}|^3 \Rightarrow \kappa = \dfrac{|\mathbf{v} \times \mathbf{a}|}{|\mathbf{v}|^3}$

25. $\kappa = \dfrac{a}{a^2 + b^2} \Rightarrow \dfrac{d\kappa}{da} = \dfrac{-a^2 + b^2}{(a^2 + b^2)^2}$. If $\dfrac{d\kappa}{da} = 0$, then $-a^2 + b^2 = 0 \Rightarrow a = \pm b$. When $a = b\ (b > 0)$,
$\dfrac{d\kappa}{da} > 0$ if $a < b$ and $\dfrac{d\kappa}{da} < 0$ if $a > b$. $\therefore\ \kappa$ is at a maximum when $a = b \Rightarrow \kappa(b) = \dfrac{b}{b^2 + b^2} = \dfrac{1}{2b}$, the maximum value of $\kappa$.

## SECTION 12.5 PLANETARY MOTION AND SATELLITES

1. $\dfrac{T^2}{a^3} = \dfrac{4\pi^2}{GM} \Rightarrow T^2 = \dfrac{4\pi^2}{GM}a^3 \Rightarrow T^2 = \dfrac{4\pi^2}{(6.6720 \times 10^{-11} Nm^2kg^{-2})(5.975 \times 10^{24}\ kg)}(6\,808\,000\ m)^3 =$
$3.125 \times 10^7\ sec^2 \Rightarrow T = \sqrt{3125 \times 10^4\ sec^2} = 55.90 \times 10^2\ sec = 93.17$ minutes.

3. 92.25 minutes = 5535 seconds. $\dfrac{T^2}{a^3} = \dfrac{4\pi^2}{GM} \Rightarrow a^3 = \dfrac{GM}{4\pi^2}T^2 \Rightarrow a^3 =$
$\dfrac{(6.6720 \times 10^{-11} Nm^2kg^{-2})(5.975 \times 10^{24}kg)}{4\pi^2}(5535\ s)^2 = 3.094 \times 10^{20}\ m^3 \Rightarrow a = \sqrt[3]{3.094 \times 10^{20}\ m^3}$
$= 6.764 \times 10^6\ m = 6764$ km. Mean distance from center of the Earth $= \dfrac{12758\ km + 183\ km + 589\ km}{2}$
$= 6765$ km

5. $a = 22030\ km = 2.203 \times 10^7\ m$. $T^2 = \dfrac{4\pi^2}{GM}a^3 \Rightarrow T^2 =$
$\dfrac{4\pi^2}{(6.670 \times 10^{-11} Nm^2kg^{-2})(6.418 \times 10^{23}\ kg)}(2.203 \times 10^7 s)^3 = 9.857 \times 10^9\ sec^2 \Rightarrow T =$
$\sqrt{9.857 \times 10^8\ sec^2} = 9.928 \times 10^4\ sec = 1655$ minutes.

7. T = 1477.4 minutes = 88644 seconds. $a^3 = \dfrac{GMT^2}{4\pi^2} =$
$\dfrac{(6.6720 \times 10^{-11} Nm^2kg^{-2})(6.418 \times 10^{23}kg)(88644\ s)^2}{4\pi^2} = 8.523 \times 10^{21}\ m^3 \Rightarrow$
$a = \sqrt[3]{8.523 \times 10^{21}\ m^3} = 2.043 \times 10^7\ m = 20430$ km

9. $r = \dfrac{GM}{v^2} \Rightarrow v^2 = \dfrac{GM}{r} \Rightarrow |v| = \sqrt{\dfrac{GM}{r}} = \sqrt{\dfrac{(6.6720 \times 10^{-11}\ Nm^2kg^{-2})(5.975 \times 10^{24}\ kg)}{r}} =$
$1.9966 \times 10^7\ r^{-1/2}$ m/s

11. $e = \dfrac{r_0 v_0^2}{GM} - 1 \Rightarrow v_0^2 = \dfrac{GM(e + 1)}{r_0} \Rightarrow v_0 = \sqrt{\dfrac{GM(e + 1)}{r_0}}$

Circle: $e = 0 \Rightarrow v_0 = \sqrt{\dfrac{GM}{r_0}}$

Ellipse: $0 < e < 1 \Rightarrow \sqrt{\dfrac{GM}{r_0}} < v_0 < \sqrt{\dfrac{2GM}{r_0}}$

11. (Continued)

Hyperbola: $e > 1 \Rightarrow v_0 > \sqrt{\frac{2GM}{r_0}}$

13. $\frac{d\theta}{dt} = 2$, $\theta = 2t$, $r = \cosh\theta = \cosh(2t)$. $\mathbf{a} = \left(\frac{d^2r}{dt^2} - r\left(\frac{d\theta}{dt}\right)^2\right)\mathbf{u}_r + \left(r\left(\frac{d^2\theta}{dt^2}\right) + 2\left(\frac{dr}{dt}\right)\left(\frac{d\theta}{dt}\right)\right)\mathbf{u}_\theta \Rightarrow$ the $\mathbf{u}_r$ component of acceleration $= \left(\frac{d^2r}{dt^2} - r\left(\frac{d\theta}{dt}\right)^2\right)$. $\frac{dr}{dt} = 2\sinh(2t) \Rightarrow \frac{d^2r}{dt^2} = 4\cosh(2t)$.

$\therefore \left(\frac{d^2r}{dt^2} - r\left(\frac{d\theta}{dt}\right)^2\right) = 4\cosh(2t) - \cosh(2t)(2^2) = 0$

## PRACTICE EXERCISES

1. $\mathbf{r} = (4\cos t)\,\mathbf{i} + (\sqrt{2}\sin t)\,\mathbf{j} \Rightarrow x = 4\cos t \Rightarrow x^2 = 16\cos^2 t$
$y = \sqrt{2}\sin t \Rightarrow y^2 = 2\sin^2 t \Rightarrow 8y^2 = 16\sin^2 t \Rightarrow$
$x^2 + 8y^2 = 16 \Rightarrow \frac{x^2}{16} + \frac{y^2}{2} = 1$. $t = 0 \Rightarrow x = 4, y = 0$;
$t = \frac{\pi}{4} \Rightarrow x = 2\sqrt{2}$, $y = 1$. $\mathbf{v} = (-4\sin t)\,\mathbf{i} + (\sqrt{2}\cos t)\,\mathbf{j}$
$\Rightarrow \mathbf{v}(0) = \sqrt{2}\,\mathbf{j}$, $\mathbf{v}\left(\frac{\pi}{4}\right) = -2\sqrt{2}\,\mathbf{i} + \mathbf{j}$.
$\mathbf{a} = (-4\cos t)\,\mathbf{i} + (-\sqrt{2}\sin t)\,\mathbf{j} \Rightarrow \mathbf{a}(0) = -4\,\mathbf{i}$, $\mathbf{a}\left(\frac{\pi}{4}\right) =$
$-2\sqrt{2}\,\mathbf{i} - \mathbf{j}$.

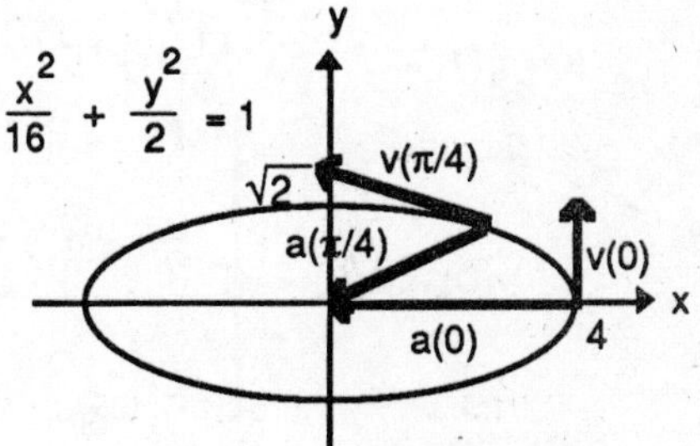

Graph 12.P.1

3. $\int_0^1 [(3 + 6t)\,\mathbf{i} + (4 + 8t)\,\mathbf{j} + (6\pi\cos\pi t)\,\mathbf{k}]\,dt = \left[3t + 3t^2\right]_0^1\mathbf{i} + \left[4t + 4t^2\right]_0^1\mathbf{j} + \left[-6\sin\pi t\right]_0^1\mathbf{j} = 6\,\mathbf{i} + 8\,\mathbf{j}$

5. $\mathbf{r} = \int ((-\sin t)\,\mathbf{i} + (\cos t)\,\mathbf{j} + \mathbf{k})\,dt = (\cos t)\,\mathbf{i} + (\sin t)\,\mathbf{j} + t\,\mathbf{k} + \mathbf{C}$. $\mathbf{r}(0) = \mathbf{j} \Rightarrow (\cos 0)\,\mathbf{i} + (\sin 0)\,\mathbf{j} + (0)\,\mathbf{k} + \mathbf{C} = \mathbf{j} \Rightarrow \mathbf{C} = \mathbf{j} - \mathbf{i} \Rightarrow \mathbf{r} = ((\cos t) - 1)\,\mathbf{i} + ((\sin t) + 1)\,\mathbf{j} + t\,\mathbf{k}$

7. $\frac{d\mathbf{r}}{dt} = \int 2\,\mathbf{j}\,dt = 2t\,\mathbf{j} + \mathbf{C}_1$. $\frac{d\mathbf{r}}{dt}(0) = \mathbf{k} \Rightarrow 2(0)\,\mathbf{j} + \mathbf{C}_1 = \mathbf{k} \Rightarrow \mathbf{C}_1 = \mathbf{k}$. $\therefore \frac{d\mathbf{r}}{dt} = 2t\,\mathbf{j} + \mathbf{k}$.
$\mathbf{r} = \int (2t\,\mathbf{j} + \mathbf{k})\,dt = t^2\,\mathbf{j} + t\,\mathbf{k} + \mathbf{C}_2$. $\mathbf{r}(0) = \mathbf{i} \Rightarrow (0^2)\,\mathbf{j} + (0)\,\mathbf{k} + \mathbf{C}_2 = \mathbf{i} \Rightarrow \mathbf{C}_2 = \mathbf{i}$
$\therefore \mathbf{r} = \mathbf{i} + t^2\,\mathbf{j} + t\,\mathbf{k}$

9. $\mathbf{r} = (2\cos t)\,\mathbf{i} + (2\sin t)\,\mathbf{j} + t^2\,\mathbf{k} \Rightarrow \mathbf{v} = (-2\sin t)\,\mathbf{i} + (2\cos t)\,\mathbf{j} + 2t\,\mathbf{k} \Rightarrow$
$|\mathbf{v}| = \sqrt{(-2\sin t)^2 + (2\cos t)^2 + (2t)^2} = 2\sqrt{1 + t^2}$. Length $= \int_0^{\pi/4} 2\sqrt{1 + t^2}\,dt =$
$\left[t\sqrt{1 + t^2} + \ln\left|t + \sqrt{1 + t^2}\right|\right]_0^{\pi/4} = \frac{\pi}{4}\sqrt{1 + \frac{\pi^2}{16}} + \ln\left(\frac{\pi}{4} + \sqrt{1 + \frac{\pi^2}{16}}\right)$

11. $\mathbf{r} = \frac{4}{9}(1+t)^{3/2}\,\mathbf{i} + \frac{4}{9}(1-t)^{3/2}\,\mathbf{j} + \frac{1}{3}t\,\mathbf{k} \Rightarrow \mathbf{v} = \frac{2}{3}(1+t)^{1/2}\,\mathbf{i} - \frac{2}{3}(1-t)^{1/2}\,\mathbf{j} + \frac{1}{3}\mathbf{k} \Rightarrow$

$|\mathbf{v}| = \sqrt{\left(\frac{2}{3}(1+t)^{1/2}\right)^2 + \left(-\frac{2}{3}(1-t)^{1/2}\right)^2 + \left(\frac{1}{3}\right)^2} = 1.$ $\mathbf{T} = \frac{2}{3}(1+t)^{1/2}\,\mathbf{i} - \frac{2}{3}(1-t)^{1/2} + \frac{1}{3}\mathbf{k} \Rightarrow$

$\mathbf{T}(0) = \frac{2}{3}\mathbf{i} - \frac{2}{3}\mathbf{j} + \frac{1}{3}\mathbf{k}.$ $\frac{d\mathbf{T}}{dt} = \frac{1}{3}(1+t)^{-1/2}\,\mathbf{i} + \frac{1}{3}(1-t)^{-1/2}\,\mathbf{j} \Rightarrow \left|\frac{d\mathbf{T}}{dt}\right| =$

$\sqrt{\left(\frac{1}{3}(1+t)^{-1/2}\right)^2 + \left(\frac{1}{3}(1-t)^{-1/2}\right)^2} = \frac{1}{3}\sqrt{\frac{2}{1-t^2}}.$ $\frac{d\mathbf{T}}{dt}(0) = \frac{1}{3}\mathbf{i} + \frac{1}{3}\mathbf{j} \Rightarrow \left|\frac{d\mathbf{T}}{dt}(0)\right| = \frac{\sqrt{2}}{3}.$

$\therefore\ \mathbf{N}(0) = \frac{1}{\sqrt{2}}\mathbf{i} + \frac{1}{\sqrt{2}}\mathbf{j}.$ $\mathbf{B}(0) = \mathbf{T}(0) \times \mathbf{N}(0) = \begin{vmatrix} \mathbf{i} & \mathbf{j} & \mathbf{k} \\ \frac{2}{3} & -\frac{2}{3} & \frac{1}{3} \\ \frac{1}{\sqrt{2}} & \frac{1}{\sqrt{2}} & 0 \end{vmatrix} = -\frac{1}{3\sqrt{2}}\mathbf{i} + \frac{1}{3\sqrt{2}}\mathbf{j} + \frac{4}{3\sqrt{2}}\mathbf{k}$

$\mathbf{a} = \frac{1}{3}(1+t)^{-1/2}\,\mathbf{i} + \frac{1}{3}(1-t)^{-1/2}\,\mathbf{j} \Rightarrow \mathbf{a}(0) = \frac{1}{3}\mathbf{i} + \frac{1}{3}\mathbf{j}.$ $\mathbf{v}(0) = \frac{2}{3}\mathbf{i} - \frac{2}{3}\mathbf{j} + \frac{1}{3}\mathbf{k}.$

$\therefore\ \mathbf{v}(0) \times \mathbf{a}(0) = \begin{vmatrix} \mathbf{i} & \mathbf{j} & \mathbf{k} \\ \frac{2}{3} & -\frac{2}{3} & \frac{1}{3} \\ \frac{1}{3} & \frac{1}{3} & 0 \end{vmatrix} = -\frac{1}{9}\mathbf{i} + \frac{1}{9}\mathbf{j} + \frac{4}{9}\mathbf{k} \Rightarrow |\mathbf{v} \times \mathbf{a}| = \frac{\sqrt{2}}{3} \Rightarrow \kappa = \frac{|\mathbf{v} \times \mathbf{a}|}{|\mathbf{v}|^3} = \frac{\sqrt{2}/3}{1^3} = \frac{\sqrt{2}}{3}.$

$\dot{\mathbf{a}} = -\frac{1}{6}(1+t)^{-3/2}\,\mathbf{i} + \frac{1}{6}(1-t)^{-3/2}\,\mathbf{j} \Rightarrow \dot{\mathbf{a}}(0) = -\frac{1}{6}\mathbf{i} + \frac{1}{6}\mathbf{j} \Rightarrow \tau = \frac{\begin{vmatrix} \frac{2}{3} & -\frac{2}{3} & \frac{1}{3} \\ \frac{1}{3} & \frac{1}{3} & 0 \\ -\frac{1}{6} & \frac{1}{6} & 0 \end{vmatrix}}{|\mathbf{v} \times \mathbf{a}|^2} = \frac{1/27}{(\sqrt{2}/3)^2} = \frac{1}{6}$

13. $\mathbf{r} = (2 + 3t + 3t^2)\,\mathbf{i} + (4t + 4t^2)\,\mathbf{j} - (6\cos t)\,\mathbf{k} \Rightarrow \mathbf{v} = (3 + 6t)\,\mathbf{i} + (4 + 8t)\,\mathbf{j} + (6\sin t)\,\mathbf{k} \Rightarrow$

$|\mathbf{v}| = \sqrt{(3+6t)^2 + (4+8t)^2 + (6\sin t)^2} = \sqrt{25 + 100t + 100t^2 + 36\sin^2 t}$

$\frac{d|\mathbf{v}|}{dt} = \frac{1}{2}(25 + 100t + 100t^2 + 36\sin^2 t)^{-1/2}(100 + 200t + 72\sin t\cos t) \Rightarrow a_T(0) = \frac{d|\mathbf{v}|}{dt}(0) = 10.$

$\mathbf{a} = 6\,\mathbf{i} + 8\,\mathbf{j} + (6\cos t)\,\mathbf{k} \Rightarrow |\mathbf{a}| = \sqrt{6^2 + 8^2 + (6\cos t)^2} = \sqrt{100 + 36\cos^2 t} \Rightarrow |\mathbf{a}(0)| = \sqrt{136} = 2\sqrt{34}$

$a_N = \sqrt{|\mathbf{a}|^2 - a_T^{\,2}} = \sqrt{(2\sqrt{34})^2 - 10^2} = \sqrt{36} = 6.$ $\therefore\ \mathbf{a}(0) = 10\,\mathbf{T} + 6\,\mathbf{N}$

15. $\mathbf{r} = \frac{1}{\sqrt{1+t^2}}\mathbf{i} + \frac{t}{\sqrt{1+t^2}}\mathbf{j} \Rightarrow \mathbf{v} = -t\left(1+t^2\right)^{-3/2}\,\mathbf{i} + (1+t^2)^{-3/2}\,\mathbf{j} \Rightarrow$

$|\mathbf{v}| = \sqrt{\left(-t\left(1+t^2\right)^{-3/2}\right)^2 + \left((1+t^2)^{-3/2}\right)^2} = \frac{1}{1+t^2}.$ Want to maximize $|\mathbf{v}|$: $\frac{d|\mathbf{v}|}{dt} = \frac{-2t}{(1+t^2)^2}$

When $\frac{d|\mathbf{v}|}{dt} = 0,\ \frac{-2t}{(1+t^2)^2} = 0 \Rightarrow -2t = 0 \Rightarrow t = 0.$ For $t < 0,\ \frac{-2t}{(1+t^2)^2} > 0$; for $t > 0,\ \frac{-2t}{(1+t^2)^2} < 0 \Rightarrow$

$|\mathbf{v}|_{max}$ occurs when $t = 0 \Rightarrow |\mathbf{v}|_{max} = 1$

17. $\mathbf{r} = \mathbf{i} + (5\cos t)\,\mathbf{j} + (3\sin t)\,\mathbf{k} \Rightarrow \mathbf{v} = (-5\sin t)\,\mathbf{j} + (3\cos t)\,\mathbf{k} \Rightarrow \mathbf{a} = (-5\cos t)\,\mathbf{j} - (3\sin t)\,\mathbf{k} \Rightarrow$

$\mathbf{v}\cdot\mathbf{a} = 25\sin t\cos t - 9\sin t\cos t = 16\sin t\cos t$. $\mathbf{v}\cdot\mathbf{a} = 0 \Rightarrow 16\sin t\cos t = 0 \Rightarrow \sin t = 0$ or $\cos t = 0$

$\Rightarrow t = 0, \pi$ or $t = \frac{\pi}{2}$

19. $y = \left(v_0 \sin\alpha\right)t - \frac{1}{2}gt^2 \Rightarrow y = (44 \text{ ft/sec})(\sin 45°) - \frac{1}{2}(32 \text{ ft/sec}^2)(3 \text{ sec})^2 = 66\sqrt{2} - 144 \text{ ft} = -50.66 \text{ ft}$

$y + 8 \text{ ft} = -42.66$ ft. The shot put is on the ground.

21. $R = \frac{v_0^2}{g}\sin 2\alpha \Rightarrow v_0 = \sqrt{\frac{Rg}{\sin 2\alpha}}$

For 4325 yards: 4325 yards = 12975 ft $\Rightarrow v_0 = \sqrt{\frac{(12975 \text{ ft})(32 \text{ ft/sec}^2)}{\sin 90°}} = 644.36$ ft/sec

For 4752 yards: 4752 yards = 14256 ft $\Rightarrow v_0 = \sqrt{\frac{(14256 \text{ ft})(32 \text{ ft/sec}^2)}{\sin 90°}} = 675.42$ ft/sec

23. $\mathbf{r} = e^t\,\mathbf{i} + (\sin t)\,\mathbf{j} + (\ln(1-t))\,\mathbf{k} \Rightarrow \mathbf{v} = e^t\,\mathbf{i} + (\cos t)\,\mathbf{j} - \frac{1}{1-t}\mathbf{k} \Rightarrow \mathbf{v}(0) = \mathbf{i} + \mathbf{j} - \mathbf{k}$. $\mathbf{r}(0) = \mathbf{i} \Rightarrow (1,0,0)$ is on the line. $\therefore$ $x = 1 + t$, $y = t$, $z = -t$ are the parametric equations of the line.

25. $\mathbf{v} = 3\,\mathbf{i} + 4\,\mathbf{j}$, $\mathbf{a} = 5\,\mathbf{i} + 15\,\mathbf{j} \Rightarrow \mathbf{v} \times \mathbf{a} = \begin{vmatrix} \mathbf{i} & \mathbf{j} & \mathbf{k} \\ 3 & 4 & 0 \\ 5 & 15 & 0 \end{vmatrix} = 25\,\mathbf{k} \Rightarrow |\mathbf{v} \times \mathbf{a}| = \sqrt{25^2} = 25$. $|\mathbf{v}| = \sqrt{3^2 + 4^2}$

$= 5$. $\therefore$ $\kappa = \frac{|\mathbf{v} \times \mathbf{a}|}{|\mathbf{v}|^3} = \frac{25}{5^3} = \frac{1}{5}$.

27. $\mathbf{r} = (a\cos t)\,\mathbf{i} + (a\sin t)\,\mathbf{j} + bt\,\mathbf{k} \Rightarrow \mathbf{v} = (-a\sin t)\,\mathbf{i} + (a\cos t)\,\mathbf{j} + b\,\mathbf{k} \Rightarrow \mathbf{a} = (-a\cos t)\,\mathbf{i} - (a\sin t)\,\mathbf{j}$

$|\mathbf{v}| = \sqrt{(-a\sin t)^2 + (a\cos t)^2 + b^2} = \sqrt{a^2 + b^2}$. $\mathbf{v} \times \mathbf{a} = \begin{vmatrix} \mathbf{i} & \mathbf{j} & \mathbf{k} \\ -a\sin t & a\cos t & b \\ -a\cos t & -a\sin t & 0 \end{vmatrix} =$

$(ab\sin t)\,\mathbf{i} + (ab\cos t)\,\mathbf{j} + a^2\,\mathbf{k} \Rightarrow |\mathbf{v} \times \mathbf{a}| = \sqrt{(ab\sin t)^2 + (ab\cos t)^2 + (a^2)^2} = \sqrt{a^2b^2 + a^4}$

$\dot{\mathbf{a}} = (a\sin t)\,\mathbf{i} - (a\cos t)\,\mathbf{j} \Rightarrow \tau = \frac{\begin{vmatrix} -a\sin t & a\cos t & b \\ -a\cos t & -a\sin t & 0 \\ a\sin t & -a\cos t & 0 \end{vmatrix}}{|\mathbf{v} \times \mathbf{a}|^2} = \frac{b(a^2\cos^2 t + a^2\sin^2 t)}{a^2b^2 + a^4} = \frac{b}{a^2 + b^2}$

$\tau'(b) = \frac{a^2 - b^2}{(a^2 + b^2)^2}$. $\tau'(b) = 0 \Rightarrow \frac{a^2 - b^2}{(a^2 + b^2)^2} = 0 \Rightarrow a^2 - b^2 = 0 \Rightarrow b = \pm a$. Use $b = a$. Then $b < a \Rightarrow$

$\tau' > 0$; $b > a \Rightarrow \tau' < 0$. $\therefore$ $\tau_{max}$ occurs when $b = a$. $\tau_{max} = \frac{a}{a^2 + a^2} = \frac{1}{2a}$

29. $\Delta\ ACB \approx \Delta\ BCD \Rightarrow \frac{DC}{BC} = \frac{BC}{AC} \Rightarrow$

$$\frac{y_0}{6380 \text{ miles}} = \frac{6380 \text{ miles}}{(6380 + 437) \text{ miles}} \Rightarrow$$

$$y_0 = \frac{6380^2 \text{ miles}^2}{6817 \text{ miles}} = 5971 \text{ miles}$$

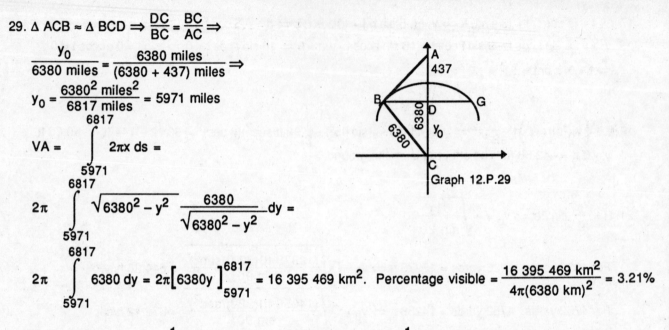

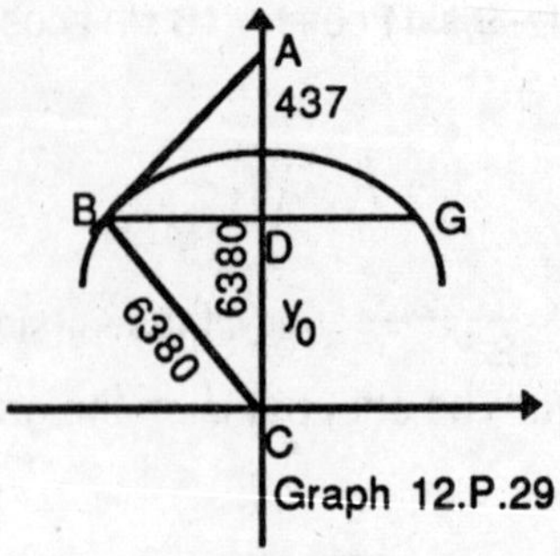

Graph 12.P.29

$$VA = \int_{5971}^{6817} 2\pi x \, ds =$$

$$2\pi \int_{5971}^{6817} \sqrt{6380^2 - y^2} \frac{6380}{\sqrt{6380^2 - y^2}} dy =$$

$$2\pi \int_{5971}^{6817} 6380 \, dy = 2\pi\Big[6380y\Big]_{5971}^{6817} = 16\,395\,469 \text{ km}^2.$$ Percentage visible $= \frac{16\,395\,469 \text{ km}^2}{4\pi(6380 \text{ km})^2} = 3.21\%$

31. a) Given $f(x) = x - 1 - \frac{1}{2}\sin x = 0$, $f(0) = -1$ and $f(2) = 2 - 1 - \frac{1}{2}\sin 2 = 0.54535$ Since f is continuous on [0,2], the intermediate value theorem implies there is a root between 0 and 2.

b) Root $\approx 1.49870113$

# CHAPTER 13

# FUNCTIONS OF TWO OR MORE VARIABLES AND THEIR DERIVATIVES

## 13.1 FUNCTIONS OF TWO OR MORE INDEPENDENT VARIABLES

1. Domain: All points in the xy–plane; Range: All Real Numbers
Level curves are straight lines parallel to the line $y = x$.

3. Domain: Set of all $(x,y)$ so that $(x,y) \neq (0,0)$; Range: All Real Numbers
Level curves are circles with center $(0,0)$ and radii $\geq 0$.

5. Domain: All points in the xy–plane; Range: All Real Numbers
Level curves are hyperbolas with the x and y axes as asymptotes when $f(x,y) \neq 0$, and the x and y axes when $f(x,y) = 0$.

7. Domain: All points in the xy–plane; Range: $z \geq 0$
Level curves are ellipses with center $(0,0)$ and major and minor axes along the x and y axes.

9. a)

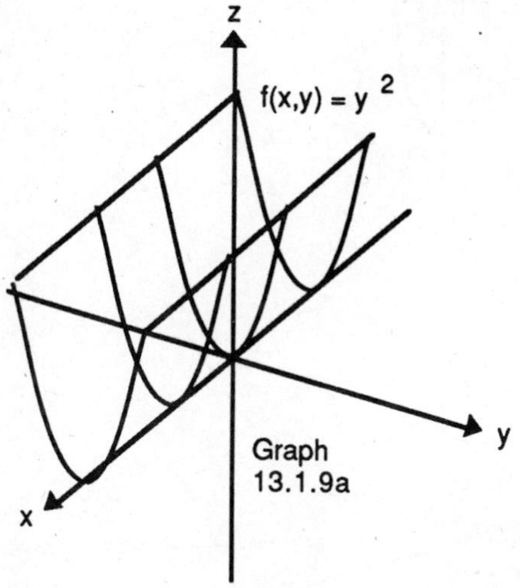

Graph 13.1.9a

b)

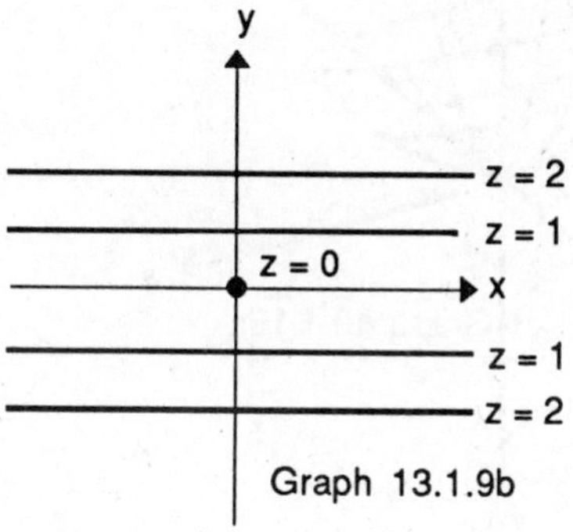

Graph 13.1.9b

11. a)

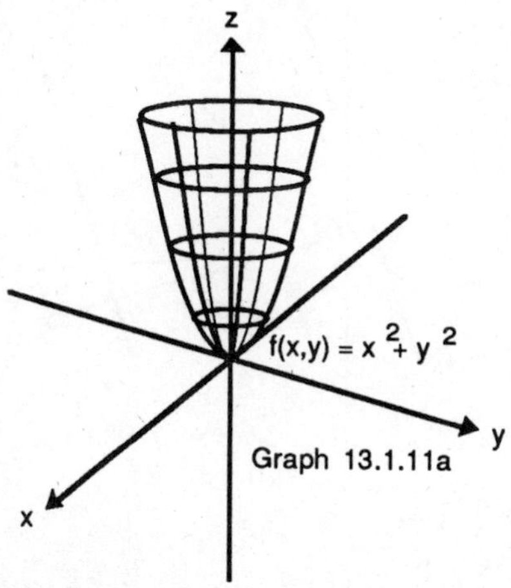

Graph 13.1.11a

b)

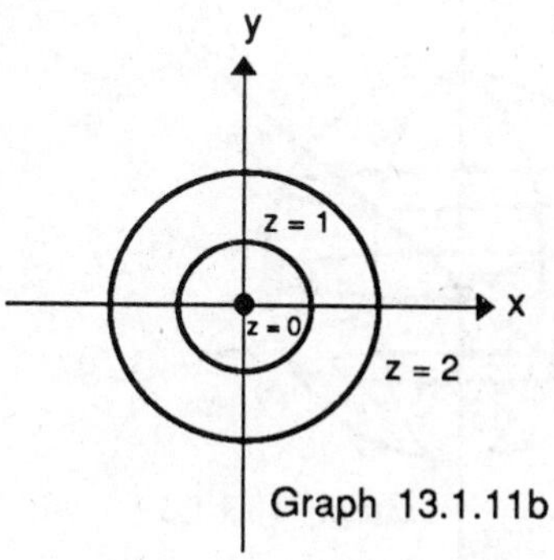

Graph 13.1.11b

13. a)

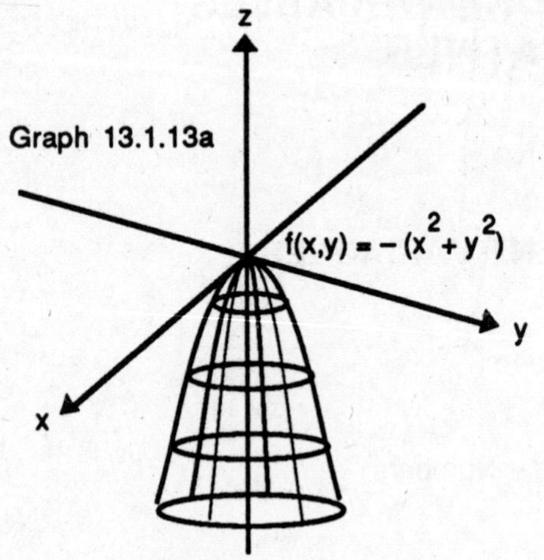


b)

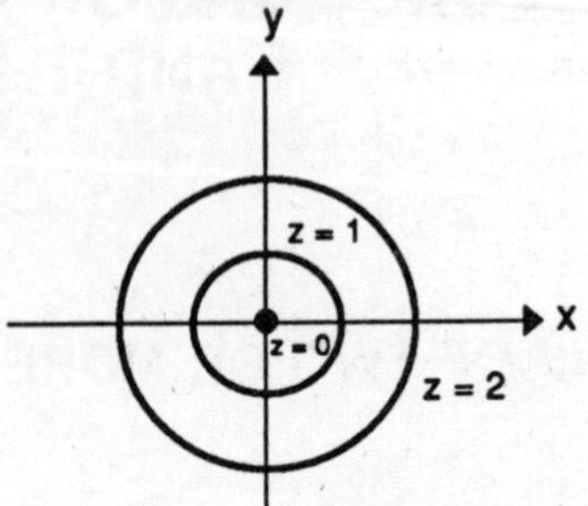


Graph 13.1.13b

15. a)

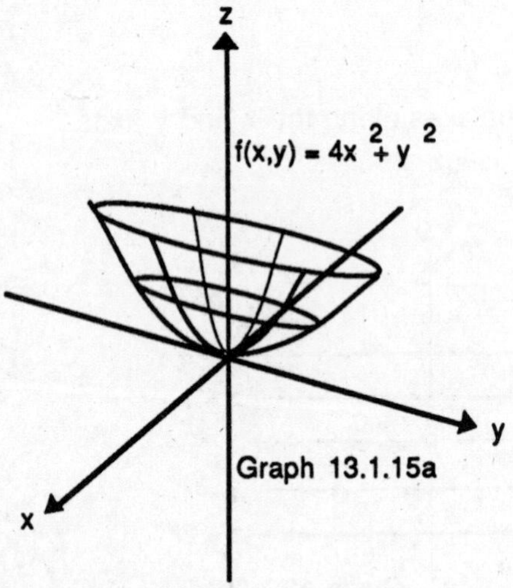


Graph 13.1.15a

b)

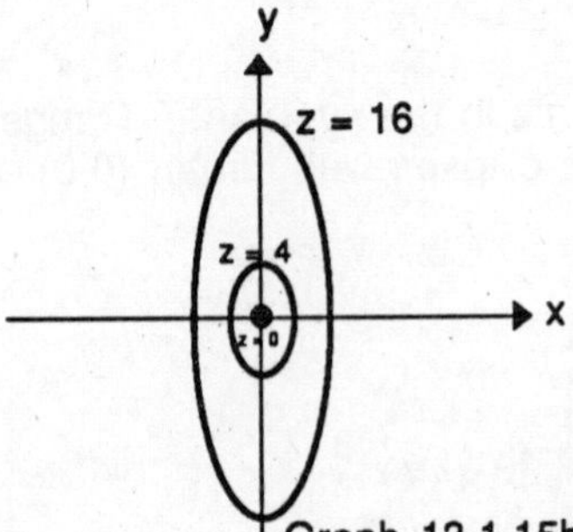


Graph 13.1.15b

17. f

19. a

21. d

23.

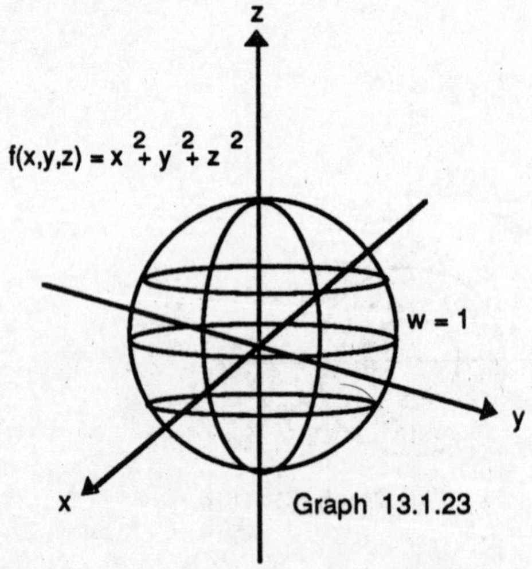


Graph 13.1.23

25.

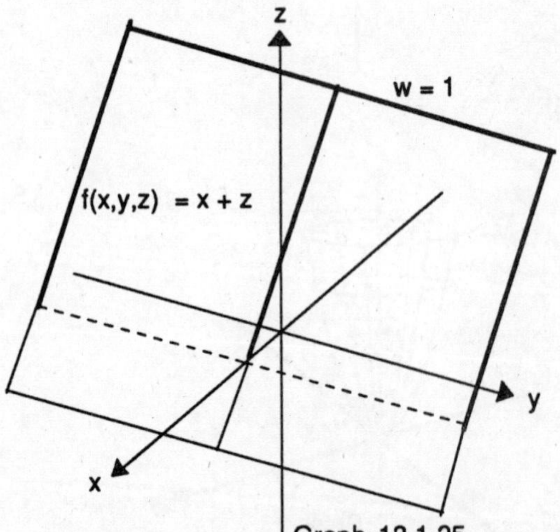


Graph 13.1.25

27.

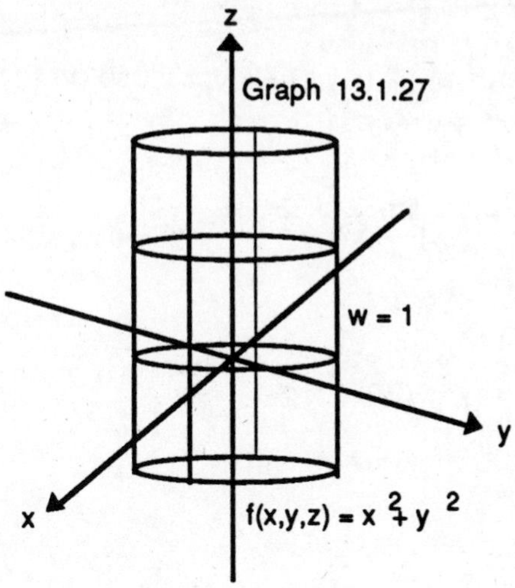


29.

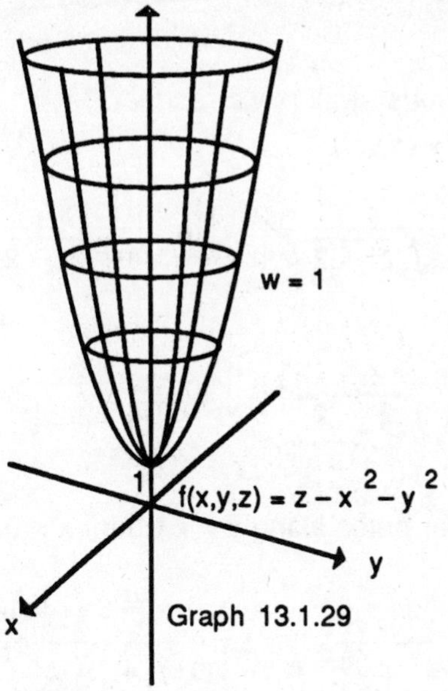


31. $f(x,y) = 16 - x^2 - y^2$ and $(2\sqrt{2},\sqrt{2}) \Rightarrow z = 16 - \left(2\sqrt{2}\right)^2 - \left(\sqrt{2}\right)^2 = 6 \Rightarrow 6 = 16 - x^2 - y^2 \Rightarrow x^2 + y^2 = 10$

33. $w = 4\left(\frac{Th}{d}\right)^{1/2} = 4\left(\frac{(290\text{ k})(16.8\text{ km})}{5\text{ k/km}}\right)^{1/2} = 124.86$ km. $\therefore$ must be 62.43 km south of Nantucket.

## SECTION 13.2 LIMITS AND CONTINUITY

1. $\lim\limits_{(x,y)\to(0,0)} \frac{3x^2 - y^2 + 5}{x^2 + y^2 + 2} = \frac{5}{2}$

3. $\lim\limits_{(x,y)\to(0,\ln 2)} e^{x-y} = \frac{1}{2}$

5. $\lim\limits_{P\to(1,3,4)} \sqrt{x^2 + y^2 + z^2 - 1} = 5$

7. $\lim\limits_{(x,y)\to(0,\pi/4)} \sec x \tan y = 1$

9. $\lim\limits_{(x,y)\to(1,1)} \cos\left(\sqrt[3]{|xy| - 1}\right) = 1$

11. $\lim\limits_{(x,y)\to(0,0)} \frac{e^y \sin x}{x} = 1$

13. $\lim\limits_{\substack{(x,y)\to(1,1)\\ x\neq y}} \frac{x^2 - 2xy + y^2}{x - y} = \lim\limits_{(x,y)\to(1,1)} \frac{(x-y)^2}{x-y} = \lim\limits_{(x,y)\to(1,1)} (x - y) = 0$

15. $\lim\limits_{\substack{(x,y)\to(1,1)\\ x\neq 1}} \frac{xy - y - 2x + 2}{x - 1} = \lim\limits_{(x,y)\to(1,1)} \frac{(x-1)(y-2)}{x-1} = \lim\limits_{(x,y)\to(1,1)} (y - 2) = -1$

17. $\lim\limits_{P\to(2,3,-6)} \sqrt{x^2 + y^2 + z^2} = 7$

19. $\lim\limits_{P\to(3,3,0)} \left(\sin^2 x + \cos^2 y + \sec^2 z\right) =$
$\sin^2 3 + \cos^2 3 + \sec^2 0 = 2$

21. $\lim\limits_{P\to(-1/4,\pi/2,2)} \tan^{-1}(xyz) = \tan^{-1}\left(-\frac{\pi}{4}\right)$

23. a) Continuous at all (x,y)
b) Continuous at all (x,y) except (0,0)

25. a) Continuous at all (x,y) except where x = 0 or y = 0
b) Continuous at all (x,y)

27. a) Continuous at all (x,y,z)

b) Continuous at all (x,y,z) except the interior of the cylinder $x^2 + y^2 = 1$

29. a) Continuous at all (x,y,z) so that $(x,y,z) \neq (x,y,0)$

b) Continuous at all (x,y,z) except those on the sphere $x^2 + y^2 + z^2 = 1$

31. $\lim\limits_{\substack{(x,y)\to(0,0)\\ \text{along } y = x,\\ x>0}} \frac{x}{\sqrt{x^2+y^2}} = \lim\limits_{(x,y)\to(0,0)} \frac{x}{\sqrt{x^2+x^2}} = \lim\limits_{(x,y)\to(0,0)} \frac{x}{\sqrt{2}\,|x|} = \lim\limits_{(x,y)\to(0,0)} \frac{x}{\sqrt{2}\,x} = \lim\limits_{(x,y)\to(0,0)} \frac{1}{\sqrt{2}} = \frac{1}{\sqrt{2}}$

$\lim\limits_{\substack{(x,y)\to(0,0)\\ \text{along } y = x,\\ x<0}} \frac{x}{\sqrt{x^2+y^2}} = \lim\limits_{(x,y)\to(0,0)} \frac{x}{\sqrt{2}\,|x|} = \lim\limits_{(x,y)\to(0,0)} \frac{x}{\sqrt{2}(-x)} = \lim\limits_{(x,y)\to(0,0)} -\frac{1}{\sqrt{2}} = -\frac{1}{\sqrt{2}}$

$\therefore$ consider paths along y = x where x > 0 or x < 0.

33. $\lim\limits_{\substack{(x,y)\to(0,0)\\ \text{along } y = kx^2}} \frac{x^4-y^2}{x^4+y^2} = \lim\limits_{(x,y)\to(0,0)} \frac{x^4-(kx^2)^2}{x^4+(kx^2)^2} = \lim\limits_{(x,y)\to(0,0)} \frac{x^4-k^2x^4}{x^4+k^2x^4} = \frac{1-k^2}{1+k^2} \Rightarrow$ different limits for different values of k. $\therefore$ consider paths along $y = kx^2$, k a constant.

35. $\lim\limits_{\substack{(x,y)\to(0,0)\\ \text{along } y = kx,\ k\neq -1}} \frac{x-y}{x+y} = \lim\limits_{(x,y)\to(0,0)} \frac{x-kx}{x+kx} = \frac{1-k}{1+k} \Rightarrow$ different limits for different values of k. $\therefore$ consider paths along y = kx, k a constant, $k \neq -1$.

37. $\lim\limits_{\substack{(x,y)\to(0,0)\\ \text{along } y = kx^2,\ k\neq 0}} \frac{x^2+y}{y} = \lim\limits_{(x,y)\to(0,0)} \frac{x^2+kx^2}{kx^2} = \frac{1+k}{k} \Rightarrow$ different limits for different values of k. $\therefore$ consider paths along $y = kx^2$, k a constant, $k \neq 0$.

39. a) $f(x,y)\big|_{y=mx} = \frac{2m}{1+m^2} = \frac{2\tan\theta}{1+\tan^2\theta} = \sin 2\theta$. The value of f(x,y) is $\sin 2\theta$ where $\tan\theta = m$ along y = mx.

b) Since $f(x,y)\big|_{y=mx} = \sin 2\theta$ and since $-1 \le \sin 2\theta \le 1$ for every $\theta$, $\lim\limits_{(x,y)\to(0,0)} f(x,y)$ varies from −1 to 1 along y = mx.

## SECTION 13.3 PARTIAL DERIVATIVES

1. $\frac{\partial f}{\partial x} = 2, \frac{\partial f}{\partial y} = 0$

3. $\frac{\partial f}{\partial x} = 0, \frac{\partial f}{\partial y} = 0$

5. $\frac{\partial f}{\partial x} = y - 1, \frac{\partial f}{\partial y} = x$

7. $\frac{\partial f}{\partial x} = 2x - y, \frac{\partial f}{\partial y} = -x + 2y$

9. $\frac{\partial f}{\partial x} = 5y - 14x + 3,$
$\frac{\partial f}{\partial y} = 5x - 2y - 6$

11. $\frac{\partial f}{\partial x} = x(x^2 + y^2)^{-1/2},$
$\frac{\partial f}{\partial y} = y(x^2 + y^2)^{-1/2}$

13. $\frac{\partial f}{\partial x} = \frac{-y^2 - 1}{(xy - 1)^2}, \frac{\partial f}{\partial y} = \frac{-x^2 - 1}{(xy - 1)^2}$

15. $\frac{\partial f}{\partial x} = e^x \ln y, \frac{\partial f}{\partial y} = \frac{e^x}{y}$

17. $\frac{\partial f}{\partial x} = e^x \sin(y + 1), \frac{\partial f}{\partial y} = e^x \cos(y + 1)$

19. $f_x(x,y,z) = y + z, f_y(x,y,z) = x + z, f_z(x,y,z) = y + x$

21. $f_x(x,y,z) = 0, f_y(x,y,z) = 2y, f_z(x,y,z) = 4z$

23. $f_x(x,y,z) = \cos(x + yz), f_y(x,y,z) = z \cos(x + yz), f_z(x,y,z) = y \cos(x + yz)$

25. $\frac{\partial f}{\partial t} = -2\pi \sin(2\pi t - \alpha), \frac{\partial f}{\partial \alpha} = \sin(2\pi t - \alpha)$

27. $\frac{\partial h}{\partial \rho} = \sin \phi \cos \theta, \frac{\partial h}{\partial \phi} = \rho \cos \phi \cos \theta, \frac{\partial h}{\partial \theta} = -\rho \sin \phi \sin \theta$

29. $\frac{\partial W}{\partial P} = V, \frac{\partial W}{\partial V} = P + \frac{\rho v^2}{2g}, \frac{\partial W}{\partial \rho} = \frac{Vv^2}{2g}, \frac{\partial W}{\partial v} = \frac{V\rho v}{g}, \frac{\partial W}{\partial g} = -\frac{V\rho v^2}{2g^2}$

31. $\frac{\partial f}{\partial x} = y + z, \frac{\partial f}{\partial y} = x + z, \frac{\partial^2 f}{\partial x^2} = 0, \frac{\partial^2 f}{\partial y^2} = 0, \frac{\partial^2 f}{\partial y \partial x} = \frac{\partial^2 f}{\partial x \partial y} = 1$

33. $\frac{\partial g}{\partial x} = 2xy + y \cos x, \frac{\partial g}{\partial y} = x^2 - \sin y + \sin x, \frac{\partial^2 g}{\partial x^2} = 2y - y \sin x, \frac{\partial^2 g}{\partial y^2} = -\cos y, \frac{\partial^2 g}{\partial y \partial x} = \frac{\partial^2 g}{\partial x \partial y} = 2x + \cos x$

35. $\frac{\partial r}{\partial x} = \frac{1}{x + y}, \frac{\partial r}{\partial y} = \frac{1}{x + y}, \frac{\partial^2 r}{\partial x^2} = \frac{-1}{(x + y)^2}, \frac{\partial^2 r}{\partial y^2} = \frac{-1}{(x + y)^2}, \frac{\partial^2 r}{\partial y \partial x} = \frac{\partial^2 r}{\partial x \partial y} = \frac{-1}{(x + y)^2}$

37. $\frac{\partial w}{\partial x} = \frac{2}{2x + 3y}, \frac{\partial w}{\partial y} = \frac{3}{2x + 3y}, \frac{\partial^2 w}{\partial y \partial x} = \frac{-6}{(2x + 3y)^2}$ and $\frac{\partial^2 w}{\partial x \partial y} = \frac{-6}{(2x + 3y)^2}$

39. $\frac{\partial w}{\partial x} = y^2 + 2xy^3 + 3x^2y^4, \frac{\partial w}{\partial y} = 2xy + 3x^2y^2 + 4x^3y^3, \frac{\partial^2 w}{\partial y \partial x} = 2y + 6xy^2 + 12x^2y^3$ and
$\frac{\partial^2 w}{\partial x \partial y} = 2y + 6xy^2 + 12x^2y^3$

41. a) x first b) y first c) x first d) x first e) y first f) y first

43. $xy + z^3x - 2yz = 0 \Rightarrow y + 3z^2x \frac{\partial z}{\partial x} - 2y \frac{\partial z}{\partial x} = 0 \Rightarrow (3z^2x - 2y) \frac{\partial z}{\partial x} = -y - z^3 \Rightarrow \frac{\partial z}{\partial x} = \frac{-y - z^3}{3z^2x - 2y} \Rightarrow$
$\frac{\partial z}{\partial x}(1,1,1) = -2$

45. $\frac{\partial f}{\partial x} = 2x, \frac{\partial f}{\partial y} = 2y, \frac{\partial f}{\partial z} = -4z \Rightarrow \frac{\partial^2 f}{\partial x^2} = 2, \frac{\partial^2 f}{\partial y^2} = 2, \frac{\partial^2 f}{\partial z^2} = -4 \Rightarrow \frac{\partial^2 f}{\partial x^2} + \frac{\partial^2 f}{\partial y^2} + \frac{\partial^2 f}{\partial z^2} = 2 + 2 + (-4) = 0$

47. $\frac{\partial f}{\partial x} = -2e^{-2y} \sin 2x, \frac{\partial f}{\partial y} = -2e^{-2y} \cos 2x, \frac{\partial^2 f}{\partial x^2} = -4e^{-2y} \cos 2x, \frac{\partial^2 f}{\partial y^2} = 4e^{-2y} \cos 2x$

$\therefore \frac{\partial^2 f}{\partial x^2} + \frac{\partial^2 f}{\partial y^2} = -4e^{-2y} \cos 2x + 4e^{-2y} \cos 2x = 0$

49. $\frac{\partial f}{\partial x} = -\frac{1}{2}(x^2 + y^2 + z^2)^{-3/2}(2x) = -x(x^2 + y^2 + x^2)^{-3/2}, \frac{\partial f}{\partial y} = -\frac{1}{2}(x^2 + y^2 + z^2)^{-3/2}(2y) = -y(x^2 + y^2 + z^2)^{-3/2}$

$\frac{\partial f}{\partial z} = -\frac{1}{2}(x^2 + y^2 + z^2)^{-3/2}(2z) = -z(x^2 + y^2 + z^2)^{-3/2}. \frac{\partial^2 f}{\partial x^2} = -(x^2 + y^2 + z^2)^{-3/2} + 3x^2(x^2 + y^2 + z^2)^{-5/2}$

$\frac{\partial^2 f}{\partial y^2} = -(x^2 + y^2 + z^2)^{-3/2} + 3y^2(x^2 + y^2 + z^2)^{-5/2}, \frac{\partial^2 f}{\partial z^2} = -(x^2 + y^2 + z^2)^{-3/2} + 3z^2(x^2 + y^2 + z^2)^{-5/2}$

$\therefore \frac{\partial^2 f}{\partial x^2} + \frac{\partial^2 f}{\partial y^2} + \frac{\partial^2 f}{\partial z^2} = \left(-(x^2 + y^2 + z^2)^{-3/2} + 3x^2(x^2 + y^2 + z^2)^{-5/2}\right) +$

$\left(-(x^2 + y^2 + z^2)^{-3/2} + 3y^2(x^2 + y^2 + z^2)^{-5/2}\right) + \left(-(x^2 + y^2 + z^2)^{-3/2} + 3z^2(x^2 + y^2 + z^2)^{-5/2}\right) =$

$-3(x^2 + y^2 + z^2)^{-3/2} + (3x^2 + 3y^2 + 3z^2)(x^2 + y^2 + z^2)^{-5/2} = 0$

51. $\frac{\partial w}{\partial x} = \cos(x + ct), \frac{\partial w}{\partial t} = c \cos(x + ct). \frac{\partial^2 w}{\partial x^2} = -\sin(x + ct), \frac{\partial^2 w}{\partial t^2} = -c^2 \sin(x + ct)$

$\therefore \frac{\partial^2 w}{\partial t^2} = c^2(-\sin(x + ct)) = c^2 \frac{\partial^2 w}{\partial x^2}$

53. $\frac{\partial w}{\partial x} = \cos(x + ct) - 2 \sin(2x + 2ct), \frac{\partial w}{\partial t} = c \cos(x + ct) - 2c \sin(2x + 2ct). \frac{\partial^2 w}{\partial x^2} = -\sin(x + ct) -$

$4 \cos(2x + 2ct), \frac{\partial^2 w}{\partial t^2} = -c^2 \sin(x + ct) - 4c^2 \cos(2x + 2ct) \quad \therefore \quad \frac{\partial^2 w}{\partial t^2} = c^2(-\sin(x + ct) - 4 \cos(2x + 2ct)) =$

$c^2 \frac{\partial^2 w}{\partial x^2}$

55. $\frac{\partial w}{\partial x} = 2 \sec^2(2x - 2ct), \frac{\partial w}{\partial t} = -2c \sec^2(2x - 2ct). \frac{\partial^2 w}{\partial x^2} = 8 \sec^2(2x - 2ct) \tan(2x - 2ct)$ ,

$\frac{\partial^2 w}{\partial t^2} = 8c^2 \sec^2(2x - 2ct) \tan(2x - 2ct) \quad \therefore \frac{\partial^2 w}{\partial t^2} = c^2\left(8 \sec^2(2x - 2ct) \tan(2x - 2ct)\right) = c^2 \frac{\partial^2 w}{\partial x^2}$

## SECTION 13.4 THE CHAIN RULE

1. $\frac{\partial w}{\partial x} = 2x, \frac{\partial w}{\partial y} = 2y, \frac{dx}{dt} = -\sin t, \frac{dy}{dt} = \cos t \Rightarrow \frac{dw}{dt} = -2x \sin t + 2y \cos t = -2 \cos t \sin t + 2 \sin t \cos t = 0$

$\Rightarrow \frac{dw}{dt}(\pi) = 0$

3. $\frac{\partial w}{\partial x} = \frac{1}{z}, \frac{\partial w}{\partial y} = \frac{1}{z}, \frac{\partial w}{\partial z} = \frac{-(x + y)}{z^2}, \frac{dx}{dt} = -2 \cos t \sin t, \frac{dy}{dt} = 2 \sin t \cos t, \frac{dz}{dt} = -\frac{1}{t^2} \Rightarrow$

$\frac{dw}{dt} = -\frac{2}{z} \cos t \sin t + \frac{2}{z} \sin t \cos t + \frac{x + y}{z^2 t^2} = \frac{\cos^2 t + \sin^2 t}{\frac{1}{t^2}(t^2)} = 1 \Rightarrow \frac{dw}{dt}(3) = 1$

5. $\frac{\partial w}{\partial x} = 2ye^x$, $\frac{\partial w}{\partial y} = 2e^x$, $\frac{\partial w}{\partial z} = -\frac{1}{z}$, $\frac{dx}{dt} = \frac{2t}{t^2+1}$, $\frac{dy}{dt} = \frac{1}{t^2+1}$, $\frac{dz}{dt} = e^t \Rightarrow \frac{dw}{dt} = \frac{4yte^x}{t^2+1} + \frac{2e^x}{t^2+1} - \frac{e^t}{z} =$
$\frac{4t\tan^{-1} t\, e^{\ln(t^2+1)}}{t^2+1} + \frac{2(t^2+1)}{t^2+1} - \frac{e^t}{e^t} = 4t\tan^{-1} t + 1 \Rightarrow \frac{dw}{dt}(1) = \pi + 1$

7. $\frac{dz}{dt} = \frac{\partial z}{\partial x}\frac{dx}{dt} + \frac{\partial z}{\partial y}\frac{dy}{dt}$

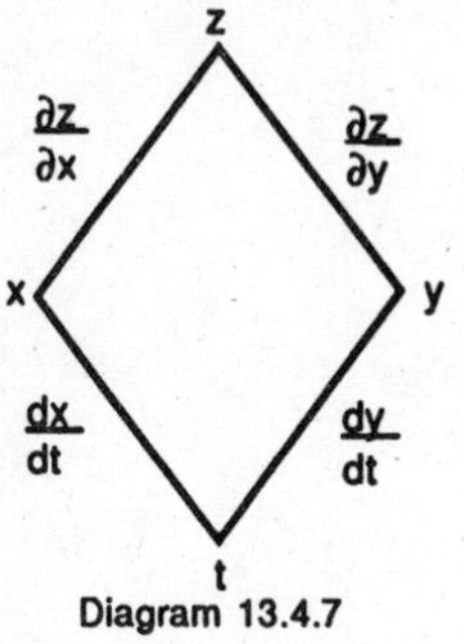

Diagram 13.4.7

9. $\frac{\partial w}{\partial u} = \frac{\partial w}{\partial x}\frac{\partial x}{\partial u} + \frac{\partial w}{\partial y}\frac{\partial y}{\partial u} + \frac{\partial w}{\partial z}\frac{\partial z}{\partial u}$

$\frac{\partial w}{\partial v} = \frac{\partial w}{\partial x}\frac{\partial x}{\partial v} + \frac{\partial w}{\partial y}\frac{\partial y}{\partial v} + \frac{\partial w}{\partial z}\frac{\partial z}{\partial v}$

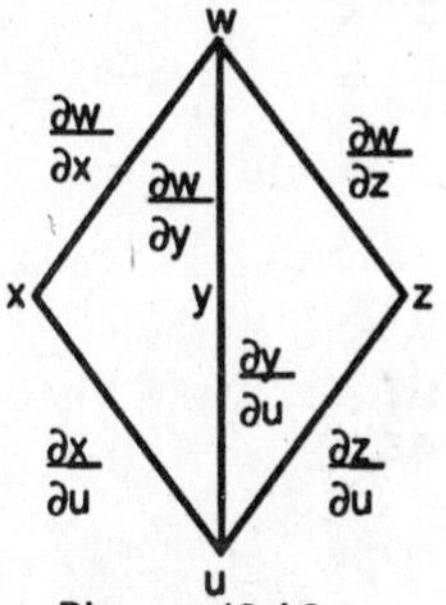

Diagram 13.4.9 a

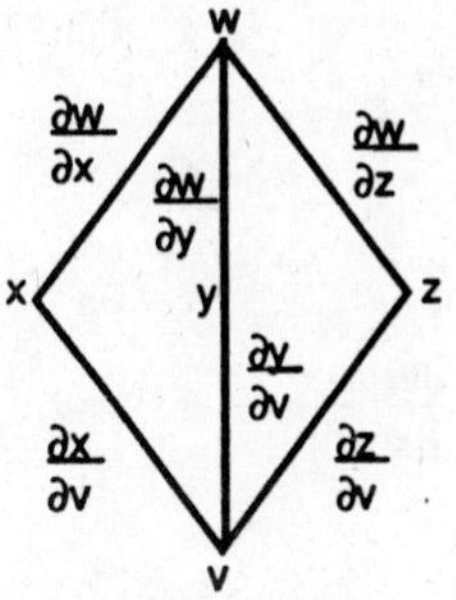

Diagram 13.4.9 b

11. $\frac{\partial w}{\partial u} = \frac{\partial w}{\partial x}\frac{\partial x}{\partial u} + \frac{\partial w}{\partial y}\frac{\partial y}{\partial u}$

$\frac{\partial w}{\partial v} = \frac{\partial w}{\partial x}\frac{\partial x}{\partial v} + \frac{\partial w}{\partial y}\frac{\partial y}{\partial v}$

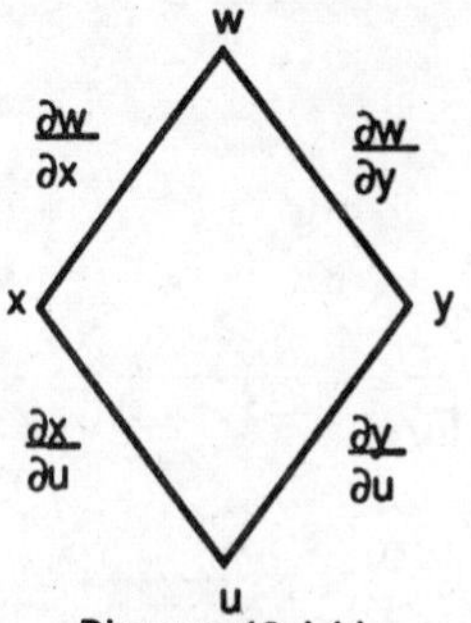

Diagram 13.4.11 a

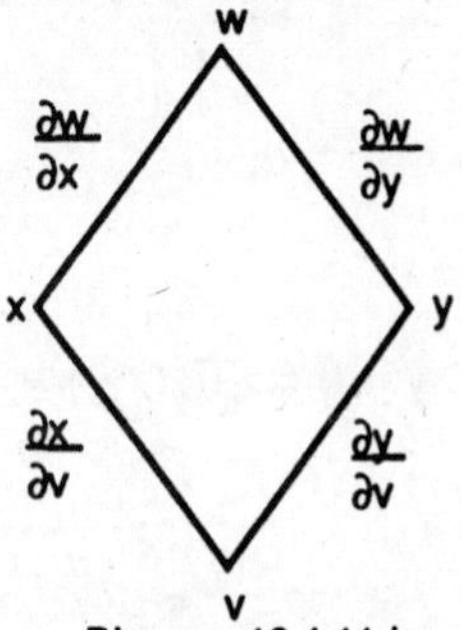

Diagram 13.4.11 b

13. $\frac{\partial z}{\partial t} = \frac{\partial z}{\partial x}\frac{\partial x}{\partial t} + \frac{\partial z}{\partial y}\frac{\partial y}{\partial t}$ $\frac{\partial z}{\partial s} = \frac{\partial z}{\partial x}\frac{\partial x}{\partial s} + \frac{\partial z}{\partial y}\frac{\partial y}{\partial s}$

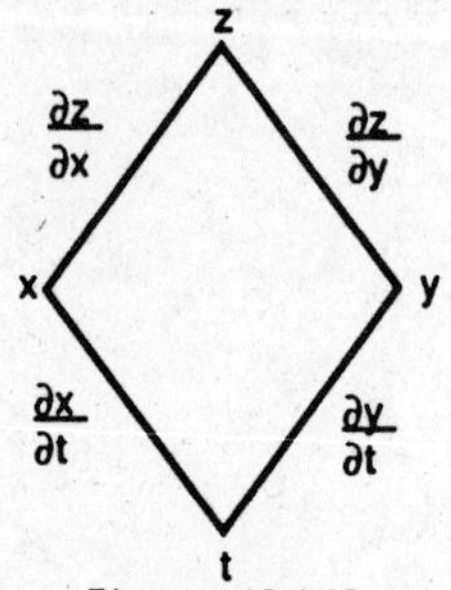

Diagram 13.4.13 a

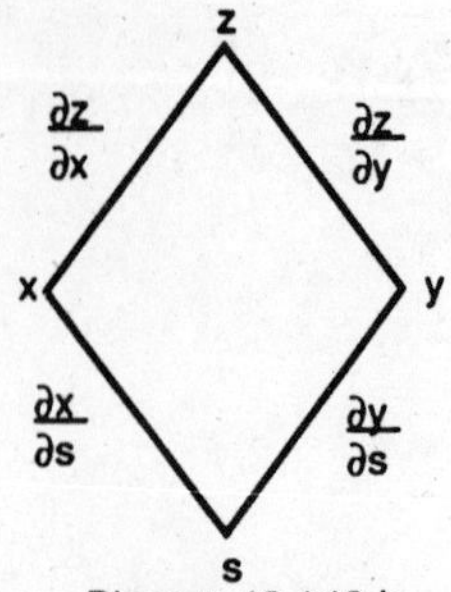

Diagram 13.4.13 b

15. $\frac{\partial w}{\partial s} = \frac{dw}{du}\frac{\partial u}{\partial s}$ $\frac{\partial w}{\partial t} = \frac{dw}{du}\frac{\partial u}{\partial t}$

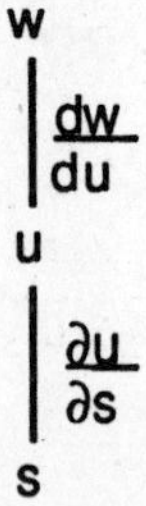

Diagram 13.4.15 a

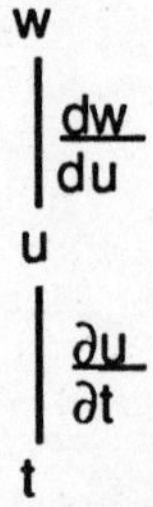

Diagram 13.4.15 b

17. $\frac{\partial w}{\partial r} = \frac{\partial w}{\partial x}\frac{dx}{dr} + \frac{\partial w}{\partial y}\frac{dy}{dr} = \frac{\partial w}{\partial x}\frac{dx}{dr}$ since $\frac{dy}{dr} = 0$

$\frac{\partial w}{\partial s} = \frac{\partial w}{\partial x}\frac{dx}{ds} + \frac{\partial w}{\partial y}\frac{dy}{ds} = \frac{\partial w}{\partial y}\frac{dy}{ds}$ since $\frac{dx}{ds} = 0$

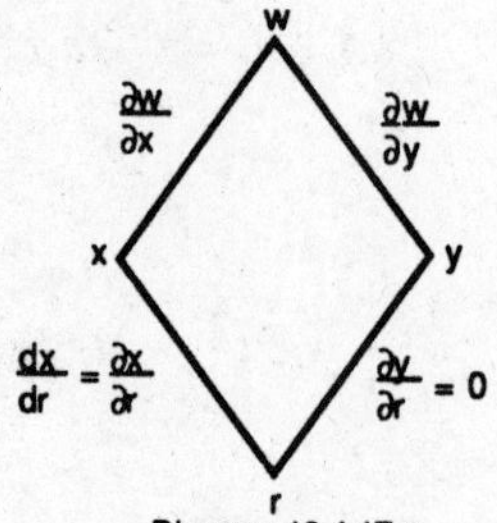

Diagram 13.4.17 a

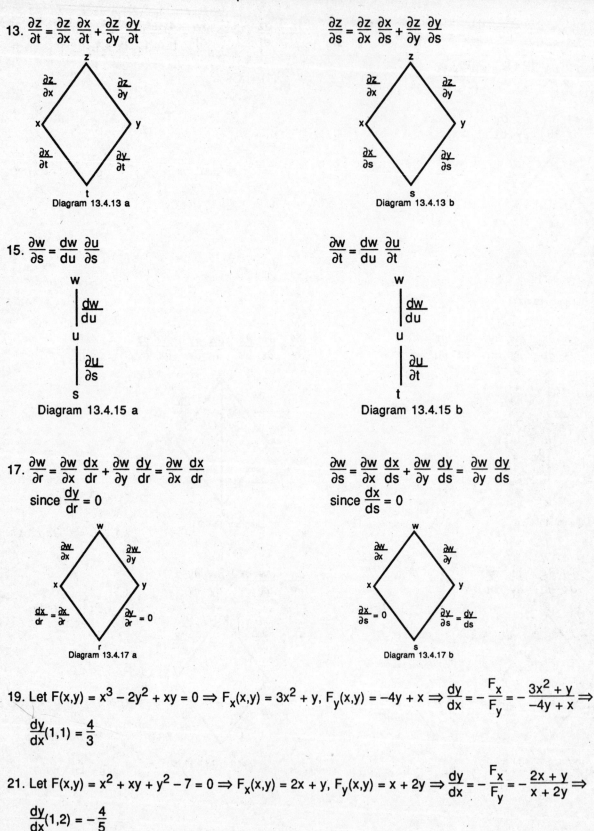

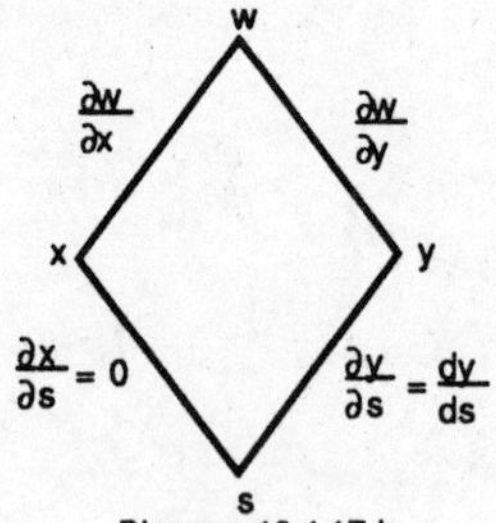

Diagram 13.4.17 b

19. Let $F(x,y) = x^3 - 2y^2 + xy = 0 \Rightarrow F_x(x,y) = 3x^2 + y,\ F_y(x,y) = -4y + x \Rightarrow \frac{dy}{dx} = -\frac{F_x}{F_y} = -\frac{3x^2 + y}{-4y + x} \Rightarrow$
$\frac{dy}{dx}(1,1) = \frac{4}{3}$

21. Let $F(x,y) = x^2 + xy + y^2 - 7 = 0 \Rightarrow F_x(x,y) = 2x + y,\ F_y(x,y) = x + 2y \Rightarrow \frac{dy}{dx} = -\frac{F_x}{F_y} = -\frac{2x + y}{x + 2y} \Rightarrow$
$\frac{dy}{dx}(1,2) = -\frac{4}{5}$

23. Let $F(x,y,z)) = z^3 - xy + yz + y^3 - 2 = 0 \Rightarrow F_x(x,y,z) = -y,\ F_y(x,y,z) = -x + z + 3y^2,\ F_z(x,y,z) = 3z^2 + y \Rightarrow$
$\frac{\partial z}{\partial x} = -\frac{F_x}{F_z} = -\frac{-y}{3z^2+y} = \frac{y}{3z^2+y} \Rightarrow \frac{\partial z}{\partial x}(1,1,1) = \frac{1}{4}$. $\frac{\partial z}{\partial y} = -\frac{F_y}{F_z} = -\frac{-x+z+3y^2}{3z^2+y} = \frac{x-z-3y^2}{3z^2+y} \Rightarrow$
$\frac{\partial z}{\partial y}(1,1,1) = -\frac{3}{4}$

25. Let $F(x,y,z) = \sin(x+y) + \sin(y+z) + \sin(x+z) = 0 \Rightarrow F_x(x,y,z) = \cos(x+y) + \cos(x+z)$,
$F_y(x,y,z) = \cos(x+y) + \cos(y+z),\ F_z(x,y,z) = \cos(y+z) + \cos(x+z) \Rightarrow \frac{\partial z}{\partial x} = -\frac{F_x}{F_z} = -\frac{\cos(x+y)+\cos(x+z)}{\cos(y+z)+\cos(x+z)}$
$\Rightarrow \frac{\partial z}{\partial x}(\pi,\pi,\pi) = -1$. $\frac{\partial z}{\partial y} = -\frac{F_y}{F_z} = -\frac{\cos(x+y)+\cos(y+z)}{\cos(y+z)+\cos(x+z)} \Rightarrow \frac{\partial z}{\partial y}(\pi,\pi,\pi) = -1$

27. $\frac{\partial w}{\partial r} = \frac{\partial w}{\partial x}\frac{\partial x}{\partial r} + \frac{\partial w}{\partial y}\frac{\partial y}{\partial r} + \frac{\partial w}{\partial z}\frac{\partial z}{\partial r} = 2(x+y+z)(1) + 2(x+y+z)(-\sin(r+s)) + 2(x+y+z)(\cos(r+s)) =$
$2(x+y+z)(1-\sin(r+s)+\cos(r+s)) = 2(r-s+\cos(r+s)+\sin(r+s))(1-\sin(r+s)+\cos(r+s)) \Rightarrow$
$\left.\frac{\partial w}{\partial r}\right|_{r=1,s=-1} = 12$

29. $\frac{\partial w}{\partial v} = \frac{\partial w}{\partial x}\frac{\partial x}{\partial v} + \frac{\partial w}{\partial y}\frac{\partial y}{\partial v} = \left(2x - \frac{y}{x^2}\right)(-2) + \frac{1}{x}(1) = \left(2(u-2v+1) - \frac{2u+v-2}{(u-2v+1)^2}\right)(-2) + \frac{1}{u-2v+1} \Rightarrow$
$\left.\frac{\partial w}{\partial v}\right|_{u=0,v=0} = -7$

31. $\frac{\partial z}{\partial u} = \frac{dz}{dx}\frac{\partial x}{\partial u} = \frac{5}{1+x^2}\left(e^u + \ln v\right) = \frac{5}{1+\left(e^u+\ln v\right)^2}\left(e^u + \ln v\right) \Rightarrow \left.\frac{\partial z}{\partial u}\right|_{u=\ln 2,v=1} = 2$
$\frac{\partial z}{\partial v} = \frac{dz}{dx}\frac{\partial x}{\partial v} = \frac{5}{1+x^2}\left(\frac{1}{v}\right) = \frac{5}{1+\left(e^u+\ln v\right)^2}\left(\frac{1}{v}\right) \Rightarrow \left.\frac{\partial z}{\partial v}\right|_{u=\ln 2,v=1} = 5$

33. $\frac{dV}{dt} = \frac{\partial V}{\partial I}\frac{dI}{dt} + \frac{\partial V}{\partial R}\frac{dR}{dt}$. $V = IR \Rightarrow \frac{\partial V}{\partial I} = R,\ \frac{\partial V}{\partial R} = I \Rightarrow \frac{dV}{dt} = R\frac{dI}{dt} + I\frac{dR}{dt} \Rightarrow -0.01$ volts/sec = (600 ohms)$\frac{dI}{dt}$ +
(0.04 amps)(0.5 ohms/sec) $\Rightarrow \frac{dI}{dt} = -0.00005$ amps/sec.

35. $f_x(x,y,z) = \cos t,\ f_y(x,y,z) = \sin t,\ f_z(x,y,z) = t^2 + t - 2$. $\frac{df}{dt} = \frac{\partial f}{\partial x}\frac{dx}{dt} + \frac{\partial f}{\partial y}\frac{dy}{dt} + \frac{\partial f}{\partial z}\frac{dz}{dt} = (\cos t)(-\sin t) +$
$(\sin t)(\cos t) + (t^2+t-2)(1) = t^2 + t - 2$. $\frac{df}{dt} = 0 \Rightarrow t^2 + t - 2 = 0 \Rightarrow t = -2$ or $t = 1$
$t = -2 \Rightarrow x = \cos(-2),\ y = \sin(-2),\ z = -2$; $t = 1 \Rightarrow x = \cos 1,\ y = \sin 1,\ z = 1$

37. a) $\frac{\partial T}{\partial x} = 8x - 4y,\ \frac{\partial T}{\partial y} = 8y - 4x$. $\frac{dT}{dt} = \frac{\partial T}{\partial x}\frac{dx}{dt} + \frac{\partial T}{\partial y}\frac{dy}{dt} = (8x-4y)(-\sin t) + (8y-4x)(\cos t) =$
$(8\cos t - 4\sin t)(-\sin t) + (8\sin t - 4\cos t)(\cos t) = 4\sin^2 t - 4\cos^2 t \Rightarrow \frac{d^2T}{dt^2} = 16\sin t\cos t$
$\frac{dT}{dt} = 0 \Rightarrow 4\sin^2 t - 4\cos^2 t = 0 \Rightarrow \sin^2 t = \cos^2 t \Rightarrow \sin t = \cos t$ or $\sin t = -\cos t \Rightarrow$
$t = \frac{\pi}{4}, \frac{5\pi}{4}$ or $\frac{3\pi}{4}, \frac{7\pi}{4}$ on the interval $0 \le t \le 2\pi$.
$\left.\frac{d^2T}{dt^2}\right|_{t=\pi/4} = 16\sin\frac{\pi}{4}\cos\frac{\pi}{4} > 0 \Rightarrow$ T has a minimum at $t = \frac{\pi}{4}$
$\left.\frac{d^2T}{dt^2}\right|_{t=3\pi/4} = 16\sin\frac{3\pi}{4}\cos\frac{3\pi}{4} < 0 \Rightarrow$ T has a maximum at $t = \frac{3\pi}{4}$
$\left.\frac{d^2T}{dt^2}\right|_{t=5\pi/4} = 16\sin\frac{5\pi}{4}\cos\frac{5\pi}{4} > 0 \Rightarrow$ T has a minimum at $t = \frac{5\pi}{4}$

37. (Continued)

$\left.\frac{d^2T}{dt^2}\right|_{t=7\pi/4} = 16 \sin\frac{7\pi}{4}\cos\frac{7\pi}{4} < 0 \Rightarrow$ T has a maximum at $t = \frac{7\pi}{4}$

b) $T = 4x^2 - 4xy + 4y^2 \Rightarrow \frac{\partial T}{\partial x} = 8x - 4y,\ \frac{\partial T}{\partial y} = 8y - 4x$ (See part a above.)

$t = \frac{\pi}{4} \Rightarrow x = \cos\frac{\pi}{4} = \frac{\sqrt{2}}{2},\ y = \sin\frac{\pi}{4} = \frac{\sqrt{2}}{2} \Rightarrow T\left(\frac{\pi}{4}\right) = 2$

$t = \frac{3\pi}{4} \Rightarrow x = \cos\frac{3\pi}{4} = -\frac{\sqrt{2}}{2},\ y = \sin\frac{3\pi}{4} = \frac{\sqrt{2}}{2} \Rightarrow T\left(\frac{3\pi}{4}\right) = 6$

$t = \frac{5\pi}{4} \Rightarrow x = \cos\frac{5\pi}{4} = -\frac{\sqrt{2}}{2},\ y = \sin\frac{5\pi}{4} = -\frac{\sqrt{2}}{2} \Rightarrow T\left(\frac{5\pi}{4}\right) = 2$

$t = \frac{7\pi}{4} \Rightarrow x = \cos\frac{7\pi}{4} = \frac{\sqrt{2}}{2},\ y = \sin\frac{7\pi}{4} = -\frac{\sqrt{2}}{2} \Rightarrow T\left(\frac{7\pi}{4}\right) = 6$

$\therefore\ T_{max} = 6$ and $T_{min} = 2$.

## SECTION 13.5 DIRECTIONAL DERIVATIVES AND GRADIENT VECTORS

1. $\frac{\partial f}{\partial x} = 2x \Rightarrow \frac{\partial f}{\partial x}(1,1,1) = 2.\ \frac{\partial f}{\partial y} = 2y \Rightarrow \frac{\partial f}{\partial y}(1,1,1) = 2.\ \frac{\partial f}{\partial z} = -4z \Rightarrow \frac{\partial f}{\partial z}(1,1,1) = -4 \Rightarrow \nabla f = 2\,\mathbf{i} + 2\,\mathbf{j} - 4\,\mathbf{k}$

3. $\frac{\partial f}{\partial x} = -x(x^2+y^2+z^2)^{-3/2} \Rightarrow \frac{\partial f}{\partial x}(1,2,-2) = -\frac{1}{27}.\ \frac{\partial f}{\partial y} = -y(x^2+y^2+z^2)^{-3/2} \Rightarrow \frac{\partial f}{\partial y}(1,2,-2) = -\frac{2}{27}.$
$\frac{\partial f}{\partial z} = -z(x^2+y^2+z^2)^{-3/2} \Rightarrow \frac{\partial f}{\partial z}(1,2,-2) = \frac{2}{27} \Rightarrow \nabla f = -\frac{1}{27}\mathbf{i} - \frac{2}{27}\mathbf{j} + \frac{2}{27}\mathbf{k}$

5. $\frac{\partial f}{\partial x} = -1,\ \frac{\partial f}{\partial y} = 1 \Rightarrow \nabla f = -\mathbf{i} + \mathbf{j}$
$-1 = y - x$ is the level curve.

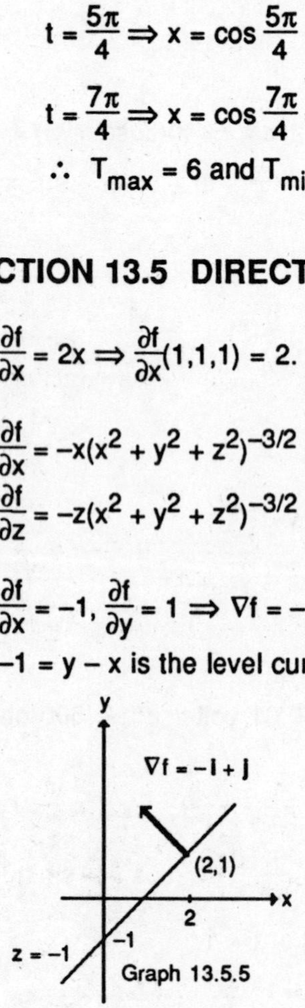

Graph 13.5.5

7. $\frac{\partial f}{\partial x} = -2x \Rightarrow \frac{\partial f}{\partial x}(-1,0) = 2.\ \frac{\partial f}{\partial y} = 1 \Rightarrow$
$\nabla f = 2\,\mathbf{i} + \mathbf{j}$. $-1 = y - x^2$ is the level curve.

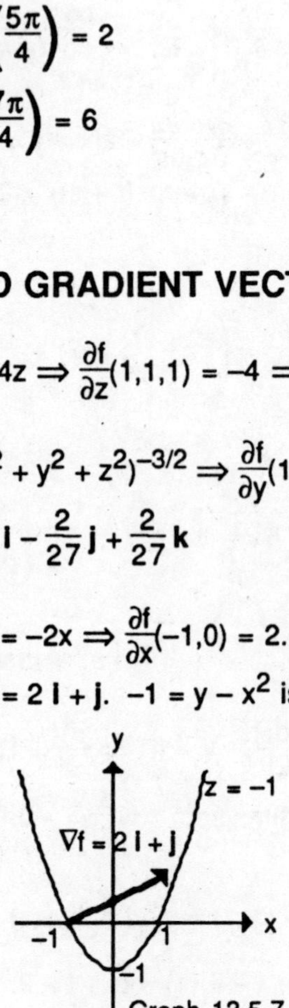

Graph 13.5.7

9. $\mathbf{u} = \frac{\mathbf{A}}{|\mathbf{A}|} = \frac{3\,\mathbf{i} + 6\,\mathbf{j} - 2\,\mathbf{k}}{\sqrt{3^2 + 6^2 + (-2)^2}} = \frac{3}{7}\mathbf{i} + \frac{6}{7}\mathbf{j} - \frac{2}{7}\mathbf{k}$. $f_x(x,y,z) = y + z \Rightarrow f_x(1,-1,2) = 1,\ f_y(x,y,z) = x + z \Rightarrow$
$f_y(1,-1,2) = 3,\ f_z(x,y,z) = y + x \Rightarrow f_z(1,-1,2) = 0 \Rightarrow \nabla f = \mathbf{i} + 3\,\mathbf{j} \Rightarrow \left(D_{\mathbf{u}}f\right)_{P_0} = \nabla f \cdot \mathbf{u} = \frac{3}{7} + \frac{18}{7} = 3$

11. $\mathbf{u} = \frac{\mathbf{A}}{|\mathbf{A}|} = \frac{2\,\mathbf{i} + \mathbf{j} - 2\mathbf{k}}{\sqrt{2^2 + 1^2 + (-2)^2}} = \frac{2}{3}\mathbf{i} + \frac{1}{3}\mathbf{j} - \frac{2}{3}\mathbf{k}$. $f_x(x,y,z) = 3e^x \cos yz \Rightarrow f_x(0,0,0) = 3,\ f_y(x,y,z) =$
$-3ze^x \sin yz \Rightarrow f_y(0,0,0) = 0,\ f_z(x,y,z) = -3ye^x \sin yz \Rightarrow f_z(0,0,0) = 0 \Rightarrow \nabla f = 3\,\mathbf{i} \Rightarrow \left(D_{\mathbf{u}}f\right)_{P_0} = \nabla f \cdot \mathbf{u} = 2$

13. $\mathbf{u} = \frac{\mathbf{A}}{|\mathbf{A}|} = \frac{12\,\mathbf{i} + 5\,\mathbf{j}}{\sqrt{12^2 + 5^2}} = \frac{12}{13}\mathbf{i} + \frac{5}{13}\mathbf{j}$. $f_x(x,y,z) = 1 + \frac{y^2}{x^2} \Rightarrow f_x(1,1) = 2$, $f_y(x,y,z) = -\frac{2y}{x} \Rightarrow f_y(1,1) = -2 \Rightarrow$
$\nabla f = 2\,\mathbf{i} - 2\,\mathbf{j} \Rightarrow \left(D_{\mathbf{u}}f\right)_{P_0} = \nabla f \cdot \mathbf{u} = \frac{24}{13} - \frac{10}{13} = \frac{14}{13}$

15. $\mathbf{u} = \frac{\mathbf{A}}{|\mathbf{A}|} = \frac{3\,\mathbf{i} + 4\,\mathbf{j}}{\sqrt{3^2 + 4^2}} = \frac{3}{5}\mathbf{i} + \frac{4}{5}\mathbf{j}$. $f_x(x,y,z) = 2x + 2y \Rightarrow f_x(1,1) = 4$, $f_y(x,y,z) = 2x = 6y \Rightarrow f_y(1,1) = -4 \Rightarrow$
$\nabla f = 4\,\mathbf{i} - 4\,\mathbf{j} \Rightarrow \left(D_{\mathbf{u}}f\right)_{P_0} = \nabla f \cdot \mathbf{u} = \frac{12}{5} - \frac{16}{5} = -\frac{4}{5}$

17. $\nabla f = 2x\,\mathbf{i} + \cos y\,\mathbf{j} \Rightarrow \nabla f(1,0) = 2\,\mathbf{i} + \mathbf{j} \Rightarrow \mathbf{u} = \frac{\nabla f}{|\nabla f|} = \frac{2\,\mathbf{i} + \mathbf{j}}{\sqrt{2^2 + 1^2}} = \frac{2}{\sqrt{5}}\mathbf{i} + \frac{1}{\sqrt{5}}\mathbf{j}$. f increases most rapidly in the direction of $\mathbf{u} = \frac{2}{\sqrt{5}}\mathbf{i} + \frac{1}{\sqrt{5}}\mathbf{j}$; decreases most rapidly in the direction $-\mathbf{u} = -\frac{2}{\sqrt{5}}\mathbf{i} - \frac{1}{\sqrt{5}}\mathbf{j}$.
$\left(D_{\mathbf{u}}f\right)_{P_0} = \nabla f \cdot \mathbf{u} = \sqrt{5}$, $\left(D_{-\mathbf{u}}f\right)_{P_0} = -\sqrt{5}$.

19. $\nabla f = e^y\,\mathbf{i} + xe^y\,\mathbf{j} + 2z\,\mathbf{k} \Rightarrow \nabla f\left(1, \ln 2, \frac{1}{2}\right) = 2\,\mathbf{i} + 2\,\mathbf{j} + \mathbf{k} \Rightarrow \mathbf{u} = \frac{\nabla f}{|\nabla f|} = \frac{2\,\mathbf{i} + 2\,\mathbf{j} + \mathbf{k}}{\sqrt{2^2 + 2^2 + 1^2}} = \frac{2}{3}\mathbf{i} + \frac{2}{3}\mathbf{j} + \frac{1}{3}\mathbf{k}$.
f increases most rapidly in the direction $\mathbf{u} = \frac{2}{3}\mathbf{i} + \frac{2}{3}\mathbf{j} + \frac{1}{3}\mathbf{k}$; decreases most rapidly in the direction $-\mathbf{u} = -\frac{2}{3}\mathbf{i} - \frac{2}{3}\mathbf{j} - \frac{1}{3}\mathbf{k}$. $\left(D_{\mathbf{u}}f\right)_{P_0} = \nabla f \cdot \mathbf{u} = 3$; $\left(D_{-\mathbf{u}}f\right)_{P_0} = -3$

21. $\nabla f = (\cos x)\,\mathbf{i} + (\cos y)\,\mathbf{j} + (\cos z)\,\mathbf{k} \Rightarrow \nabla f(0,0,0) = \mathbf{i} + \mathbf{j} + \mathbf{k} \Rightarrow \mathbf{u} = \frac{\nabla f}{|\nabla f|} = \frac{\mathbf{i} + \mathbf{j} + \mathbf{k}}{\sqrt{1^2 + 1^2 + 1^2}} = \frac{1}{\sqrt{3}}\mathbf{i} + \frac{1}{\sqrt{3}}\mathbf{j} + \frac{1}{\sqrt{3}}\mathbf{k}$. f increases most rapidly in the direction $\mathbf{u} = \frac{1}{\sqrt{3}}\mathbf{i} + \frac{1}{\sqrt{3}}\mathbf{j} + \frac{1}{\sqrt{3}}\mathbf{k}$; decreases most rapidly in the direction $-\mathbf{u} = -\frac{1}{\sqrt{3}}\mathbf{i} - \frac{1}{\sqrt{3}}\mathbf{j} - \frac{1}{\sqrt{3}}\mathbf{k}$. $\left(D_{\mathbf{u}}f\right)_{P_0} = \nabla f \cdot \mathbf{u} = \sqrt{3}$, $\left(D_{-\mathbf{u}}f\right)_{P_0} = -\sqrt{3}$

23. $\nabla f = (-\pi y \sin(\pi xy) + y^2)\,\mathbf{i} + (-\pi x \sin(\pi xy) + 2xy)\,\mathbf{j} \Rightarrow \nabla f(-1,-1) = \mathbf{i} + 2\,\mathbf{j}$. $\mathbf{u} = \frac{\mathbf{A}}{|\mathbf{A}|} = \frac{\mathbf{i} + \mathbf{j}}{\sqrt{1^2 + 1^2}} = \frac{1}{\sqrt{2}}\mathbf{i} + \frac{1}{\sqrt{2}}\mathbf{j}$
$\nabla f \cdot \mathbf{u} = \frac{1}{\sqrt{2}} + \frac{2}{\sqrt{2}} = \frac{3}{\sqrt{2}}$. $\therefore$ $df = (\nabla f \cdot \mathbf{u})ds = \frac{3}{\sqrt{2}}(0.1) = 0.15\sqrt{2} \approx 0.212132$

25. $\nabla f = \left(e^x \cos yz\right)\mathbf{i} - \left(ze^x \sin yz\right)\mathbf{j} - \left(ye^x \sin yz\right)\mathbf{k} \Rightarrow \nabla f(0,0,0) = \mathbf{i}$. $\mathbf{u} = \frac{\mathbf{A}}{|\mathbf{A}|} = \frac{2\,\mathbf{i} + \mathbf{j} - 2\,\mathbf{k}}{\sqrt{2^2 + 1^2 + (-2)^2}} =$
$\frac{2}{3}\mathbf{i} + \frac{1}{3}\mathbf{j} - \frac{2}{3}\mathbf{k} \Rightarrow \nabla f \cdot \mathbf{u} = \frac{2}{3}$. $\therefore$ $df = (\nabla f \cdot \mathbf{u})ds = \frac{2}{3}(0.1) = \frac{1}{15}$ or 0.067

27. $\nabla f = y\,\mathbf{i} + (x + 2y)\,\mathbf{j} \Rightarrow \nabla f(3,2) = 2\,\mathbf{i} + 7\,\mathbf{j}$. $\mathbf{A}$, orthogonal to $\nabla f$, is $\mathbf{A} = 7\,\mathbf{i} - 2\,\mathbf{j} \Rightarrow \mathbf{u} = \frac{\mathbf{A}}{|\mathbf{A}|} = \frac{7\,\mathbf{i} - 2\,\mathbf{j}}{\sqrt{7^2 + (-2)^2}} =$
$\frac{7}{\sqrt{53}}\mathbf{i} - \frac{2}{\sqrt{53}}\mathbf{j} \Rightarrow -\mathbf{u} = -\frac{7}{\sqrt{53}}\mathbf{i} + \frac{2}{\sqrt{53}}\mathbf{j}$

29. $\nabla f = 2x\,\mathbf{i} + 2y\,\mathbf{j} + 2z\,\mathbf{k} = (2\cos t)\,\mathbf{i} + (2\sin t)\,\mathbf{j} + 2t\,\mathbf{k}$. $\mathbf{v} = (-\sin t)\,\mathbf{i} + (\cos t)\,\mathbf{j} + \mathbf{k} \Rightarrow \mathbf{T} = \frac{\mathbf{v}}{|\mathbf{v}|} =$
$\frac{(-\sin t)\,\mathbf{i} + (\cos t)\,\mathbf{j} + \mathbf{k}}{\sqrt{(\sin t)^2 + (\cos t)^2 + 1^2}} = \left(\frac{-\sin t}{\sqrt{2}}\right)\mathbf{i} + \left(\frac{\cos t}{\sqrt{2}}\right)\mathbf{j} + \frac{1}{\sqrt{2}}\mathbf{k}$.
$\left(D_{\mathbf{T}}f\right)_{P_0} = \nabla f \cdot \mathbf{T} = (2\cos t)\left(\frac{-\sin t}{\sqrt{2}}\right) + (2\sin t)\left(\frac{\cos t}{\sqrt{2}}\right) + 2t\left(\frac{1}{\sqrt{2}}\right) = \frac{2t}{\sqrt{2}} \Rightarrow \left(D_{\mathbf{T}}f\right)\left(\frac{-\pi}{4}\right) = \frac{-\pi}{2\sqrt{2}}$,
$\left(D_{\mathbf{T}}f\right)(0) = 0$, $\left(D_{\mathbf{T}}f\right)\left(\frac{\pi}{4}\right) = \frac{\pi}{2\sqrt{2}}$

31. $\nabla f = f_x(1,2)\,\mathbf{i} + f_y(1,2)\,\mathbf{j}$. $\mathbf{u}_1 = \frac{\mathbf{i}+\mathbf{j}}{\sqrt{1^2+1^2}} = \frac{1}{\sqrt{2}}\mathbf{i} + \frac{1}{\sqrt{2}}\mathbf{j}$. $\left(D_{\mathbf{u}_1}f\right)(1,2) =$
$f_x(1,2)\left(\frac{1}{\sqrt{2}}\right) + f_y(1,2)\left(\frac{1}{\sqrt{2}}\right) = 2\sqrt{2} \Rightarrow f_x(1,2) + f_y(1,2) = 4$. $\mathbf{u}_2 = -\mathbf{j}$. $\left(D_{\mathbf{u}_2}f\right)(1,2) = f_x(1,2)(0) + f_y(1,2)(-1)$
$= -3 \Rightarrow -f_y(1,2) = -3 \Rightarrow f_y(1,2) = 3$. $\therefore$ $f_x(1,2) + 3 = 4 \Rightarrow f_x(1,2) = 1$. Then $\nabla f(1,2) = \mathbf{i} + 3\,\mathbf{j}$.
$\mathbf{u} = \frac{\mathbf{A}}{|\mathbf{A}|} = \frac{-\mathbf{i} - 2\,\mathbf{j}}{\sqrt{(-1)^2 + (-2)^2}} = -\frac{1}{\sqrt{5}}\mathbf{i} - \frac{2}{\sqrt{5}}\mathbf{j} \Rightarrow \left(D_{\mathbf{u}}f\right)_{P_0} = \nabla f \cdot \mathbf{u} = -\frac{1}{\sqrt{5}} - \frac{6}{\sqrt{5}} = -\frac{7}{\sqrt{5}}$

33. $x = g(t)$, $y = h(t) \Rightarrow \mathbf{r} = g(t)\,\mathbf{i} + h(t)\,\mathbf{j} \Rightarrow \mathbf{v} = g'(t)\,\mathbf{i} + h'(t)\,\mathbf{j} \Rightarrow \mathbf{T} = \frac{\mathbf{v}}{|\mathbf{v}|} = \frac{g'(t)\,\mathbf{i} + h'(t)\,\mathbf{j}}{\sqrt{(g'(t))^2 + (h'(t))^2}}$
$z = f(x,y) \Rightarrow \frac{df}{dt} = \frac{\partial f}{\partial x}\frac{dx}{dt} + \frac{\partial f}{\partial y}\frac{dy}{dt} = \frac{\partial f}{\partial x}g'(t) + \frac{\partial f}{\partial y}h'(t)$. If $f(g(t),h(t)) = c$, then $\frac{df}{dt} = 0 \Rightarrow \frac{\partial f}{\partial x}g'(t) + \frac{\partial f}{\partial y}h'(t) = 0$.
Now $\nabla f = \frac{\partial f}{\partial x}\mathbf{i} + \frac{\partial f}{\partial y}\mathbf{j}$. $\therefore$ $\left(D_{\mathbf{T}}f\right) = \nabla f \cdot \mathbf{T} = \frac{\frac{\partial f}{\partial x}g'(t) + \frac{\partial f}{\partial y}h'(t)}{\sqrt{(g'(t))^2 + (h'(t))^2}} = 0 \Rightarrow \nabla f$ is normal to $\mathbf{T}$

## SECTION 13.6 TANGENT PLANES AND NORMAL LINES

1. $\nabla f = 2x\,\mathbf{i} + 2y\,\mathbf{j} + 2z\,\mathbf{k} \Rightarrow \nabla f(1,1,1) = 2\,\mathbf{i} + 2\,\mathbf{j} + 2\,\mathbf{k} \Rightarrow$ Tangent plane: $2(x-1) + 2(y-1) + 2(z-1) = 0$
$\Rightarrow x + y + z = 3$; Normal line: $x = 1 + 2t$, $y = 1 + 2t$, $z = 1 + 2t$

3. $\nabla f = -2x\,\mathbf{i} - 2y\,\mathbf{j} + 2z\,\mathbf{k} \Rightarrow \nabla f(3,4,-5) = -6\,\mathbf{i} - 8\,\mathbf{j} - 10\,\mathbf{k} \Rightarrow$ Tangent plane: $-6(x-3) - 8(y-4) -$
$10(z+5) = 0 \Rightarrow 3x + 4y + 5z = 0$; Normal line: $x = 3 - 6t$, $y = 4 - 8t$, $z = -5 - 10t$

5. $\nabla f = \frac{-2x}{x^2+y^2}\mathbf{i} - \frac{2y}{x^2+y^2}\mathbf{j} \Rightarrow \nabla f(1,0,0) = -2\,\mathbf{i} + \mathbf{k} \Rightarrow$ Tangent plane: $-2(x-1) + z = 0 \Rightarrow -2x + z = -2$;
Normal line: $x = 1 - 2t$, $y = 0$, $z = t$

7. $\nabla f = (2x + 2y)\,\mathbf{i} + (2x - 2y)\,\mathbf{j} + 2z\,\mathbf{k} \Rightarrow \nabla f(1,-1,3) = 4\,\mathbf{j} + 6\,\mathbf{k} \Rightarrow$ Tangent plane: $4(y+1) + 6(z-3) = 0$
$\Rightarrow 2y + 3z = 7$; Normal line: $x = 1$, $y = -1 + 4t$, $z = 3 + 6t$

9. $\nabla f = (2x - y)\,\mathbf{i} - (x + 2y)\,\mathbf{j} - \mathbf{k} \Rightarrow \nabla f(1,1,-1) = \mathbf{i} - 3\,\mathbf{j} - \mathbf{k} \Rightarrow$ Tangent plane: $1(x-1) - 3(y-1) -$
$(z+1) = 0 \Rightarrow x - 3y - z = -1$; Normal line: $x = 1 + t$, $y = 1 - 3t$, $z = -1 - t$

11. $\nabla f = (2x - 2y - 1)\,\mathbf{i} + (2y - 2x + 3)\,\mathbf{j} - \mathbf{k} \Rightarrow \nabla f(2,-3,18) = 9\,\mathbf{i} - 7\,\mathbf{j} - \mathbf{k} \Rightarrow$ Tangent plane: $9(x-2) -$
$7(y+3) - (z-18) = 0 \Rightarrow 9x - 7y - z = 21$; Normal line: $x = 2 + 9t$, $y = -3 - 7t$, $z = 18 - t$

13. $\nabla f = 2x\,\mathbf{i} + 2y\,\mathbf{j} \Rightarrow \nabla f(1,\sqrt{3},1) = 2\,\mathbf{i} + 2\sqrt{3}\,\mathbf{j}$

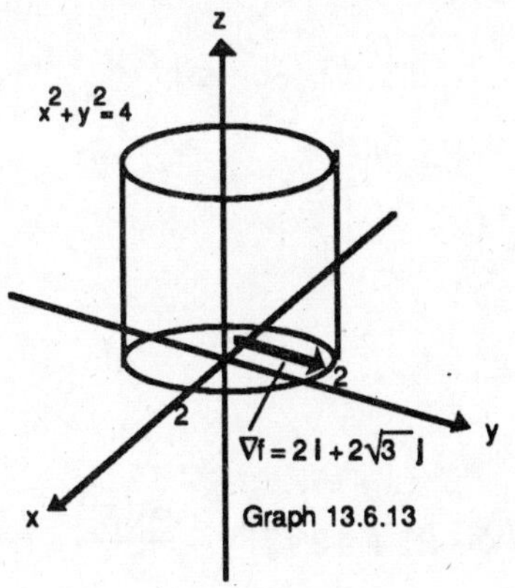


Graph 13.6.13

15. $\nabla f = 2x\,\mathbf{i} + 2y\,\mathbf{j} - \mathbf{k} \Rightarrow \nabla f(1,1,2) =$
$2\,\mathbf{i} + 2\,\mathbf{j} - \mathbf{k}$

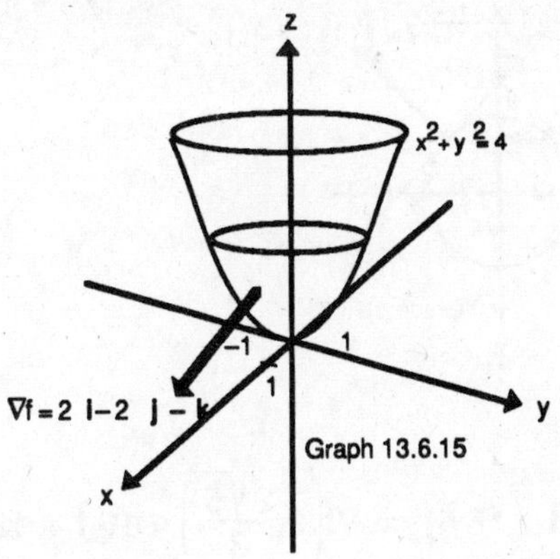


Graph 13.6.15

17. $\nabla f = 2x\,\mathbf{i} + 2y\,\mathbf{j} - 2z\,\mathbf{k} \Rightarrow \nabla f(1,2,-2\sqrt{5}) =$
$2\,\mathbf{i} + 4\,\mathbf{j} - 4\sqrt{5}\,\mathbf{k}$

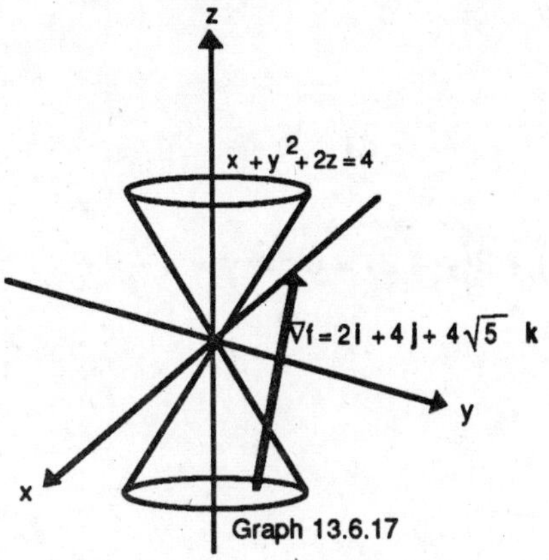


Graph 13.6.17

19. $\nabla f = \mathbf{i} + 2y\,\mathbf{j} + 2\,\mathbf{k} \Rightarrow \nabla f(1,1,1) = \mathbf{i} + 2\,\mathbf{j} + 2\,\mathbf{k}$.
$\Rightarrow \nabla f(1,1,1) = \mathbf{i} + 2\,\mathbf{j} + 2\,\mathbf{k}$. $\nabla g = \mathbf{i}$ for all points P.

$$\mathbf{v} = \nabla f \times \nabla g \Rightarrow \mathbf{v} = \begin{vmatrix} \mathbf{i} & \mathbf{j} & \mathbf{k} \\ 1 & 2 & 2 \\ 1 & 0 & 0 \end{vmatrix} \quad 2\,\mathbf{j} - 2\,\mathbf{k}$$

Tangent line: $x = 1,\ y = 1 + 2t,\ z = 1 - 2t$

21. $\nabla f = 2x\,\mathbf{i} + 2\,\mathbf{j} + 2\,\mathbf{k} \Rightarrow \nabla f(1,1,\frac{1}{2}) = 2\,\mathbf{i} + 2\,\mathbf{j} + 2\,\mathbf{k}$. $\nabla g = \mathbf{j}$ for all points P. $\mathbf{v} = \nabla f \times \nabla g \Rightarrow$

$$\mathbf{v} = \begin{vmatrix} \mathbf{i} & \mathbf{j} & \mathbf{k} \\ 2 & 2 & 2 \\ 0 & 1 & 0 \end{vmatrix} = -2\,\mathbf{i} + 2\,\mathbf{k} \Rightarrow \text{Tangent line: } x = 1 - 2t,\ y = 1,\ z = \frac{1}{2} + 2t$$

23. $\nabla f = \left(3x^2 + 6xy^2 + 4y\right)\mathbf{i} + \left(6x^2 y + 3y^2 + 4x\right)\mathbf{j} - 2z\,\mathbf{k} \Rightarrow \nabla f(1,1,3) = 13\,\mathbf{i} + 13\,\mathbf{j} - 6\,\mathbf{k}$.

$$\nabla g = 2x\,\mathbf{i} + 2y\,\mathbf{j} + 2z\,\mathbf{k} \Rightarrow \nabla g(1,1,3) = 2\,\mathbf{i} + 2\,\mathbf{j} + 6\,\mathbf{k}.\ \mathbf{v} = \nabla f \times \nabla g \Rightarrow \mathbf{v} = \begin{vmatrix} \mathbf{i} & \mathbf{j} & \mathbf{k} \\ 13 & 13 & -6 \\ 2 & 2 & 6 \end{vmatrix} =$$

$90\,\mathbf{i} - 90\,\mathbf{j} \Rightarrow$ Tangent line: $x = 1 + 90t,\ y = 1 - 90t,\ z = 3$

25. $\nabla f = 2x\,\mathbf{i} + 2y\,\mathbf{j} \Rightarrow \nabla f(\sqrt{2},\sqrt{2}) = 2\sqrt{2}\,\mathbf{i} + 2\sqrt{2}\,\mathbf{j} \Rightarrow$ Tangent line: $2\sqrt{2}\left(x - \sqrt{2}\right) + 2\sqrt{2}\left(y - \sqrt{2}\right) = 0 \Rightarrow$
$\sqrt{2}\,x + \sqrt{2}\,y = 4$

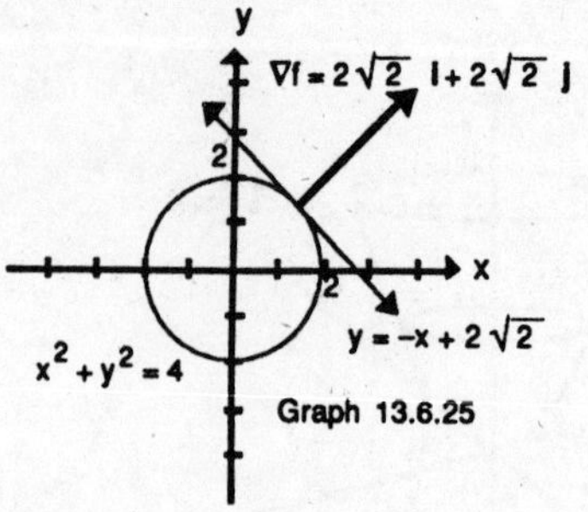

Graph 13.6.25

27. $\nabla f = 8x\,\mathbf{i} + 18y\,\mathbf{j} \Rightarrow \nabla f\left(2, \frac{2\sqrt{5}}{3}\right) = 16\,\mathbf{i} + 12\sqrt{5}\,\mathbf{j} \Rightarrow$ Tangent line: $16(x - 2) + 12\sqrt{5}\left(y - \frac{2\sqrt{5}}{3}\right) = 0 \Rightarrow$
$4x + 3\sqrt{5}\,y = 18.$

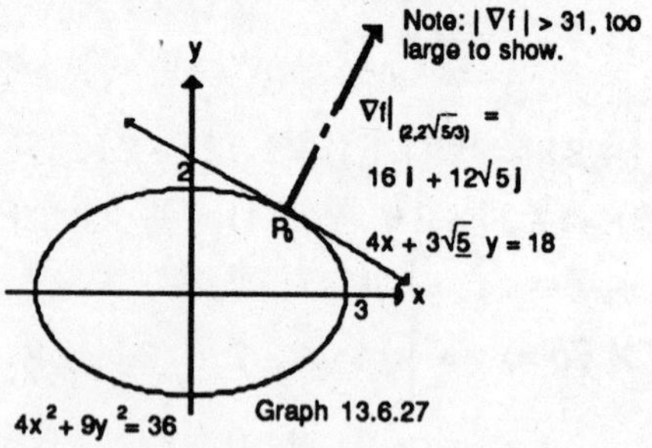

Graph 13.6.27

29. $\nabla f = y\,\mathbf{i} + x\,\mathbf{j} \Rightarrow \nabla f(2,-2) = -2\,\mathbf{i} + 2\,\mathbf{j} \Rightarrow$ Tangent line: $-2(x - 2) + 2(y + 2) = 0 \Rightarrow y = x - 4$

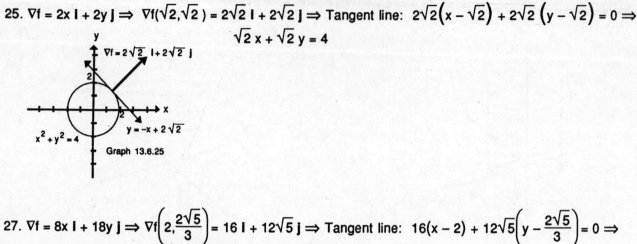

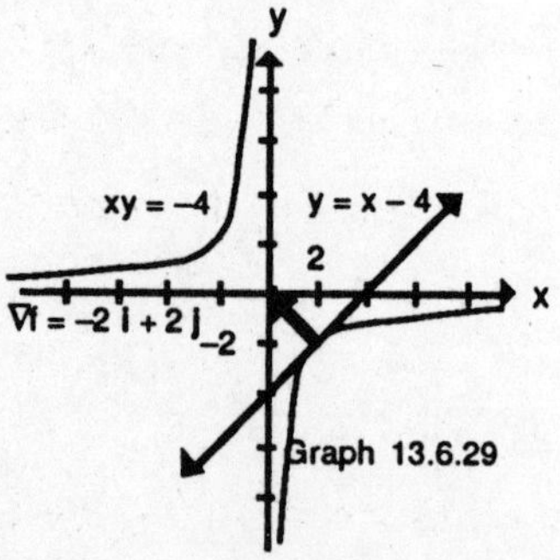

Graph 13.6.29

31. $\mathbf{r} = \sqrt{t}\,\mathbf{i} + \sqrt{t}\,\mathbf{j} - \frac{1}{4}(t + 3)\,\mathbf{k} \Rightarrow \mathbf{v} = \frac{1}{2}t^{-1/2}\,\mathbf{i} + \frac{1}{2}t^{-1/2}\,\mathbf{j} - \frac{1}{4}\mathbf{k}$. $t = 1 \Rightarrow x = 1, y = 1, z = -1 \Rightarrow P_0 = (1,1,-1)$. Also $t = 1 \Rightarrow \mathbf{v}(1) = \frac{1}{2}\mathbf{i} + \frac{1}{2}\mathbf{j} - \frac{1}{4}\mathbf{k}$. $f(x,y,z) = x^2 + y^2 - z - 3 = 0 \Rightarrow \nabla f = 2x\,\mathbf{i} + 2y\,\mathbf{j} - \mathbf{k} \Rightarrow \nabla f(1,1,-1) = 2\,\mathbf{i} + 2\,\mathbf{j} - \mathbf{k}$. $\therefore\ \mathbf{v} = \frac{1}{4}(\nabla f) \Rightarrow$ The curve is normal to the surface.

## SECTION 13.7 LINEARIZATION AND DIFFERENTIALS

1. a) $f(0,0) = 1$, $f_x(x,y) = 2x \Rightarrow f_x(0,0) = 0$, $f_y(x,y) = 2y \Rightarrow f_y(0,0) = 0 \Rightarrow L(x,y) = 1 + 0(x-0) + 0(y-0) = 1$
   b) $f(1,1) = 3$, $f_x(1,1) = 2$, $f_y(1,1) = 2 \Rightarrow L(x,y) = 3 + 2(x-1) + 2(y-1) = 2x + 2y - 1$

3. a) $f(0,0) = 1$, $f_x(x,y) = e^x \cos y \Rightarrow f_x(0,0) = 1$, $f_y(x,y) = -e^x \sin y \Rightarrow f_y(0,0) = 0 \Rightarrow L(x,y) = 1 + 1(x-0) + 0(y-0) = 1 + x$
   b) $f\left(0,\frac{\pi}{2}\right) = 0$, $f_x\left(0,\frac{\pi}{2}\right) = 0$, $f_y\left(0,\frac{\pi}{2}\right) = -1 \Rightarrow L(x,y) = 0 + 0(x-0) - 1\left(y - \frac{\pi}{2}\right) = -y + \frac{\pi}{2}$

5. a) $f(0,0) = 5$, $f_x(x,y) = 3$ for all $(x,y)$, $f_y(x,y) = -4$ for all $(x,y) \Rightarrow L(x,y) = 5 + 3(x-0) - 4(y-0) = 5 + 3x - 4y$
   b) $f(1,1) = 4$, $f_x(1,1) = 3$, $f_y(1,1) = -4 \Rightarrow L(x,y) = 4 + 3(x-1) - 4(y-1) = 3x - 4y + 5$

7. $f(2,1) = 3$, $f_x(x,y) = 2x - 3y \Rightarrow f_x(2,1) = 1$, $f_y(x,y) = -3x \Rightarrow f_y(2,1) = -6 \Rightarrow L(x,y) = 3 + 1(x-2) - 6(y-1) = 7 + x - 6y$. $f_{xx}(x,y) = 2$, $f_{yy}(x,y) = 0$, $f_{xy}(x,y) = -3 \Rightarrow M = 3$. $\therefore$ $|E(x,y)| \le \frac{1}{2}(3)(|x-2| + |y-1|)^2 \le \frac{3}{2}(0.1 + 0.1)^2 = 0.06$

9. $f(0,0) = 1$, $f_x(x,y) = \cos y \Rightarrow f_x(0,0) = 1$, $f_y(x,y) = 1 - x \sin y \Rightarrow f_y(0,0) = 1 \Rightarrow L(x,y) = 1 + 1(x-0) + 1(y-0) = x + y + 1$. $f_{xx}(x,y) = 0$, $f_{yy}(x,y) = 0$, $f_{xy}(x,y) = -\sin y \Rightarrow M = 1$.
   $\therefore$ $|E(x,y)| \le \frac{1}{2}(1)(|x| + |y|)^2 \le \frac{1}{2}(0.2 + 0.2)^2 = 0.08$

11. $f(0,0) = 1$, $f_x(x,y) = e^x \cos y \Rightarrow f_x(0,0) = 1$, $f_y(x,y) = -e^x \sin y \Rightarrow f_y(0,0) = 0 \Rightarrow L(x,y) = 1 + 1(x-0) + 0(y-0) = 1 + x$. $f_{xx}(x,y) = e^x \cos y$, $f_{yy}(x,y) = -e^x \cos y$, $f_{xy}(x,y) = -e^x \sin y$. $|x| \le 0.1 \Rightarrow -0.1 \le x \le 0.1$, $|y| \le 0.1 \Rightarrow -0.1 \le y \le 0.1. \Rightarrow$ max of $|f_{xx}(x,y)|$ on R is $e^{0.1}\cos(0.1) \le 1.11$, max of $|f_{yy}(x,y)|$ on R is $e^{0.1}\cos(0.1) \le 1.11$, max of $|f_{xy}(x,y)|$ on R is $e^{0.1}\sin(0.1) \le 0.002 \Rightarrow M = 1.11$.
    $\therefore$ $|E(x,y)| \le \frac{1}{2}(1.11)(|x| + |y|)^2 \le 0.555(0.1 + 0.1)^2 = 0.0222$

13. Let the width, w, be the long side. Then $A = lw \Rightarrow dA = A_l\,dl + A_w\,dw \Rightarrow dA = w\,dl + l\,dw$. Since $w > l$, dA is more sensitive to a change in w than l. $\therefore$ pay more attention to the width.

15. $T_x(x,y) = e^y + e^{-y}$, $T_y(x,y) = x(e^y - e^{-y}) \Rightarrow dT = T_x(x,y)\,dx + T_y(x,y)\,dy = (e^y + e^{-y})dx + x(e^y - e^{-y})dy \Rightarrow dT|_{(2,\ln 2)} = 2.5\,dx + 3.0\,dy$. If $|dx| \le 0.1$, $|dy| \le 0.02$, then the maximum possible error (estimate) $\le 2.5(0.1) + 3.0(0.02) = 0.31$ in magnitude.

17. $V_r = 2\pi rh$, $V_h = \pi r^2 \Rightarrow dV = V_r\,dr + V_h\,dh \Rightarrow dV = 2\pi rh\,dr + \pi r^2\,dh \Rightarrow dV|_{(5,12)} = 120\pi\,dr + 25\pi\,dh$. Since $|dr| \le 0.1$ cm, $|dh| \le 0.1$ cm, $dV \le 120\pi(0.1) + 25\pi(0.1) = 14.5\pi\text{ cm}^3$. $V(5,12) = 300\pi\text{ cm}^3 \Rightarrow$ Maximum percentage error $= \pm\frac{14.5\pi}{300\pi} \times 100 = \pm 4.83\%$

19. $df = f_x(x,y)\,dx + f_y(x,y)\,dy = 3x^2y^4dx + 4x^3y^3dy \Rightarrow df|_{(1,1)} = 3\,dx + 4\,dy$. Let $dx = dy \Rightarrow df = 7\,dx$. $|df| \le 0.1 \Rightarrow 7|dx| \le 0.1 \Rightarrow |dx| \le \frac{0.1}{7} \approx 0.014$. $\therefore$ for the square, let $|x-1| \le 0.014$, $|y-1| \le 0.014$

21. $dR = \left(\frac{R}{R_1}\right)^2 dR_1 + \left(\frac{R}{R_2}\right)^2 dR_2$ (See Exercise 20 above). $R_1$ changes from 20 to 20.1 ohms $\Rightarrow dR_1 =$ 0.1 ohms, $R_2$ changes from 25 to 24.9 ohms $\Rightarrow dR_2 = -$ 0.1 ohms. $\frac{1}{R} = \frac{1}{R_1} + \frac{1}{R_2} \Rightarrow R = \frac{100}{9}$ ohms.

$dR\Big|_{(20,25)} = \frac{(100/9)^2}{(20)^2}(0.1) + \frac{(100/9)^2}{(25)^2}(-0.1) = 0.011$ ohms $\Rightarrow$ Percentage change $= \frac{dR}{R}\Big|_{(20,25)} \times 100$ $= \frac{0.011}{100/9} \times 100 \approx 0.099\%$

23. a) $f(1,1,1) = 3$, $f_x(1,1,1) = y + z\big|_{(1,1,1)} = 2$, $f_y(1,1,1) = x + z\big|_{(1,1,1)} = 2$, $f_z(1,1,1) = y + x\big|_{(1,1,1)} = 2 \Rightarrow$ $L(x,y,z) = 2x + 2y + 2z - 3$

b) $f(1,0,0) = 0$, $f_x(1,0,0) = 0$, $f_y(1,0,0) = 1$, $f_z(1,0,0) = 1 \Rightarrow L(x,y,z) = y + z$

c) $f(0,0,0) = 0$, $f_x(0,0,0) = 0$, $f_y(0,0,0) = 0$, $f_z(0,0,0) = 0 \Rightarrow L(x,y,z) = 0$

25. a) $f(1,0,0) = 1$, $f_x(1,0,0) = \frac{x}{\sqrt{x^2+y^2+z^2}}\Big|_{(1,0,0)} = 1$, $f_y(1,0,0) = \frac{y}{\sqrt{x^2+y^2+z^2}}\Big|_{(1,0,0)} = 0$,

$f_z(1,0,0) = \frac{z}{\sqrt{x^2+y^2+z^2}}\Big|_{(1,0,0)} = 0 \Rightarrow L(x,y,z) = x$

b) $f(1,1,0) = \sqrt{2}$, $f_x(1,1,0) = \frac{1}{\sqrt{2}}$, $f_y(1,1,0) = \frac{1}{\sqrt{2}}$, $f_z(1,1,0) = 0 \Rightarrow L(x,y,z) = \frac{1}{\sqrt{2}}x + \frac{1}{\sqrt{2}}y$

c) $f(1,2,2) = 3$, $f_x(1,2,2) = \frac{1}{3}$, $f_y(1,2,2) = \frac{2}{3}$, $f_z(1,2,2) = \frac{2}{3} \Rightarrow L(x,y,z) = \frac{1}{3}x + \frac{2}{3}y + \frac{2}{3}z$

27. a) $f(0,0,0) = 2$, $f_x(0,0,0) = e^x\big|_{(0,0,0)} = 1$, $f_y(0,0,0) = -\sin(y+z)\big|_{(0,0,0)} = 0$, $f_z(0,0,0) = -\sin(y+z)\big|_{(0,0,0)}$ $= 0 \Rightarrow L(x,y,z) = 2 + x$

b) $f\left(0,\frac{\pi}{2},0\right) = 1$, $f_x\left(0,\frac{\pi}{2},0\right) = 1$, $f_y\left(0,\frac{\pi}{2},0\right) = -1$, $f_z\left(0,\frac{\pi}{2},0\right) = -1 \Rightarrow L(x,y,z) = x - y - z + \frac{\pi}{2} + 1$

c) $f\left(0,\frac{\pi}{4},\frac{\pi}{4}\right) = 1$, $f_x\left(0,\frac{\pi}{4},\frac{\pi}{4}\right) = 1$, $f_y\left(0,\frac{\pi}{4},\frac{\pi}{4}\right) = -1$, $f_z\left(0,\frac{\pi}{4},\frac{\pi}{4}\right) = -1 \Rightarrow L(x,y,z) = x - y - z + \frac{\pi}{2} + 1$

29. $f(a,b,c,d) = \begin{vmatrix} a & b \\ c & d \end{vmatrix} = ad - bc \Rightarrow f_a = d$, $f_b = -c$, $f_c = -b$, $f_d = a \Rightarrow df = d\,da - c\,db - b\,dc + a\,dd$.
Since $|a|$ is much greater than $|b|$, $|c|$, and $|d|$, f is most sensitive to a change in d.

31. $V = lwh \Rightarrow V_l = wh$, $V_w = lh$, $V_h = lw \Rightarrow dV = wh\,dl + lh\,dw + lw\,dh \Rightarrow dV\big|_{(5,3,2)} = 6\,dl + 10\,dw + 15\,dh$

$dl = 1$ in $= \frac{1}{12}$ ft, $dw = 1$ in $= \frac{1}{12}$ ft, $dh = \frac{1}{2}$ in $= \frac{1}{24}$ ft $\Rightarrow dV = 6\left(\frac{1}{12}\right) + 10\left(\frac{1}{12}\right) + 15\left(\frac{1}{24}\right) = \frac{47}{24}$ ft$^3$

33. $u_x = e^y$, $u_y = xe^y + \sin z$, $u_z = y\cos z \Rightarrow du = e^y\,dx + (xe^y + \sin z)dy + (y\cos z)dz \Rightarrow$

$du\big|_{(2,\ln 3,\pi/2)} = 3\,dx + 7\,dy + 0\,dz = 3\,dx + 7\,dy \Rightarrow$ magnitude of the maximum possible error $\le$ $3(0.2) + 7(0.6) = 4.8$

## SECTION 13.8 MAXIMA, MINIMA, AND SADDLE POINTS

1. $f_x(x,y) = 2x + y + 3 = 0$ and $f_y(x,y) = x + 2y - 3 = 0 \Rightarrow x = -3,\ y = 3 \Rightarrow$ critical point is $(-3,3)$. $f_{xx}(-3,3) = 2$, $f_{yy}(-3,3) = 2$, $f_{xy}(-3,3) = 1 \Rightarrow f_{xx}f_{yy} - f_{xy}^2 = 3 > 0$ and $f_{xx} > 0 \Rightarrow$ minimum. $f(-3,3) = -5$, absolute minimum.

3. $f_x(x,y) = 5y - 14x + 3 = 0$ and $f_y(x,y) = 5x - 6 = 0 \Rightarrow x = \frac{6}{5},\ y = \frac{69}{25} \Rightarrow$ critical point is $\left(\frac{6}{5}, \frac{69}{25}\right)$. $f_{xx}\left(\frac{6}{5}, \frac{69}{25}\right) = -14$, $f_{yy}\left(\frac{6}{5}, \frac{69}{25}\right) = 0$, $f_{xy}\left(\frac{6}{5}, \frac{69}{25}\right) = 5 \Rightarrow f_{xx}f_{yy} - f_{xy}^2 = -25 < 0 \Rightarrow$ saddle point.

5. $f_x(x,y) = 2x + y + 3 = 0$ and $f_y(x,y) = x + 2 = 0 \Rightarrow x = -2,\ y = 1 \Rightarrow$ critical point is $(-2,1)$. $f_{xx}(-2,1) = 2$, $f_{yy}(-2,1) = 0$, $f_{xy}(-2,1) = 1 \Rightarrow f_{xx}f_{yy} - f_{xy}^2 = -1 \Rightarrow$ saddle point.

7 $f_x(x,y) = 2y - 10x + 4 = 0$ and $f_y(x,y) = 2x - 4y = 0 \Rightarrow x = \frac{4}{9},\ y = \frac{2}{9} \Rightarrow$ critical point is $\left(\frac{4}{9}, \frac{2}{9}\right)$. $f_{xx}\left(\frac{4}{9}, \frac{2}{9}\right) = -10$, $f_{yy}\left(\frac{4}{9}, \frac{2}{9}\right) = -4$, $f_{xy}\left(\frac{4}{9}, \frac{2}{9}\right) = 2 \Rightarrow f_{xx}f_{yy} - f_{xy}^2 = 36 > 0$ and $f_{xx} < 0 \Rightarrow$ maximum (absolute). $f\left(\frac{4}{9}, \frac{2}{9}\right) = -\frac{252}{81}$

9. $f_x(x,y) = 2x - 4y = 0$ and $f_y(x,y) = -4x + 2y + 6 = 0 \Rightarrow x = 2,\ y = 1 \Rightarrow$ critical point is $(2,1)$. $f_{xx}(2,1) = 2$, $f_{yy}(2,1) = 2$, $f_{xy}(2,1) = -4 \Rightarrow f_{xx}f_{yy} - f_{xy}^2 = -12 \Rightarrow$ saddle point.

11. $f_x(x,y) = 4x + 3y - 5 = 0$ and $f_y(x,y) = 3x + 8y + 2 = 0 \Rightarrow x = 2,\ y = -1 \Rightarrow$ critical point is $(2,-1)$. $f_{xx}(2,-1) = 4$, $f_{yy}(2,-1) = 8$, $f_{xy}(2,-1) = 3 \Rightarrow f_{xx}f_{yy} - f_{xy}^2 = 29 > 0$ and $f_{xx} > 0 \Rightarrow$ minimum (absolute). $f(2,-1) = -6$.

13. $f_x(x,y) = 2x - 4y + 5 = 0$ and $f_y(x,y) = -4x + 2y - 2 = 0 \Rightarrow x = \frac{1}{6},\ y = \frac{4}{3} \Rightarrow$ critical point is $\left(\frac{1}{6}, \frac{4}{3}\right)$. $f_{xx}\left(\frac{1}{6}, \frac{4}{3}\right) = 2$, $f_{yy}\left(\frac{1}{6}, \frac{4}{3}\right) = 2$, $f_{xy}\left(\frac{1}{6}, \frac{4}{3}\right) = -4 \Rightarrow f_{xx}f_{yy} - f_{xy}^2 = -12 \Rightarrow$ saddle point.

15. $f_x(x,y) = 2x - 2y - 2 = 0$ and $f_y(x,y) = -2x + 4y + 2 = 0 \Rightarrow x = 1,\ y = 0 \Rightarrow$ critical point is $(1,0)$. $f_{xx}(1,0) = 2$, $f_{yy}(1,0) = 4$, $f_{xy}(1,0) = -2 \Rightarrow f_{xx}f_{yy} - f_{xy}^2 = 4 > 0$ and $f_{xx} > 0 \Rightarrow$ minimum (absolute). $f(1,0) = 0$

17. $f_x(x,y) = 2 - 4x - 2y = 0$ and $f_y(x,y) = 2 - 2x - 2y = 0 \Rightarrow x = 0,\ y = 1 \Rightarrow$ critical point is $(0,1)$. $f_{xx}(0,1) = -4$, $f_{yy}(0,1) = -2$, $f_{xy}(0,1) = -2 \Rightarrow f_{xx}f_{yy} - f_{xy}^2 = 4 > 0$ and $f_{xx} < 0 \Rightarrow$ maximum (absolute). $f(0,1) = 4$

19. $f_x(x,y) = 3x^2 - 2y = 0$ and $f_y(x,y) = -3y^2 - 2x = 0 \Rightarrow x = 0,\ y = 0$ or $x = -\frac{2}{3},\ y = \frac{2}{3} \Rightarrow$ critical points are $(0,0)$ and $\left(-\frac{2}{3}, \frac{2}{3}\right)$. For $(0,0)$: $f_{xx}(0,0) = 6x\big|_{(0,0)} = 0$, $f_{yy}(0,0) = -6y\big|_{(0,0)} = 0$, $f_{xy}(0,0) = -2 \Rightarrow f_{xx}f_{yy} - f_{xy}^2 = -4 \Rightarrow$ saddle point. For $\left(-\frac{2}{3}, \frac{2}{3}\right)$: $f_{xx}\left(-\frac{2}{3}, \frac{2}{3}\right) = -4$, $f_{yy}\left(-\frac{2}{3}, \frac{2}{3}\right) = -4$, $f_{xy}\left(-\frac{2}{3}, \frac{2}{3}\right) = -2$ $\Rightarrow f_{xx}f_{yy} - f_{xy}^2 = 12 > 0$ and $f_{xx} < 0 \Rightarrow$ local maximum. $f\left(-\frac{2}{3}, \frac{2}{3}\right) = \frac{170}{27}$

21. $f_x(x,y) = 12x - 6x^2 + 6y = 0$ and $f_y(x,y) = 6y + 6x = 0 \Rightarrow x = 0,\ y = 0$ or $x = 1,\ y = -1 \Rightarrow$ critical points are (0,0) and (1,–1). For (0,0): $f_{xx}(0,0) = 12 - 12x\big|_{(0,0)} = 12$, $f_{yy}(0,0) = 6$, $f_{xy}(0,0) = 6 \Rightarrow$ $f_{xx}f_{yy} - f_{xy}{}^2 = 36 > 0$ and $f_{xx} > 0 \Rightarrow$ local minimum. $f(0,0) = 0$. For (1,–1): $f_{xx}(1,-1) = 0$, $f_{yy}(1,-1) = 6$, $f_{xy}(1,-1) = 6 \Rightarrow f_{xx}f_{yy} - f_{xy}{}^2 = -36 \Rightarrow$ saddle point.

23. $f_x(x,y) = 3x^2 + 3y = 0$ and $f_y(x,y) = 3x + 3y^2 = 0 \Rightarrow x = 0,\ y = 0$ or $x = -1,\ y = -1 \Rightarrow$ critical points are (0,0) and (–1,–1). For (0,0): $f_{xx}(0,0) = 6x\big|_{(0,0)} = 0$, $f_{yy}(0,0) = 6y\big|_{(0,0)} = 0$, $f_{xy}(0,0) = 3 \Rightarrow$ $f_{xx}f_{yy} - f_{xy}{}^2 = -9 \Rightarrow$ saddle point. For (–1,–1): $f_{xx}(-1,-1) = -6$. $f_{yy}(-1,-1) = -6$, $f_{xy}(-1,-1) = 3 \Rightarrow$ $f_{xx}f_{yy} - f_{xy}{}^2 = 27 > 0$ and $f_{xx} < 0 \Rightarrow$ local maximum. $f(-1,-1) = 1$

25.

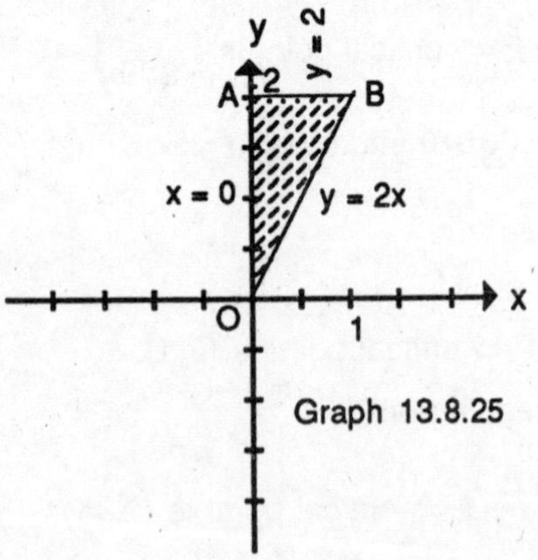


Graph 13.8.25

1. On OA, $f(x,y) = y^2 - 4y + 1 = f(0,y)$ on $0 \le y \le 2$. $y = 0 \Rightarrow f(0,0) = 1$. $y = 2 \Rightarrow f(0,2) = -3$. $f'(0,y) = 2y - 4 = 0 \Rightarrow y = 2 \Rightarrow f(0,2) = -3$.
2. On AB, $f(x,y) = 2x^2 - 4x - 3 = f(x,2)$ on $0 \le x \le 2$. $x = 0 \Rightarrow f(0,2) = -3$. $x = 2 \Rightarrow f(2,2) = -3$. $f'(x,2) = 4x - 4 = 0 \Rightarrow x = 1 \Rightarrow f(1,2) = -5$.
3. On OB, $f(x,y) = 6x^2 - 12x + 1 = f(x,2x)$ on $0 \le x \le 1$. Endpoint values have been found above. $f'(x,2x) = 12x - 12 = 0 \Rightarrow x = 1,\ y = 2 \Rightarrow (1,2)$, not an interior point of OB.
4. For interior points of the triangular region, $f_x(x,y) = 4x - 4 = 0$ and $f_y(x,y) = 2y - 4 = 0 \Rightarrow x = 1,\ y = 2 \Rightarrow (1,2)$, not an interior point of the region.

∴ absolute maximum is 1 at (0,0); absolute minimum is –5 at (1,2)

27.

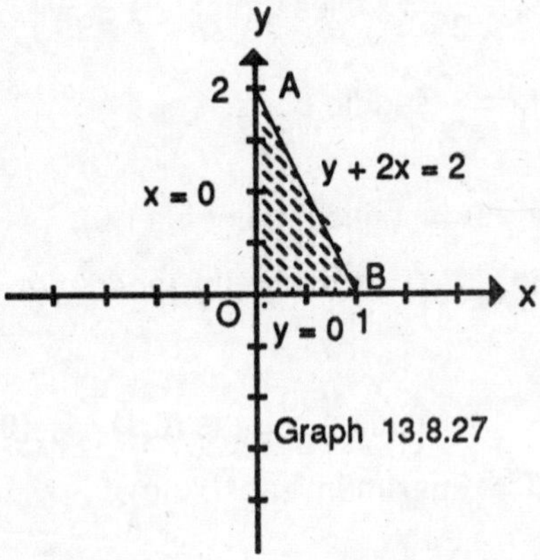


Graph 13.8.27

1. On OA, $f(x,y) = y^2 = f(0,y)$ on $0 \le y \le 2$. $f(0,0) = 0$. $f(0,2) = 4$. $f'(0,y) = 2y = 0 \Rightarrow y = 0,\ x = 0 \Rightarrow (0,0)$
2. On OB, $f(x,y) = x^2 = f(x,0)$ on $0 \le x \le 1$. $f(1,0) = 1$. $f'(x,0) = 2x = 0 \Rightarrow x = 0,\ y = 0 \Rightarrow (0,0)$
3. On AB, $f(x,y) = 5x^2 - 8x + 4 = f(x,-2x+2)$ on $0 \le x \le 1$. $f(0,2) =$ 4. $f'(x,-2x+2) = 10x - 8 = 0 \Rightarrow x = \frac{4}{5},\ y = \frac{2}{5}$. $f\left(\frac{4}{5},\frac{2}{5}\right) = \frac{4}{5}$
4. For interior points of the triangular region, $f_x(x,y) = 2x = 0$ and $f_y(x,y) = 2y = 0 \Rightarrow x = 0,\ y = 0 \Rightarrow (0,0)$, not an interior point of the region.

∴ absolute maximum is 4 at (0,2); absolute minimum is 0 at (0,0)

29.

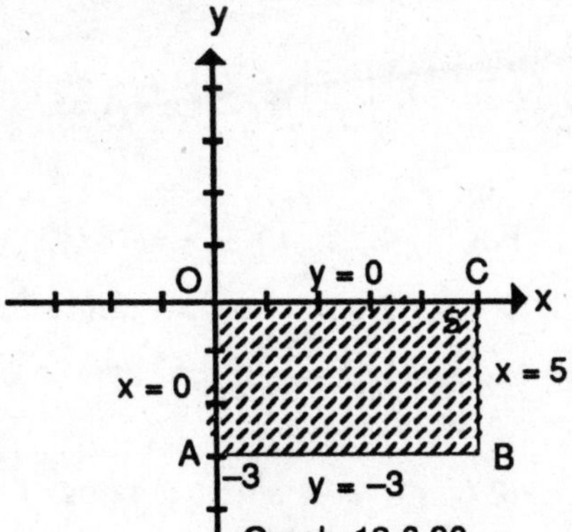

Graph 13.8.29

1. On OC, $T(x,y) = x^2 - 6x + 2 = T(x,0)$ on $0 \le x \le 5$. $T(0,0) = 2$. $T(5,0) = -3$. $T'(x,0) = 2x - 6 = 0 \Rightarrow x = 3, y = 0$. $T(3,0) = -7$
2. On CB, $T(x,y) = y^2 + 5y - 3 = T(5,y)$ on $-3 \le y \le 0$. $T(5,-3) = -9$ $T'(5,y) = 2y + 5 = 0 \Rightarrow y = -\frac{5}{2}, x = 5$. $T\left(5,-\frac{5}{2}\right) = -\frac{37}{4}$
3. On AB, $T(x,y) = x^2 - 9x + 11 = T(x,-3)$ on $0 \le x \le 5$. $T(0,-3) = 11$. $T'(x,-3) = 2x - 9 = 0 \Rightarrow x = \frac{9}{2}, y = -3$. $T\left(\frac{9}{2},-3\right) = -\frac{37}{4}$
4. On AO, $T(x,y) = y^2 + 2 = T(0,y)$ on $-3 \le y \le 0$. $T'(0,y) = 2y = 0 \Rightarrow y = 0, x = 0$. (0,0) not an interior point of AO.
5. For interior points of the rectangular region, $T_x(x,y) = 2x + y - 6 = 0$ and $T_y(x,y) = x + 2y = 0 \Rightarrow x = 4, y = -2 \Rightarrow T(4,-2) = -10$.

$\therefore$ absolute maximum is 11 at (0,–3); absolute minimum is –10 at (4,–2).

31.

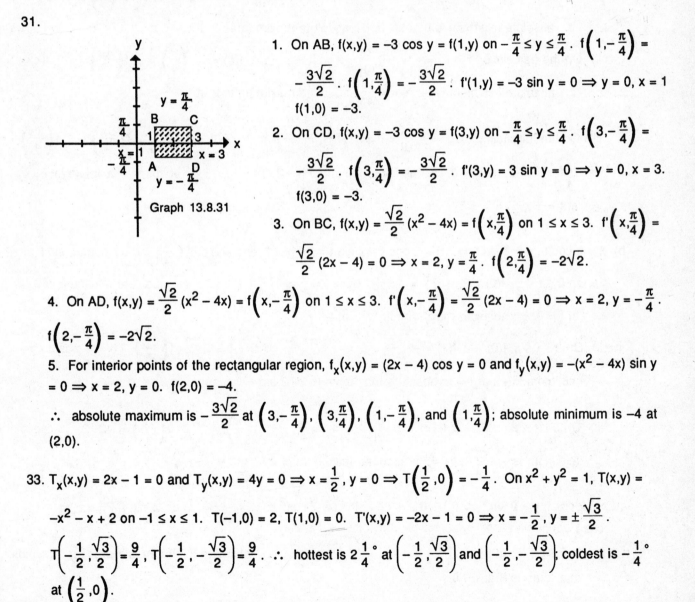

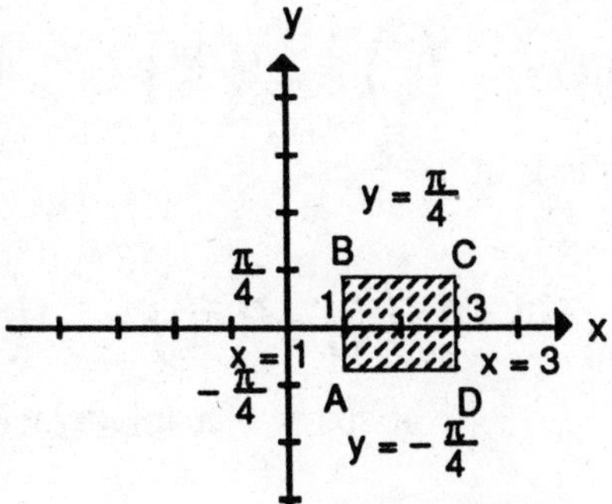

Graph 13.8.31

1. On AB, $f(x,y) = -3\cos y = f(1,y)$ on $-\frac{\pi}{4} \le y \le \frac{\pi}{4}$. $f\left(1,-\frac{\pi}{4}\right) = -\frac{3\sqrt{2}}{2}$. $f\left(1,\frac{\pi}{4}\right) = -\frac{3\sqrt{2}}{2}$. $f'(1,y) = -3\sin y = 0 \Rightarrow y = 0, x = 1$ $f(1,0) = -3$.
2. On CD, $f(x,y) = -3\cos y = f(3,y)$ on $-\frac{\pi}{4} \le y \le \frac{\pi}{4}$. $f\left(3,-\frac{\pi}{4}\right) = -\frac{3\sqrt{2}}{2}$. $f\left(3,\frac{\pi}{4}\right) = -\frac{3\sqrt{2}}{2}$. $f'(3,y) = 3\sin y = 0 \Rightarrow y = 0, x = 3$. $f(3,0) = -3$.
3. On BC, $f(x,y) = \frac{\sqrt{2}}{2}(x^2 - 4x) = f\left(x,\frac{\pi}{4}\right)$ on $1 \le x \le 3$. $f'\left(x,\frac{\pi}{4}\right) = \frac{\sqrt{2}}{2}(2x - 4) = 0 \Rightarrow x = 2, y = \frac{\pi}{4}$. $f\left(2,\frac{\pi}{4}\right) = -2\sqrt{2}$.
4. On AD, $f(x,y) = \frac{\sqrt{2}}{2}(x^2 - 4x) = f\left(x,-\frac{\pi}{4}\right)$ on $1 \le x \le 3$. $f'\left(x,-\frac{\pi}{4}\right) = \frac{\sqrt{2}}{2}(2x - 4) = 0 \Rightarrow x = 2, y = -\frac{\pi}{4}$. $f\left(2,-\frac{\pi}{4}\right) = -2\sqrt{2}$.
5. For interior points of the rectangular region, $f_x(x,y) = (2x - 4)\cos y = 0$ and $f_y(x,y) = -(x^2 - 4x)\sin y = 0 \Rightarrow x = 2, y = 0$. $f(2,0) = -4$.

$\therefore$ absolute maximum is $-\frac{3\sqrt{2}}{2}$ at $\left(3,-\frac{\pi}{4}\right)$, $\left(3,\frac{\pi}{4}\right)$, $\left(1,-\frac{\pi}{4}\right)$, and $\left(1,\frac{\pi}{4}\right)$; absolute minimum is –4 at (2,0).

33. $T_x(x,y) = 2x - 1 = 0$ and $T_y(x,y) = 4y = 0 \Rightarrow x = \frac{1}{2}, y = 0 \Rightarrow T\left(\frac{1}{2},0\right) = -\frac{1}{4}$. On $x^2 + y^2 = 1$, $T(x,y) = -x^2 - x + 2$ on $-1 \le x \le 1$. $T(-1,0) = 2$, $T(1,0) = 0$. $T'(x,y) = -2x - 1 = 0 \Rightarrow x = -\frac{1}{2}, y = \pm\frac{\sqrt{3}}{2}$. $T\left(-\frac{1}{2},\frac{\sqrt{3}}{2}\right) = \frac{9}{4}$, $T\left(-\frac{1}{2},-\frac{\sqrt{3}}{2}\right) = \frac{9}{4}$. $\therefore$ hottest is $2\frac{1}{4}°$ at $\left(-\frac{1}{2},\frac{\sqrt{3}}{2}\right)$ and $\left(-\frac{1}{2},-\frac{\sqrt{3}}{2}\right)$; coldest is $-\frac{1}{4}°$ at $\left(\frac{1}{2},0\right)$.

35. a) $f_x(x,y) = 2x - 4y = 0$ and $f_y(x,y) = 2y - 4x = 0 \Rightarrow x = 0, y = 0$. $f_{xx}(0,0) = 2$, $f_{yy}(0,0) = -4$, $f_{xy}(0,0) = -4$ $\Rightarrow f_{xx}f_{yy} - f_{xy}^2 = -24 \Rightarrow$ saddle point.

b) $f_x(x,y) = 2x - 2 = 0$ and $f_y(x,y) = 2y - 4 = 0 \Rightarrow x = 1, y = 2$. $f_{xx}(1,2) = 2$, $f_{yy}(1,2) = 2$, $f_{xy}(1,2) = 0 \Rightarrow$ $f_{xx}f_{yy} - f_{xy}^2 = 4 > 0$ and $f_{xx} > 0 \Rightarrow$ local minimum at (1,2).

c) $f_x(x,y) = 9x^2 - 9 = 0$ and $f_y(x,y) = 2y + 4 = 0 \Rightarrow x = \pm 1, y = -2$. For (1,–2), $f_{xx}(1,-2) = 18x\big|_{(1,-2)} =$ 18, $f_{yy}(1,-2) = 2$, $f_{xy}(1,-2) = 0 \Rightarrow f_{xx}f_{yy} - f_{xy}^2 = 36 > 0$ and $f_{xx} > 0 \Rightarrow$ local minimum at (1,–2). For (–1,–2), $f_{xx}(-1,-2) = -18$, $f_{yy}(-1,-2) = 2$, $f_{xy}(-1,-2) = 0 \Rightarrow f_{xx}f_{yy} - f_{xy}^2 = -36 \Rightarrow$ saddle point.

37. a) $x = 2\cos t$, $y = 2\sin t \Rightarrow f(t) = 4\cos t \sin t \Rightarrow \frac{df}{dt} = y(-2\sin t) + x(2\cos t) = -4\sin^2 t + 4\cos^2 t$.

$\frac{df}{dt} = 0 \Rightarrow 4\cos^2 t - 4\sin^2 t = 0 \Rightarrow \cos t = \sin t$ or $\cos t = -\sin t$

i) On the quarter circle $x^2 + y^2 = 4$ in the first quadrant, $\frac{df}{dt} = 0$ at $t = \frac{\pi}{4}$. $f(0) = 0$, $f\left(\frac{\pi}{2}\right) = 0$, $f\left(\frac{\pi}{4}\right) = 2$

$\therefore$ absolute minimum is 0 at $t = 0, \frac{\pi}{2}$; absolute maximum is 2 at $t = \frac{\pi}{4}$.

ii) On the half circle $x^2 + y^2 = 4$, $y \ge 0$, $\frac{df}{dt} = 0 \Rightarrow t = \frac{\pi}{4}, \frac{3\pi}{4}$. $f(0) = 0$, $f\left(\frac{\pi}{4}\right) = 2$, $f\left(\frac{3\pi}{4}\right) = -2$, $f(\pi) = 0$

$\therefore$ absolute minimum is –2 at $t = \frac{3\pi}{4}$; absolute maximum is 2 at $t = \frac{\pi}{4}$.

iii) On the full circle $x^2 + y^2 = 4$, $\frac{df}{dt} = 0 \Rightarrow t = \frac{\pi}{4}, \frac{3\pi}{4}, \frac{5\pi}{4}, \frac{7\pi}{4}$. $f(0) = 0$, $f\left(\frac{\pi}{4}\right) = f\left(\frac{5\pi}{4}\right) = 2$, $f\left(\frac{3\pi}{4}\right) =$ $f\left(\frac{7\pi}{4}\right) = -2$, $f(2\pi) = 0$. $\therefore$ absolute minimum is –2 at $t = \frac{3\pi}{4}, \frac{7\pi}{4}$; absolute maximum is 2 at $t = \frac{\pi}{4}, \frac{5\pi}{4}$.

b) $x = 2\cos t$, $y = 2\sin t \Rightarrow f(t) = 2\cos t + 2\sin t \Rightarrow \frac{df}{dt} = -2\sin t + 2\cos t$. $\frac{df}{dt} = 0 \Rightarrow \cos t = \sin t$

i) On $0 \le t \le \frac{\pi}{2}$, $f(0) = 2$, $f\left(\frac{\pi}{2}\right) = 2$. $\frac{df}{dt} = 0 \Rightarrow t = \frac{\pi}{4} \Rightarrow f\left(\frac{\pi}{4}\right) = 2\sqrt{2}$. $\therefore$ absolute minimum is 2 at $t = 0, \frac{\pi}{2}$; absolute maximum is $2\sqrt{2}$ at $t = \frac{\pi}{4}$.

ii) On $0 \le t \le \pi$, $f(0) = 2$, $f(\pi) = -2$. $\frac{df}{dt} = 0 \Rightarrow t = \frac{\pi}{4}, \frac{3\pi}{4} \Rightarrow f\left(\frac{\pi}{4}\right) = 2\sqrt{2}$, $f\left(\frac{3\pi}{4}\right) = 0$ $\therefore$ absolute minimum is –2 at $t = \pi$; absolute maximum is $2\sqrt{2}$ at $t = \frac{\pi}{4}$.

iii) On $0 \le t \le 2\pi$, $f(0) = 2$, $f(2\pi) = 2$. $\frac{df}{dt} = 0 \Rightarrow t = \frac{\pi}{4}, \frac{5\pi}{4}$. $f\left(\frac{\pi}{4}\right) = 2\sqrt{2}$, $f\left(\frac{5\pi}{4}\right) = -2\sqrt{2}$. $\therefore$ absolute minimum is $-2\sqrt{2}$ at $t = \frac{5\pi}{4}$; absolute maximum is $2\sqrt{2}$ at $t = \frac{\pi}{4}$.

c) $x = 2\cos t$, $y = 2\sin t \Rightarrow f(t) = 8\cos^2 t + \sin^2 t = 7\cos^2 t + 1 \Rightarrow \frac{df}{dt} = -8\cos t \sin t$. $\frac{df}{dt} = 0 \Rightarrow$ $\cos t = 0$ or $\sin t = 0$

i) On $0 \le t \le \frac{\pi}{2}$, $f(0) = 8$, $f\left(\frac{\pi}{2}\right) = 1$. $\frac{df}{dt} = 0 \Rightarrow t = 0, \frac{\pi}{2}$. $\therefore$ absolute minimum is 1 at $t = \frac{\pi}{2}$; absolute maximum is 8 at $t = 0$.

37. (Continued)

ii) On $0 \le t \le \pi$, $f(0) = 8$, $f(\pi) = 8$. $\frac{df}{dt} = 0 \Rightarrow t = 0, \pi, \frac{\pi}{2}$. $f\left(\frac{\pi}{2}\right) = 1$. $\therefore$ absolute minimum is 1 at $t = \frac{\pi}{2}$; absolute maximum is 8 at $t = 0, \pi$.

iii) On $0 \le t \le 2\pi$, $f(0) = 8$, $f(2\pi) = 8$. $\frac{df}{dt} = 0 \Rightarrow t = 0, \frac{\pi}{2}, \pi, \frac{3\pi}{2}, 2\pi \Rightarrow f\left(\frac{\pi}{2}\right) = 1$, $f(\pi) = 8$, $f\left(\frac{3\pi}{2}\right) = 1$.

$\therefore$ absolute minimum is 1 at $t = \frac{\pi}{2}, \frac{3\pi}{2}$; absolute maximum is 8 at $t = 0, \pi, 2\pi$.

39. a) $x = 3\cos t$, $y = 2\sin t \Rightarrow f(t) = 9 + 3\sin^2 t \Rightarrow \frac{df}{dt} = 6\sin t\cos t$. $\frac{df}{dt} = 0 \Rightarrow \sin t = 0$ or $\cos t = 0$

i) On the quarter ellipse, $\frac{x^2}{9} + \frac{y^2}{4} = 1$, in the first quadrant, $f(0) = 9$, $f\left(\frac{\pi}{2}\right) = 12$. $\frac{df}{dt} = 0 \Rightarrow t = 0, \frac{\pi}{2}$

$\therefore$ absolute maximum is 12 at $t = \frac{\pi}{2}$; absolute minimum is 9 at $t = 0$.

ii) On the half ellipse, $\frac{x^2}{9} + \frac{y^2}{4} = 1$, $y \ge 0$, $f(0) = 9$, $f(\pi) = 9$. $\frac{df}{dt} = 0 \Rightarrow t = 0, \frac{\pi}{2}, \pi$. $f\left(\frac{\pi}{2}\right) = 12$.

$\therefore$ absolute minimum is 9 at $t = 0, \pi$; absolute maximum is 12 at $t = \frac{\pi}{2}$.

iii) On the full ellipse, $\frac{x^2}{9} + \frac{y^2}{4} = 1$, $f(0) = 9$, $f(2\pi) = 9$. $\frac{df}{dt} = 0 \Rightarrow t = 0, \frac{\pi}{2}, \pi, \frac{3\pi}{2}, 2\pi$. $f\left(\frac{\pi}{2}\right) = 12$, $f(\pi) = 9$, $f\left(\frac{3\pi}{2}\right) = 12$. $\therefore$ absolute maximum is 12 at $t = \frac{\pi}{2}, \frac{3\pi}{2}$; absolute minimum is 9 at $t = 0, \pi, 2\pi$.

b) $x = 3\cos t$, $y = 2\sin t \Rightarrow f(t) = 6\cos t + 6\sin t \Rightarrow \frac{df}{dt} = -6\sin t + 6\cos t$. $\frac{df}{dt} = 0 \Rightarrow \cos t = \sin t$.

i) On the quarter ellipse, $\frac{x^2}{9} + \frac{y^2}{4} = 1$, $f(0) = 6$, $f\left(\frac{\pi}{2}\right) = 6$. $\frac{df}{dt} = 0 \Rightarrow t = \frac{\pi}{4} \Rightarrow f\left(\frac{\pi}{4}\right) = 6\sqrt{2} \Rightarrow$ absolute maximum is $6\sqrt{2}$ at $t = \frac{\pi}{4}$; absolute minimum is 6 at $t = 0, \frac{\pi}{2}$.

ii) On the half ellipse, $\frac{x^2}{9} + \frac{y^2}{4} = 1$, $y \ge 0$, $f(0) = 6$, $f(\pi) = -6$. $\frac{df}{dt} = 0 \Rightarrow t = \frac{\pi}{4} \Rightarrow f\left(\frac{\pi}{4}\right) = 6\sqrt{2} \Rightarrow$ absolute maximum is $6\sqrt{2}$; absolute minimum is $-6$ at $t = \pi$.

iii) On the full ellipse, $\frac{x^2}{9} + \frac{y^2}{4} = 1$, $f(0) = 6$, $f(2\pi) = 6$. $\frac{df}{dt} = 0 \Rightarrow t = \frac{\pi}{4}, \frac{5\pi}{4} \Rightarrow f\left(\frac{\pi}{4}\right) = 6\sqrt{2}$, $f\left(\frac{5\pi}{4}\right) = -6\sqrt{2} \Rightarrow$ absolute maximum is $6\sqrt{2}$ at $t = \frac{\pi}{4}$; absolute minimum is $-6\sqrt{2}$ at $t = \frac{5\pi}{4}$.

41.

| k | $x_k$ | $y_k$ | $x_k^2$ | $x_k y_k$ |
|---|---|---|---|---|
| 1 | −1 | 2 | 1 | −2 |
| 2 | 0 | 1 | 0 | 0 |
| 3 | 3 | −4 | 9 | −12 |
| Σ | 2 | −1 | 10 | −14 |

$m = \frac{2(-1) - 3(-14)}{2^2 - 3(10)} \approx -1.5$, $b = \frac{1}{3}(-1 - (-1.5)2) \approx 0.7$

$\therefore\ y = -1.5x + 0.7$. $y\big|_{x=4} = -5.3$

43.

| k | $x_k$ | $y_k$ | $x_k^2$ | $x_k y_k$ |
|---|---|---|---|---|
| 1 | 0 | 0 | 0 | 0 |
| 2 | 1 | 2 | 1 | 2 |
| 3 | 2 | 3 | 4 | 6 |
| Σ | 3 | 5 | 5 | 8 |

$m = \dfrac{3(5) - 3(8)}{3^2 - 3(5)} = 1.5$, $b = \frac{1}{3}(5 - 1.5(3)) \approx 0.2$

$\therefore\ y = 1.5x + 0.2.\ y\big|_{x=4} = 6.2$

45.

| k | $x_k$ | $y_k$ | $x_k^2$ | $x_k y_k$ |
|---|---|---|---|---|
| 1 | 12 | 5.27 | 144 | 63.24 |
| 2 | 18 | 5.68 | 324 | 102.24 |
| 3 | 24 | 6.25 | 576 | 150 |
| 4 | 30 | 7.21 | 900 | 216.3 |
| 5 | 36 | 8.20 | 1296 | 295.2 |
| 6 | 42 | 8.71 | 1764 | 365.82 |
| Σ | 162 | 41.32 | 5004 | 1192.8 |

$m = \dfrac{162(41.32) - 6(1192.8)}{162^2 - 6(5004)} \approx 0.122$,

$b = \frac{1}{6}(41.32 - (0.122)(162)) \approx 3.59$

$\therefore\ y = 0.122x + 3.59$

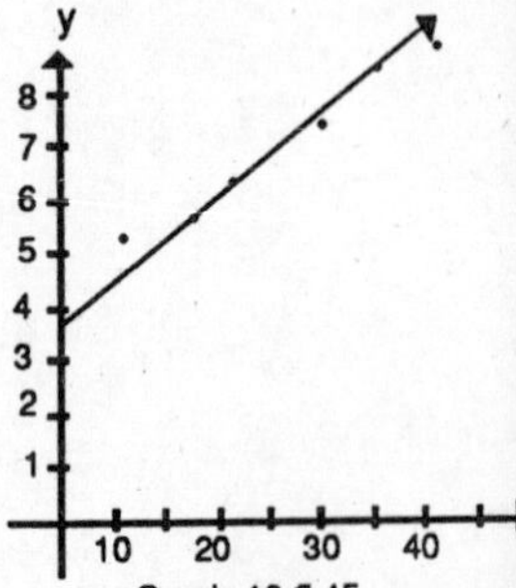


Graph 13.8.45

## SECTION 13.9 LAGRANGE MULTIPLIERS

1. $\nabla f = y\,\mathbf{i} + x\,\mathbf{j}$, $\nabla g = 2x\,\mathbf{i} + 4y\,\mathbf{j}$. $\nabla f = \lambda \nabla g \Rightarrow y\,\mathbf{i} + x\,\mathbf{j} = \lambda(2x\,\mathbf{i} + 4y\,\mathbf{j}) \Rightarrow y = 2x\lambda$ and $x = 4y\lambda \Rightarrow$ $\lambda = \pm\frac{\sqrt{2}}{4}$ or $x = 0$. CASE 1: If $x = 0$, then $y = 0$ but $(0,0)$ not on the ellipse. $\therefore\ x \neq 0$.

CASE 2: $x \neq 0 \Rightarrow \lambda = \pm\frac{\sqrt{2}}{4} \Rightarrow x = \pm\sqrt{2}\,y \Rightarrow (\pm\sqrt{2}\,y)^2 + 2y^2 = 1 \Rightarrow y = \pm\frac{1}{2}$. $\therefore$ f takes on its extreme values at $\left(\pm\sqrt{2}, \frac{1}{2}\right)$ and $\left(\pm\sqrt{2}, \frac{1}{2}\right) \Rightarrow$ the extreme values of f are $\pm\frac{\sqrt{2}}{2}$.

3. $\nabla f = -2x\,\mathbf{i} - 2y\,\mathbf{j}$, $\nabla g = \mathbf{i} + 3\,\mathbf{j}$. $\nabla f = \lambda\nabla g \Rightarrow -2x\,\mathbf{i} - 2y\,\mathbf{j} = \lambda(\mathbf{i} + 3\,\mathbf{j}) \Rightarrow x = -\frac{\lambda}{2}$ and $y = -\frac{3\lambda}{2} \Rightarrow \lambda = -2$ $\Rightarrow x = 1$ and $y = 3 \Rightarrow$ f takes on its extreme value at $(1,3) \Rightarrow$ the extreme value of f is 39.

5. $\nabla f = 2xy\,\mathbf{i} + x^2\,\mathbf{j}$, $\nabla g = \mathbf{i} + \mathbf{j}$. $\nabla f = \lambda\nabla g \Rightarrow 2xy\,\mathbf{i} + x^2\,\mathbf{j} = \lambda(\mathbf{i} + \mathbf{j}) \Rightarrow 2xy = \lambda$ and $x^2 = \lambda \Rightarrow 2xy = x^2 \Rightarrow$ $x = 0, y = 3$ or $x = 2, y = 1$. $\therefore$ f takes on its extreme values at $(0,3)$ and $(2,1)$. $\therefore$ the extreme values of f are $f(0,3) = 0$ and $f(2,1) = 4$.

7. a) $\nabla f = \mathbf{i} + \mathbf{j}$, $\nabla g = y\,\mathbf{i} + x\,\mathbf{j}$. $\nabla f = \lambda\nabla g \Rightarrow \mathbf{i} + \mathbf{j} = \lambda(y\,\mathbf{i} + x\,\mathbf{j}) \Rightarrow 1 = \lambda y$ and $1 = \lambda x \Rightarrow y = \frac{1}{\lambda}$ and $x = \frac{1}{\lambda}$ $\Rightarrow \frac{1}{\lambda^2} = 16 \Rightarrow \lambda = \pm 4$. Use $\lambda = 4$ since $x > 0$, $y > 0$. Then $x = 4$, $y = 4 \Rightarrow$ the minimum value is 8 at $x = 4$, $y = 4$.

   b) $\nabla f = y\,\mathbf{i} + x\,\mathbf{j}$, $\nabla g = \mathbf{i} + \mathbf{j}$. $\nabla f = \lambda\nabla g \Rightarrow y\,\mathbf{i} + x\,\mathbf{j} = \lambda(\mathbf{i} + \mathbf{j}) \Rightarrow y = \lambda = x \Rightarrow y = x \Rightarrow y + y = 16 \Rightarrow y = 8$ $\Rightarrow x = 8 \Rightarrow f(8,8) = 64$ is the maximum value.

9. $V = \pi r^2 h \Rightarrow 16\pi = \pi r^2 h \Rightarrow 16 = r^2 h \Rightarrow g(r,h) = r^2 h - 16$. $\nabla S = (2\pi h + 4\pi r)\,\mathbf{i} + 2\pi r\,\mathbf{j}$, $\nabla g = 2rh\,\mathbf{i} + r^2\,\mathbf{j}$. $\nabla S = \lambda\nabla g \Rightarrow (2\pi rh + 4\pi r)\,\mathbf{i} + 2\pi r\,\mathbf{j} = \lambda(2rh\,\mathbf{i} + r^2\,\mathbf{j}) \Rightarrow 2\pi h + 4\pi r = 2rh\lambda$ and $2\pi r = \lambda r^2 \Rightarrow 0 = \lambda r^2 - 2\pi r$ $\Rightarrow r = 0$ or $\lambda = \frac{2\pi}{r}$. Now $r \neq 0 \Rightarrow \lambda = \frac{2\pi}{r} \Rightarrow 2\pi h + 4\pi r = 2rh\left(\frac{2\pi}{r}\right) \Rightarrow 2r = h \Rightarrow 16 = r^2(2r) \Rightarrow r = 2 \Rightarrow$ $h = 4$. $\therefore$ r = 2 cm, h = 4 cm give the smallest surface area.

11. $\nabla T = (8x - 4y)\,\mathbf{i} + (-4x + 2y)\,\mathbf{j}$, $\nabla g = 2x\,\mathbf{i} + 2y\,\mathbf{j}$. $\nabla T = \lambda \nabla g \Rightarrow (8x - 4y)\,\mathbf{i} + (-4x + 2y)\,\mathbf{j} = \lambda(2x\,\mathbf{i} + 2y\,\mathbf{j}) \Rightarrow$ $8x - 4y = 2\lambda x$ and $-4x + 2y = 2\lambda y \Rightarrow y = \frac{-2x}{\lambda - 1}$, $\lambda \neq 1 \Rightarrow 8x - 4\left(\frac{-2x}{\lambda - 1}\right) = 2\lambda x \Rightarrow x = 0$ or $\lambda = 0$ or $\lambda = 5$. $x = 0 \Rightarrow y = 0$. But $(0,0)$ not on $x^2 + y^2 = 25$. $\therefore\ x \neq 0 \Rightarrow \lambda = 0$ or $\lambda = 5$. $\lambda = 0 \Rightarrow y = 2x \Rightarrow$ $x^2 + (2x)^2 = 25 \Rightarrow x = \pm\sqrt{5} \Rightarrow y = \pm 2\sqrt{5}$. $\lambda = 5 \Rightarrow y = \frac{-2x}{4} = -\frac{1}{2}x \Rightarrow x^2 + \left(-\frac{1}{2}x\right)^2 = 25 \Rightarrow x = \pm 2\sqrt{5}$. $x = 2\sqrt{5} \Rightarrow y = -\sqrt{5}$, $x = -2\sqrt{5} \Rightarrow y = \sqrt{5}$. $T(\sqrt{5}, 2\sqrt{5}) = 0° = T(-\sqrt{5}, -2\sqrt{5})$, the minimum value; $T(2\sqrt{5}, -\sqrt{5}) = 125° = T(-2\sqrt{5}, \sqrt{5})$, the maximum value.

13. $\nabla f = 2x\,\mathbf{i} + 2y\,\mathbf{j}$, $\nabla g = (2x - 2)\,\mathbf{i} + (2y - 4)\,\mathbf{j}$. $\nabla f = \lambda \nabla g \Rightarrow 2x\,\mathbf{i} + 2y\,\mathbf{j} = \lambda((2x - 2)\,\mathbf{i} + (2y - 4)\,\mathbf{j}) \Rightarrow$ $2x = \lambda(2x - 2)$ and $2y = \lambda(2y - 4) \Rightarrow x = \frac{\lambda}{\lambda - 1}$ and $y = \frac{2\lambda}{\lambda - 1}$, $\lambda \neq 1 \Rightarrow y = 2x \Rightarrow x^2 - 2x + (2x)^2 - 4(2x)$ $= 0 \Rightarrow x = 0, y = 0$ or $x = 2, y = 4$. $\therefore$ $f(0,0) = 0$ is the minimum value, $f(2,4) = 20$ is the maximum value.

15. $\nabla f = \mathbf{i} - 2\,\mathbf{j} + 5\,\mathbf{k}$, $\nabla g = 2x\,\mathbf{i} + 2y\,\mathbf{j} + 2z\,\mathbf{k}$. $\nabla f = \lambda \nabla g \Rightarrow \mathbf{i} - 2\,\mathbf{j} + 5\,\mathbf{k} = \lambda(2x\,\mathbf{i} + 2y\,\mathbf{j} + 2z\,\mathbf{k}) \Rightarrow 1 = 2x\lambda$, $-2 = 2y\lambda$, and $5 = 2z\lambda \Rightarrow x = \frac{1}{2\lambda}$, $y = -\frac{1}{\lambda} = -2x$, $z = \frac{5}{2\lambda} = 5x \Rightarrow x^2 + (-2x)^2 + (5x)^2 = 30 \Rightarrow x = \pm 1$. $x = 1 \Rightarrow y = -2, z = 5$. $x = -1 \Rightarrow y = 2, z = -5$. $f(1,-2,5) = 30$, the maximum value; $f(-1,2,-5) = -30$, the minimum value.

17. Let $f(x,y,z) = x^2 + y^2 + z^2$ be the square of the distance to the origin. Then $\nabla f = 2x\,\mathbf{i} + 2y\,\mathbf{j} + 2z\,\mathbf{k}$, $\nabla g = y\,\mathbf{i} + x\,\mathbf{j} - \mathbf{k}$. $\nabla f = \lambda \nabla g \Rightarrow 2x\,\mathbf{i} + 2y\,\mathbf{j} + 2z\,\mathbf{k} = \lambda(y\,\mathbf{i} + x\,\mathbf{j} - \mathbf{k}) \Rightarrow 2x = \lambda y$, $2y = \lambda x$, and $2z = -\lambda$ $\Rightarrow x = \frac{\lambda y}{2} \Rightarrow 2y = \lambda\left(\frac{\lambda y}{2}\right) \Rightarrow y = 0$ or $\lambda = \pm 2$. $y = 0 \Rightarrow x = 0 \Rightarrow -z + 1 = 0 \Rightarrow z = 1$. $\lambda = 2 \Rightarrow x = y$, $z = -1 \Rightarrow x^2 - (-1) + 1 = 0 \Rightarrow x^2 + 2 = 0$, no solution. $\lambda = -2 \Rightarrow x = -y$, $z = 1 \Rightarrow (-y)y - 1 + 1 = 0 \Rightarrow$ $y = 0$, again. $\therefore$ $(0,0,1)$ is the point on the surface closest to the origin.

19. $\nabla f = \mathbf{i} + 2\,\mathbf{j} + 3\,\mathbf{k}$, $\nabla g = 2x\,\mathbf{i} + 2y\,\mathbf{j} + 2z\,\mathbf{k}$. $\nabla f = \lambda \nabla g \Rightarrow \mathbf{i} + 2\,\mathbf{j} + 3\,\mathbf{k} = \lambda(2x\,\mathbf{i} + 2y\,\mathbf{j} + 2z\,\mathbf{k}) \Rightarrow 1 = 2x\lambda$, $2 = 2y\lambda$, and $3 = 2z\lambda \Rightarrow x = \frac{1}{2\lambda}$, $y = \frac{1}{\lambda}$, and $z = \frac{3}{2\lambda} \Rightarrow y = 2x$ and $z = 3x \Rightarrow x^2 + (2x)^2 + (3x)^2 = 25 \Rightarrow$ $x = \pm\frac{5}{\sqrt{14}}$. $x = \frac{5}{\sqrt{14}} \Rightarrow y = \frac{10}{\sqrt{14}}$, $z = \frac{15}{\sqrt{14}}$. $x = -\frac{5}{\sqrt{14}} \Rightarrow y = -\frac{10}{\sqrt{14}}$, $z = -\frac{15}{\sqrt{14}}$. $f\left(\frac{5}{\sqrt{14}}, \frac{10}{\sqrt{14}}, \frac{15}{\sqrt{14}}\right) =$ $5\sqrt{14}$, the maximum value; $f\left(-\frac{5}{\sqrt{14}}, -\frac{10}{\sqrt{14}}, -\frac{15}{\sqrt{14}}\right) = -5\sqrt{14}$, the minimum value.

21. $\nabla f = yz\,\mathbf{i} + xz\,\mathbf{j} + xy\,\mathbf{k}$, $\nabla g = \mathbf{i} + \mathbf{j} + 2z\,\mathbf{k}$. $\nabla f = \lambda \nabla g \Rightarrow yz\,\mathbf{i} + xz\,\mathbf{j} + xy\,\mathbf{k} = \lambda(\mathbf{i} + \mathbf{j} + 2z\,\mathbf{k}) \Rightarrow yz = \lambda$, $xz = \lambda$, and $xy = \lambda \Rightarrow yz = xz \Rightarrow z = 0$ or $y = x$. But $z > 0 \Rightarrow y = x \Rightarrow x^2 = 2z\lambda$ and $xz = \lambda$. Then $x^2 = 2z(xz)$ $\Rightarrow x = 0$ or $x = 2z^2$. But $x > 0 \Rightarrow x = 2z^2 \Rightarrow y = 2z^2 \Rightarrow 2z^2 + 2z^2 + z^2 = 16 \Rightarrow z = \pm\frac{4}{\sqrt{5}}$. Use $z = \frac{4}{\sqrt{5}}$ since $z > 0 \Rightarrow x = \frac{32}{5}$, $y = \frac{32}{5}$. $f\left(\frac{32}{5}, \frac{32}{5}, \frac{4}{\sqrt{5}}\right) = \frac{4096}{25\sqrt{5}}$

23. $\nabla U = (y + 2)\,\mathbf{i} + x\,\mathbf{j}$, $\nabla g = 2\,\mathbf{i} + \mathbf{j}$. $\nabla U = \lambda \nabla g \Rightarrow (y + 2)\,\mathbf{i} + x\,\mathbf{j} = \lambda(2\,\mathbf{i} + \mathbf{j}) \Rightarrow y + 2 = 2\lambda$ and $x = \lambda \Rightarrow$ $y + 2 = 2x \Rightarrow y = 2x - 2 \Rightarrow 2x + 2x - 2 = 30 \Rightarrow x = 8 \Rightarrow y = 14$. $\therefore$ $U(8,14) = \$128$, the maximum value of U under the constraint.

25. $\nabla f = \mathbf{i} + \mathbf{j}$, $\nabla g = y\,\mathbf{i} + x\,\mathbf{j}$. $\nabla f = \lambda \nabla g \Rightarrow \mathbf{i} + \mathbf{j} = \lambda(y\,\mathbf{i} + x\,\mathbf{j}) \Rightarrow 1 = y\lambda$ and $1 = x\lambda \Rightarrow y = x \Rightarrow y^2 = 16 \Rightarrow$ $y = \pm 4 \Rightarrow x = \pm 4$. But as $x \to \infty$, $y \to 0$ and $f(x,y) \to \infty$; as $x \to -\infty$, $y \to 0$ and $f(x,y) \to -\infty$.

## PRACTICE EXERCISES

1.

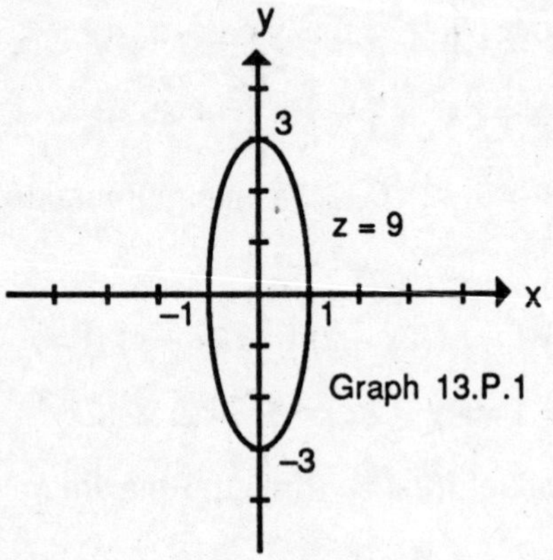

Graph 13.P.1

Domain: All points in the xy–plane
Range: f(x,y) ≥ 0

Level curves are ellipses with major axis along the y–axis and minor axis along the x–axis.

3.

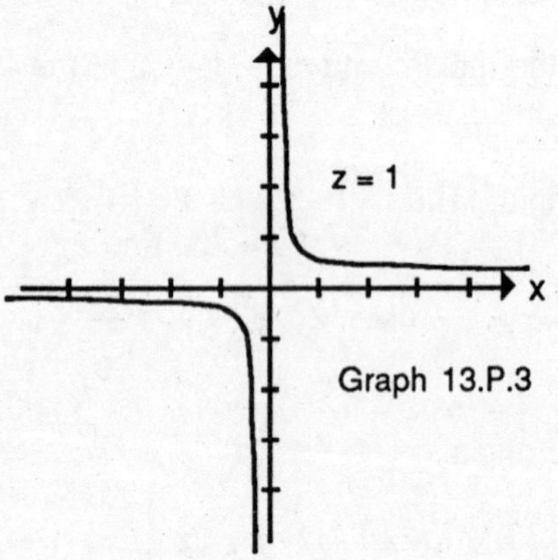

Graph 13.P.3

Domain: All (x,y) such that x ≠ 0 or y ≠ 0
Range: f(x,y) ≠ 0

Level curves are hyperbolas rotated 45° or 135°.

5.

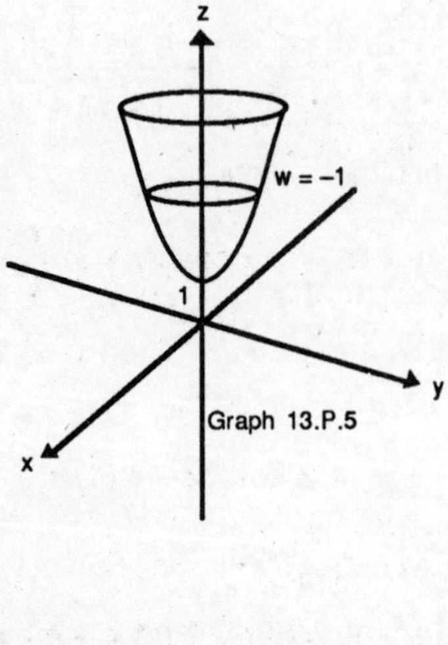

Graph 13.P.5

Domain: All (x,y,z) such (x,y,z ≠ (0,0,0)
Range: All Real Numbers

Level surfaces are paraboloids of revolution with the z–axis as axis.

7.

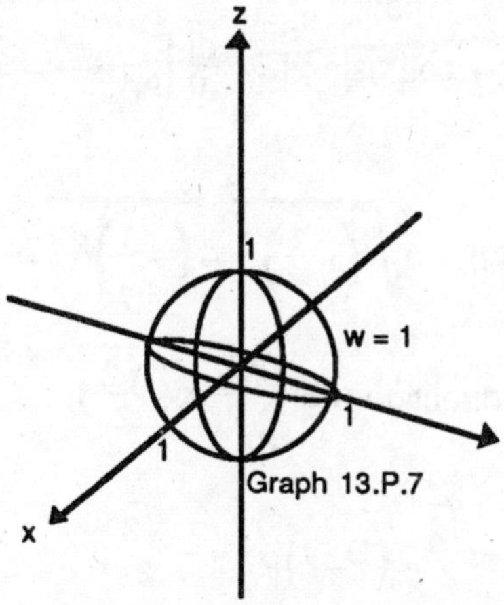


Domain: All (x,y,z) such that $(x,y,z) \neq (0,0,0)$
Range: $f(x,y,z) > 0$

Level surfaces are spheres with center (0,0,0) and radius $r > 0$.

9. $\lim_{(x,y)\to(\pi,\ln 2)} e^y \cos x = e^{\ln 2} \cos \pi = -2$

11. $\lim_{(x,y)\to(1,1)} \frac{x^2 - y^2}{x - y} = \lim_{(x,y)\to(1,1)} (x + y) = 2$

13. Let $y = kx^2$, $k \neq 1$. Then $\lim_{\substack{(x,y)\to(0,0) \\ y \neq x^2}} \frac{y}{x^2 - y} = \lim_{(x,kx^2)\to(0,0)} \frac{kx^2}{x^2 - kx^2} = \frac{k}{1 - k^2}$ which gives different limits for different values of k. ∴ the limit doesn't exist.

15. $\frac{\partial g}{\partial r} = \cos\theta + \sin\theta$, $\frac{\partial g}{\partial \theta} = -r\sin\theta + r\cos\theta$

17. $\frac{\partial f}{\partial R_1} = -\frac{1}{R_1{}^2}$, $\frac{\partial f}{\partial R_2} = -\frac{1}{R_2{}^2}$, $\frac{\partial f}{\partial R_3} = -\frac{1}{R_3{}^2}$

19. $\frac{\partial P}{\partial n} = \frac{RT}{V}$, $\frac{\partial P}{\partial R} = \frac{nT}{V}$, $\frac{\partial P}{\partial T} = \frac{nR}{V}$, $\frac{\partial P}{\partial V} = -\frac{nRT}{V^2}$

21. $\frac{\partial f}{\partial x} = \frac{1}{y}$, $\frac{\partial f}{\partial y} = 1 - \frac{x}{y^2} \Rightarrow \frac{\partial^2 f}{\partial x^2} = 0$, $\frac{\partial^2 f}{\partial y^2} = \frac{2x}{y^3}$, $\frac{\partial^2 f}{\partial y \partial x} = \frac{\partial^2 f}{\partial x \partial y} = -\frac{1}{y^2}$

23. $\frac{\partial f}{\partial x} = 1 + y - 15x^2 + \frac{2x}{x^2 + 1}$, $\frac{\partial f}{\partial y} = x \Rightarrow \frac{\partial^2 f}{\partial x^2} = -30x + \frac{2 - 2x^2}{(x^2 + 1)^2}$, $\frac{\partial^2 f}{\partial y^2} = 0$, $\frac{\partial^2 f}{\partial y \partial x} = \frac{\partial^2 f}{\partial x \partial y} = 1$

25. $\frac{\partial w}{\partial x} = y\cos(xy + \pi)$, $\frac{\partial w}{\partial y} = x\cos(xy + \pi)$, $\frac{dx}{dt} = e^t$, $\frac{dy}{dt} = \frac{1}{t + 1} \Rightarrow \frac{dw}{dt} = y\cos(xy + \pi)\, e^t + x\cos(xy + \pi)\left(\frac{1}{t + 1}\right)$
$= e^t \ln(t + 1)\cos(e^t \ln(t + 1) + \pi) + \frac{e^t}{t + 1}\cos(e^t \ln(t + 1) + \pi) \Rightarrow \left.\frac{dw}{dt}\right|_{t=0} = -1$

27. $\frac{\partial w}{\partial x} = 2\cos(2x - y)$, $\frac{\partial w}{\partial y} = -\cos(2x - y)$, $\frac{\partial x}{\partial r} = 1$, $\frac{\partial x}{\partial s} = \cos s$, $\frac{\partial y}{\partial r} = s$, $\frac{\partial y}{\partial s} = r \Rightarrow \frac{\partial w}{\partial r} = 2\cos(2x - y)(1) +$
$(-\cos(2x - y)(s)) = 2\cos(2r + 2\sin s - rs) - s\cos(2r + 2\sin s - rs) \Rightarrow \left.\frac{\partial w}{\partial r}\right|_{r=\pi, s=0} = 2.$
$\frac{\partial w}{\partial s} = 2\cos(2x - y)(\cos s) + (-\cos(2x - y)(r)) = 2\cos(2r + 2\sin s - rs)(\cos s) - r\cos(2r + 2\sin s - rs) \Rightarrow$
$\left.\frac{\partial w}{\partial s}\right|_{r=\pi, s=0} = 2 - \pi$

29. $F_x = -1 - y\cos xy$, $F_y = -2y - x\cos xy$. $\frac{dy}{dx} = -\frac{F_x}{F_y} = -\frac{-1 - y\cos xy}{-2y - x\cos xy} = \frac{1 + y\cos xy}{-2y - x\cos xy} \Rightarrow$
$\left.\frac{dy}{dx}\right|_{(x,y)=(0,1)} = -1$

31. $\frac{\partial f}{\partial x} = y + z,\ \frac{\partial f}{\partial y} = x + z,\ \frac{\partial f}{\partial z} = y + x,\ \frac{dx}{dt} = -\sin t,\ \frac{dy}{dt} = \cos t,\ \frac{dz}{dt} = -2 \sin 2t \Rightarrow \frac{df}{dt} = -(y + z)\sin t + (x + z)\cos t$
$- 2(y + x)\sin 2t = -(\sin t + \cos 2t)\sin t + (\cos t + \cos 2t)\cos t - 2(\sin t + \cos t)\sin 2t \Rightarrow \left.\frac{df}{dt}\right|_{t=1} =$
$-(\sin 1 + \cos 2)\sin 1 + (\cos 1 + \cos 2)\cos 1 - 2(\sin 1 + \cos 1)\sin 2$

33. $\nabla f = (-\sin x \cos y)\,\mathbf{i} - (\cos x \sin y)\,\mathbf{j} \Rightarrow \nabla f\big|_{(\pi/4,\pi/4)} = -\frac{1}{2}\mathbf{i} - \frac{1}{2}\mathbf{j} \Rightarrow |\nabla f| = \sqrt{\left(-\frac{1}{2}\right)^2 + \left(-\frac{1}{2}\right)^2} = \frac{1}{\sqrt{2}}$

$\mathbf{u} = \frac{\nabla f}{|\nabla f|} = \frac{-\frac{1}{2}\mathbf{i} - \frac{1}{2}\mathbf{j}}{\frac{1}{\sqrt{2}}} = -\frac{\sqrt{2}}{2}\mathbf{i} - \frac{\sqrt{2}}{2}\mathbf{j}$. f increases most rapidly in the direction $\mathbf{u} = -\frac{\sqrt{2}}{2}\mathbf{i} - \frac{\sqrt{2}}{2}\mathbf{j}$;

decreases most rapidly in the direction $-\mathbf{u} = \frac{\sqrt{2}}{2}\mathbf{i} + \frac{\sqrt{2}}{2}\mathbf{j}$. $\left(D_{\mathbf{u}}f\right)_{P_0} = \frac{\sqrt{2}}{2}$, $\left(D_{-\mathbf{u}}f\right)_{P_0} = -\frac{\sqrt{2}}{2}$.

$\mathbf{u}_1 = \frac{\mathbf{A}}{|\mathbf{A}|} = \frac{3\mathbf{i} + 4\mathbf{j}}{\sqrt{3^2 + 4^2}} = \frac{3}{5}\mathbf{i} + \frac{4}{5}\mathbf{j}$. $\left(D_{\mathbf{u}_1}f\right)_{P_0} = \nabla f \cdot \mathbf{u}_1 = -\frac{7}{10}$.

35. $\nabla f = \left(\frac{2}{2x + 3y + 6z}\right)\mathbf{i} + \left(\frac{3}{2x + 3y + 6z}\right)\mathbf{j} + \left(\frac{6}{2x + 3y + 6z}\right)\mathbf{k} \Rightarrow \nabla f\big|_{(-1,-1,1)} = 2\mathbf{i} + 3\mathbf{j} + 6\mathbf{k}$.
$\mathbf{u} = \frac{\nabla f}{|\nabla f|} = \frac{2\mathbf{i} + 3\mathbf{j} + 6\mathbf{k}}{\sqrt{2^2 + 3^2 + 6^2}} = \frac{2}{7}\mathbf{i} + \frac{3}{7}\mathbf{j} + \frac{6}{7}\mathbf{k}$. f increases most rapidly in the direction $\mathbf{u} = \frac{2}{7}\mathbf{i} + \frac{3}{7}\mathbf{j} + \frac{6}{7}\mathbf{k}$;

decreases most rapidly in the direction $-\mathbf{u} = -\frac{2}{7}\mathbf{i} - \frac{3}{7}\mathbf{j} - \frac{6}{7}\mathbf{k}$. $\left(D_{\mathbf{u}}f\right)_{P_0} = \nabla f \cdot \mathbf{u} = 7$,

$\left(D_{-\mathbf{u}}f\right)_{P_0} = -7$. $\mathbf{u}_1 = \frac{\mathbf{A}}{|\mathbf{A}|} = \frac{2}{7}\mathbf{i} + \frac{3}{7}\mathbf{j} + \frac{6}{7}\mathbf{k}$ since $\mathbf{A} = \nabla f$. $\Rightarrow \left(D_{\mathbf{u}_1}f\right)_{P_0} = 7$.

37.

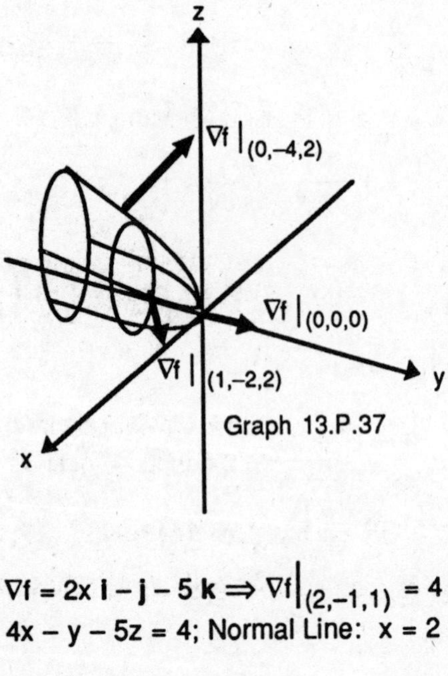

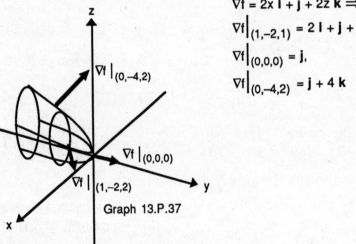

Graph 13.P.37

$\nabla f = 2x\,\mathbf{i} + \mathbf{j} + 2z\,\mathbf{k} \Rightarrow$

$\nabla f\big|_{(1,-2,1)} = 2\mathbf{i} + \mathbf{j} + 2\mathbf{k}$,

$\nabla f\big|_{(0,0,0)} = \mathbf{j}$,

$\nabla f\big|_{(0,-4,2)} = \mathbf{j} + 4\mathbf{k}$

39. $\nabla f = 2x\,\mathbf{i} - \mathbf{j} - 5\mathbf{k} \Rightarrow \nabla f\big|_{(2,-1,1)} = 4\mathbf{i} - \mathbf{j} - 5\mathbf{k} \Rightarrow$ Tangent Plane: $4(x - 2) - (y + 1) - 5(z - 1) = 0 \Rightarrow$
$4x - y - 5z = 4$; Normal Line: $x = 2 + 4t,\ y = -1 - t,\ z = 1 - 5t$

41.

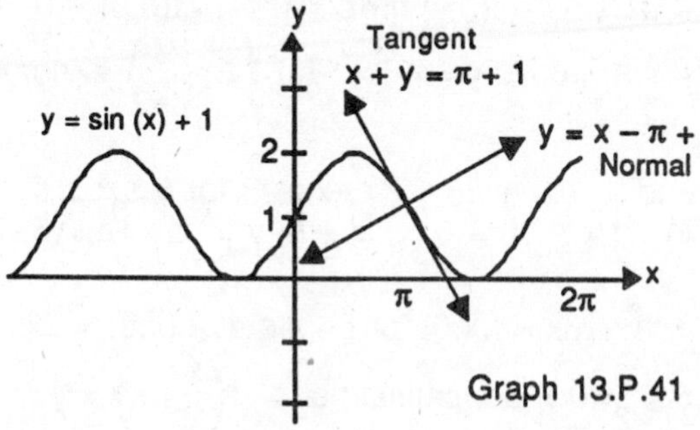

Graph 13.P.41

$\nabla f = (-\cos x)\,\mathbf{i} + \mathbf{j} \Rightarrow \nabla f\big|_{(\pi,1)} = \mathbf{i} + \mathbf{j} \Rightarrow$ Tangent Line: $(x - \pi) + (y - 1) = 0 \Rightarrow x + y = \pi + 1$; Normal Line: $y - 1 = 1(x - \pi) \Rightarrow y = x - \pi + 1$.

43. $f\left(\frac{\pi}{4},\frac{\pi}{4}\right) = \frac{1}{2}$, $f_x\left(\frac{\pi}{4},\frac{\pi}{4}\right) = \cos x \cos y\big|_{(\pi/4,\pi/4)} = \frac{1}{2}$, $f_y\left(\frac{\pi}{4},\frac{\pi}{4}\right) = -\sin x \sin y\big|_{(\pi/4,\pi/4)} = -\frac{1}{2} \Rightarrow L(x,y) = \frac{1}{2} + \frac{1}{2}\left(x - \frac{\pi}{4}\right) - \frac{1}{2}\left(y - \frac{\pi}{4}\right) = \frac{1}{2} + \frac{1}{2}x - \frac{1}{2}y$. $f_{xx}(x,y) = -\sin x \cos y$, $f_{yy}(x,y) = -\sin x \cos y$, $f_{xy}(x,y) = -\cos x \sin y$. ∴ maximum of $|f_{xx}|$, $|f_{yy}|$, and $|f_{xy}|$ is $1 \Rightarrow M = 1 \Rightarrow |E(x,y)| \le \frac{1}{2}(1)\left(\left|x - \frac{\pi}{4}\right| + \left|y - \frac{\pi}{4}\right|\right)^2 \le 0.02$.

45. a) $f(1,0,0) = 0$, $f_x(1,0,0) = y - 3z\big|_{(1,0,0)} = 0$, $f_y(1,0,0) = x + 2z\big|_{(1,0,0)} = 1$, $f_z(1,0,0) = 2y - 3x\big|_{(1,0,0)} = -3 \Rightarrow L(x,y,z) = 0(x - 1) + (y - 0) - 3(z - 0) = y - 3z$.
b) $f(1,1,0) = 1$, $f_x(1,1,0) = 1$, $f_y(1,1,0) = 1$, $f_z(1,1,0) = 2 \Rightarrow L(x,y,z) = 1 + (x - 1) + (y - 1) + 2(z - 0) = x + y + 2z - 1$

47. $dV = 2\pi rh\,dr + \pi r^2\,dh \Rightarrow dV\big|_{(1.5,5280)} = 2\pi(1.5)(5280)\,dr + \pi(1.5)^2\,dh = 15840\pi\,dr + 2.25\pi\,dh$. Be more careful with the diameter since it has a greater effect on dV.

49. $dI = \frac{1}{R}dV - \frac{V}{R^2}dR \Rightarrow dI\big|_{(24,100)} = \frac{1}{100}dV - \frac{24}{100^2}dR \Rightarrow dI\big|_{dV=-1,dR=-20} = 0.038$. % change in V $= -\frac{1}{24} = -4.17\%$; % change in R $= -\frac{20}{100} = -20\%$. $I = \frac{24}{100} = 0.24 \Rightarrow$ Estimated % change in I $= \frac{dI}{I} \times 100 = \frac{0.038}{0.24} \times 100 = 15.83\%$

51. $f_x(x,y) = 2x - y + 2 = 0$ and $f_y(x,y) = -x + 2y + 2 = 0 \Rightarrow x = -2, y = -2 \Rightarrow (-2,-2)$ is the critical point. $f_{xx}(-2,-2) = 2$, $f_{yy}(-2,-2) = 2$, $f_{xy}(-2,-2) = -1 \Rightarrow f_{xx}f_{yy} - f_{xy}^2 = 3 > 0$ and $f_{xx} > 0 \Rightarrow$ Minimum (absolute). $f(-2,-2) = -8$

53. $f_x(x,y) = 6y - 3x^2 = 0$ and $f_y(x,y) = 6x - 2y = 0 \Rightarrow x = 0, y = 0$ or $x = 6, y = 18 \Rightarrow$ critical points are (0,0) and (6,18). For (0,0): $f_{xx}(0,0) = -6x\big|_{(0,0)} = 0$, $f_{yy}(0,0) = -2$, $f_{xy}(0,0) = 6 \Rightarrow f_{xx}f_{yy} - f_{xy}^2 = -36 \Rightarrow$ Saddle Point. $f(0,0) = 0$. For (6,18): $f_{xx}(6,18) = -36$, $f_{yy}(6,18) = -2$, $f_{xy}(6,18) = 6 \Rightarrow f_{xx}f_{yy} - f_{xy}^2 = 36 > 0$ and $f_{xx} < 0 \Rightarrow$ maximum (local since $y = 0$ and $x < 0 \Rightarrow f(x,y)$ increases without bound). $f(6,18) = 108$.

55. $f_x(x,y) = 3x^2 - 3y = 0$ and $f_y(x,y) = 3y^2 - 3x = 0 \Rightarrow x = 0, y = 0$ or $x = 1, y = 1 \Rightarrow$ critical points are (0,0) and (1,1). For (0,0): $f_{xx}(0,0) = 6x\big|_{(0,0)} = 0$, $f_y(0,0) = 6y\big|_{(0,0)} = 0$, $f_{xy}(,0) = -3 \Rightarrow f_{xx}f_{yy} - f_{xy}^2 = -9 \Rightarrow$ Saddle Point. $f(0,0) = 15$. For (1,1): $f_{xx}(1,1) = 6$, $f_{yy}(1,1) = 6$, $f_{xy}(1,1) = -3 \Rightarrow f_{xx}f_{yy} - f_{xy}^2 = 27 > 0$ and $f_{xx} > 0 \Rightarrow$ Minimum (local since $y = 0$, $x < 0 \Rightarrow f(x,y)$ decreases without bound). $f(1,1) = 14$.

57.

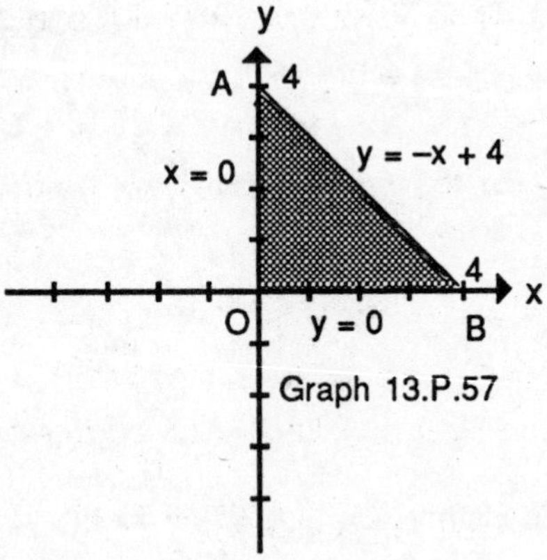


Graph 13.P.57

1. On OA, $f(x,y) = y^2 + 3y = f(0,y)$ for $0 \le y \le 4$. $f(0,0) = 0$, $f(0,4) = 28$. $f'(0,y) = 2y + 3 = 0 \Rightarrow y = -\frac{3}{2}$. But $\left(0,-\frac{3}{2}\right)$ is not in the region.
2. On AB, $f(x,y) = x^2 - 10x + 28 = f(x,-x+4)$ for $0 \le x \le 4$. $f(4,0) = 4$. $f'(x,-x+4) = 2x - 10 = 0 \Rightarrow x = 5, y = -1$. But $(5,-1)$ not in the region.
3. On OB, $f(x,y) = x^2 - 3x = f(x,0)$ for $0 \le x \le 4$. $f'(x,0) = 2x - 3 \Rightarrow x = \frac{3}{2}, y = 0 \Rightarrow \left(\frac{3}{2},0\right)$ is a critical point. $f\left(\frac{3}{2},0\right) = -\frac{9}{4}$
4. For the interior of the triangular region, $f_x(x,y) = 2x + y - 3 = 0$ and $f_y(x,y) = x + 2y + 3 = 0 \Rightarrow x = 3, y = -3$. But $(3,-3)$ is not in the region.

∴ the absolute maximum is 28 at (0,4); the absolute minimum is $-\frac{9}{4}$ at $\left(\frac{3}{2},0\right)$.

59.

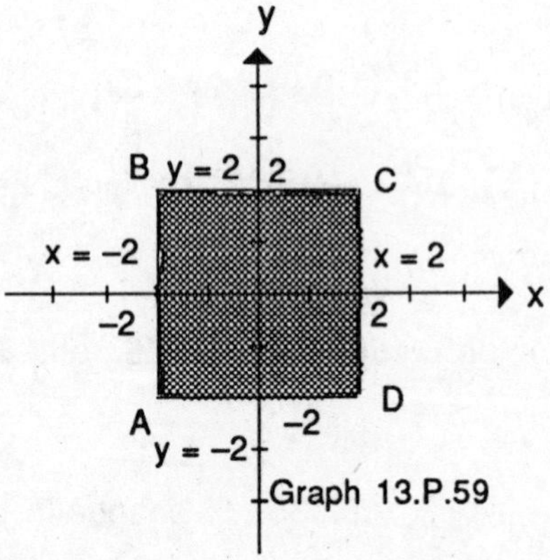


Graph 13.P.59

1. On AB, $f(x,y) = y^2 - y - 4 = f(-2,y)$ for $-2 \le y \le 2$. $f(-2,-2) = 2$, $f(-2,2) = -2$. $f'(-2,y) = 2y - 1 \Rightarrow y = \frac{1}{2}, x = -2 \Rightarrow \left(-2,\frac{1}{2}\right)$ is a critical point. $f\left(-2,\frac{1}{2}\right) = -\frac{17}{4}$.
2. On BC, $f(x,y) = -2 = f(x,2)$ for $-2 \le x \le 2$. $f(2,2) = -2$. $f'(x,2) = 0 \Rightarrow$ no critical points in the interior of BC.
3. On CD, $f(x,y) = y^2 - 5y + 4 = f(2,y)$ for $-2 \le y \le 2$. $f(2,-2) = 18$. $f'(2,y) = 2y - 5 = 0 \Rightarrow y = \frac{5}{2}, x = 2 \Rightarrow \left(2,\frac{5}{2}\right)$ which is not in the region.
4. On AD, $f(x,y) = 4x + 10 = f(x,-2)$ for $-2 \le x \le 2$. $f'(x,-2) = 4 \Rightarrow$ no critical points in the interior of AD.
5. For the interior of the square, $f_x(x,y) = -y = 2 = 0$ and $f_y(x,y) = 2y - x - 3 = 0 \Rightarrow x = 1, y = 2 \Rightarrow (1,2)$ is a critical point. $f(1,2) = -2$

∴ the absolute maximum is 18 at (2,−2); the absolute minimum is $-\frac{17}{4}$ at $\left(-2,\frac{1}{2}\right)$.

61.

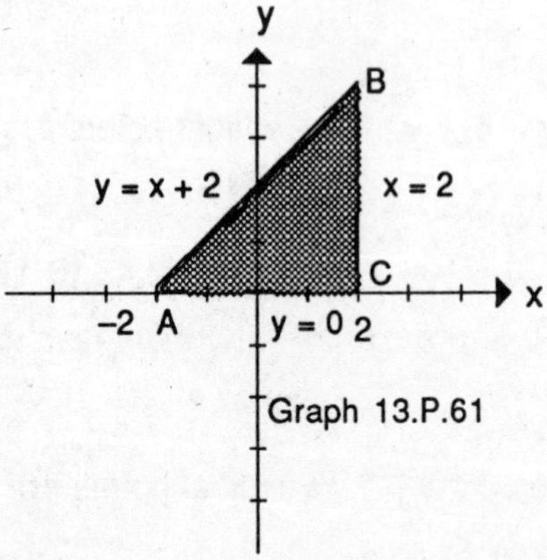


Graph 13.P.61

1. On AB, $f(x,y) = -2x + 4 = f(x,x+2)$ for $-2 \le x \le 2$. $f(-2,0) = 8$, $f(2,4) = 0$. $f'(x,x+2) = -2 \Rightarrow$ no critical points in the interior of AB.
2. On BC, $f(x,y) = -y^2 + 4y = f(2,y)$ for $0 \le y \le 4$. $f(2,0) = 0$. $f'(2,y) = -2y + 4 = 0 \Rightarrow y = 2, x = 2 \Rightarrow (2,2)$ is a critical point. $f(2,2) = 4$.
3. On AC, $f(x,y) = x^2 - 2x = f(x,0)$ for $-2 \le x \le 2$. $f'(x,0) = 2x - 2 = 0 \Rightarrow x = 1, y = 0 \Rightarrow (1,0)$ is a critical point. $f(1,0) = -1$.
4. For the interior of the triangular region, $f_x(x,y) = 2x - 2 = 0$ and $f_y(x,y) = -2y + 4 = 0 \Rightarrow x = 1, y = 2 \Rightarrow (1,2)$ is a critical point. $f(1,2) = 3$.

∴ the absolute maximum is 8 at (−2,0); the absolute minimum is −1 at (1,0).

63. Let $f(x,y) = x^2 + y^2$ be the square of the distance to the origin. $\nabla f = 2x\,\mathbf{i} + 2y\,\mathbf{j}$, $\nabla g = y^2\,\mathbf{i} + 2xy\,\mathbf{j}$. $\nabla f = \lambda \nabla g \Rightarrow 2x\,\mathbf{i} + 2y\,\mathbf{j} = \lambda(y^2\,\mathbf{i} + 2xy\,\mathbf{j}) \Rightarrow 2x = \lambda y^2$ and $2y = 2xy\lambda \Rightarrow 2y = \lambda y^2(y\lambda) \Rightarrow y = 0$ (not on $xy^2 = 54$) or $\lambda^2 y^2 - 2 = 0 \Rightarrow y^2 = \frac{2}{\lambda^2}$. $2y = 2xy\lambda \Rightarrow 1 = x\lambda$ since $y \neq 0 \Rightarrow x = \frac{1}{\lambda}$.
$\therefore \frac{1}{\lambda}\left(\frac{2}{\lambda^2}\right) = 54 \Rightarrow \lambda^3 = \frac{1}{27} \Rightarrow \lambda = \frac{1}{3} \Rightarrow x = 3,\ y^2 = 18 \Rightarrow y = \pm 3\sqrt{2} \Rightarrow$ the points nearest to the origin are $(3, \pm 3\sqrt{2})$.

65. $\nabla T = 400yz^2\,\mathbf{i} + 400xz^2\,\mathbf{j} + 800xyz\,\mathbf{k}$, $\nabla g = 2x\,\mathbf{i} + 2y\,\mathbf{j} + 2z\,\mathbf{k}$. $\nabla T = \lambda \nabla g \Rightarrow 400yz^2\,\mathbf{i} + 400xz^2\,\mathbf{j} + 800xyz\,\mathbf{k} = \lambda(2x\,\mathbf{i} + 2y\,\mathbf{j} + 2z\,\mathbf{k}) \Rightarrow 400yz^2 = 2x\lambda$, $400xz^2 = 2y\lambda$, and $800xyz = 2z\lambda$. Solving this system yields the following points: $(0,\pm 1,0)$, $(\pm 1,0,0)$, $\left(\pm\frac{1}{2}, \pm\frac{1}{2}, \pm\frac{\sqrt{2}}{2}\right)$. $T(0,\pm 1,0) = 0$, $T(\pm 1,0,0) = 0$, $T\left(\pm\frac{1}{2}, \pm\frac{1}{2}, \pm\frac{\sqrt{2}}{2}\right) = \pm 50$. $\therefore$ 50 is the maximum at $\left(\frac{1}{2}, \frac{1}{2}, \pm\frac{\sqrt{2}}{2}\right)$ and $\left(-\frac{1}{2}, -\frac{1}{2}, \pm\frac{\sqrt{2}}{2}\right)$; $-50$ is the minimum at $\left(\frac{1}{2}, -\frac{1}{2}, \pm\frac{\sqrt{2}}{2}\right)$ and $\left(-\frac{1}{2}, \frac{1}{2}, \pm\frac{\sqrt{2}}{2}\right)$.

# CHAPTER 14

# MULTIPLE INTEGRALS

## 14.1 DOUBLE INTEGRALS

1. $$\int_0^3 \int_0^2 \left(4 - y^2\right) dy\, dx = \int_0^3 \left[4y - \frac{y^3}{3}\right]_0^2 dx = \frac{16}{3}\int_0^3 dx = 16$$

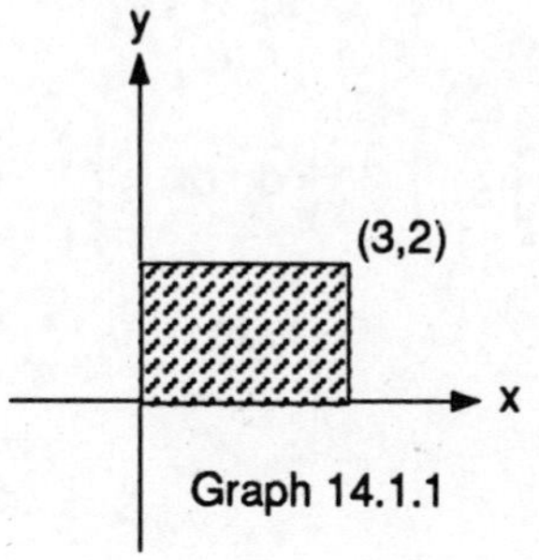

Graph 14.1.1

3. $$\int_0^3 \int_{-2}^0 \left(x^2y - 2xy\right) dy\, dx = \int_0^3 \left[\frac{x^2y^2}{2} - xy^2\right]_{-2}^0 dx =$$

$$\int_0^3 \left(4x - 2x^2\right) dx = \left[2x^2 - \frac{2x^3}{3}\right]_0^3 = 0$$

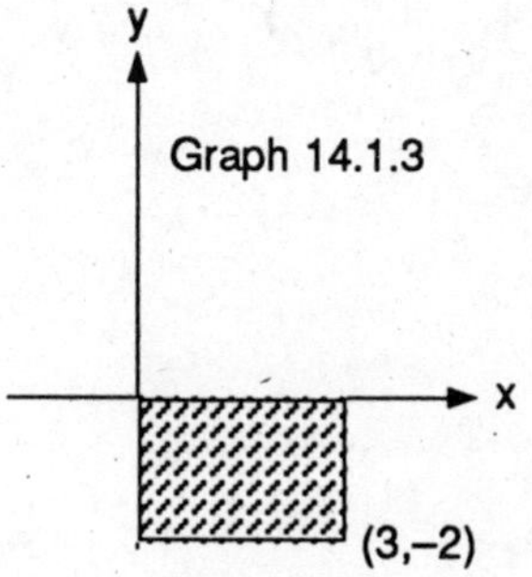

Graph 14.1.3

5. $$\int_0^\pi \int_0^x (x \sin y)\, dy\, dx = \int_0^\pi [-x \cos y]_0^x dx =$$

$$\int_0^\pi (x - x \cos x)\, dx = \frac{\pi^2}{2} + 2$$

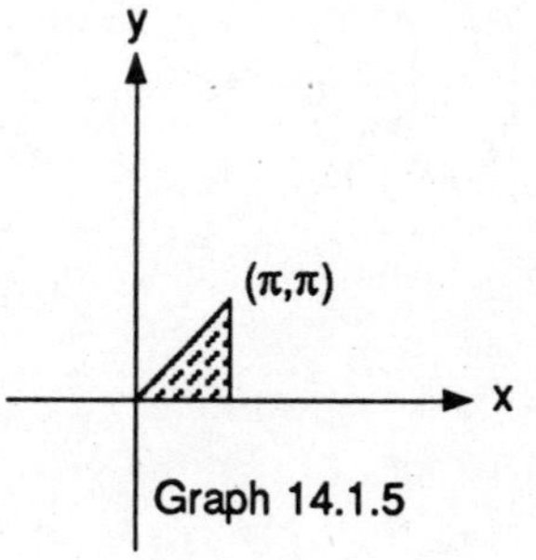

Graph 14.1.5

7. $$\int_0^\pi \int_0^{\sin x} y\, dy\, dx = \int_0^\pi \left[\frac{y^2}{2}\right]_0^{\sin x} dx =$$

$$\frac{1}{4}\int_0^\pi (1 - \cos 2x)\, dx = \frac{\pi}{4}$$

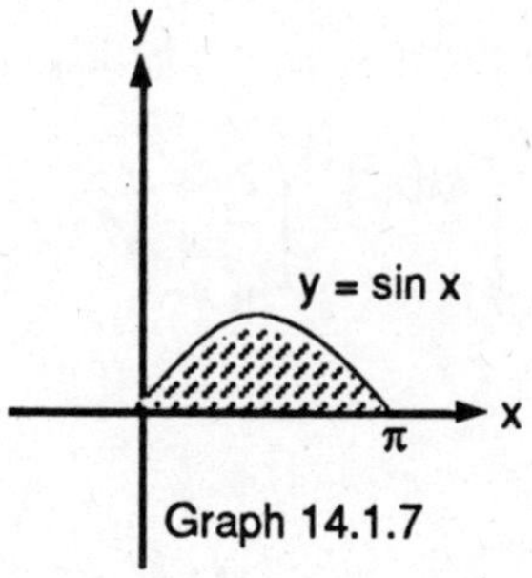

Graph 14.1.7

9. $\int_1^2 \int_x^{2x} \frac{x}{y}\, dy\, dx = \int_1^2 [x \ln y]_x^{2x}\, dx = \ln(2)\int_1^2 x\, dx = \frac{\ln 8}{2}$

11. $\int_0^1 \int_0^{1-x} y - \sqrt{x}\, dy\, dx = \int_0^1 \left[\frac{y^2}{2} - y\sqrt{x}\right]_0^{1-x} dx = \int_0^1 \frac{1 - 2x + x^2}{2} - \sqrt{x}(1 - x)\, dx = -\frac{1}{10}$

13. $\int_1^2 \int_1^2 \frac{1}{xy}\, dy\, dx = \int_1^2 \frac{1}{x}(\ln 2 - \ln 1)\, dx = \ln 2 \int_1^2 \frac{1}{x}\, dx = (\ln 2)^2$

15. $\int_0^4 \int_0^{(4-y)/2} dx\, dy$

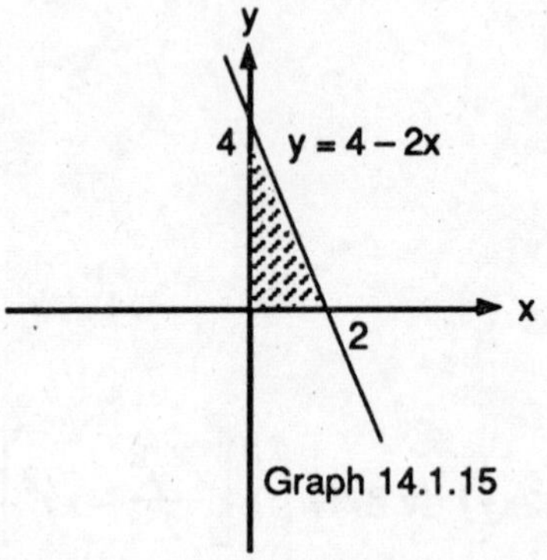

Graph 14.1.15

17. $\int_0^1 \int_{x^2}^{x} dy\, dx$

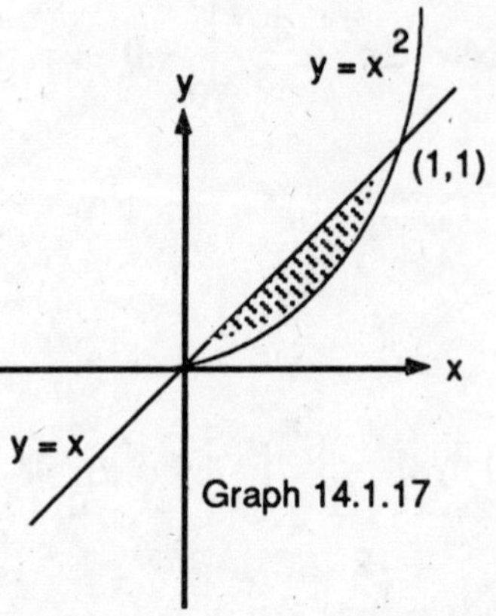

Graph 14.1.17

19. $\int_1^{\exp(2)} \int_{\ln y}^{2} dx\, dy$

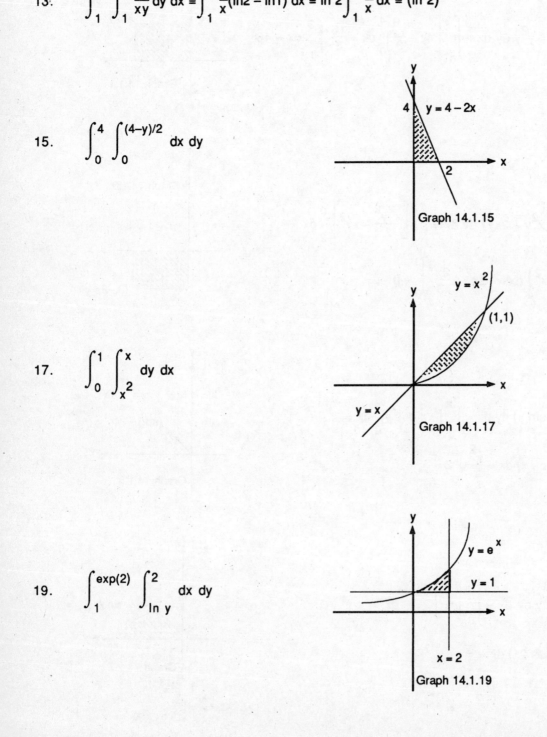

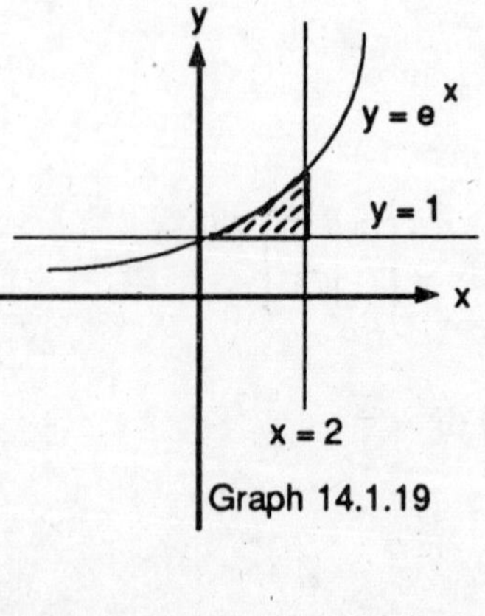

Graph 14.1.19

21. $\int_0^1 \int_1^{\exp(x)} dy\,dx = \int_1^e \int_{\ln y}^1 dx\,dy$

23. $\int_0^2 \int_0^{x^3} f(x,y)\,dy\,dx = \int_0^8 \int_{\sqrt[3]{y}}^2 f(x,y)\,dx\,dy$

25. $\int_0^\pi \int_x^\pi \frac{\sin y}{y}\,dy\,dx = \int_0^\pi \int_0^y \frac{\sin y}{y}\,dx\,dy = \int_0^\pi \sin y\,dy = 2$

27. $\int_0^2 \int_x^2 2y^2 \sin xy\,dy\,dx = \int_0^2 \int_0^y 2y^2 \sin xy\,dx\,dy = \int_0^2 [-2y \cos xy]_0^y\,dy =$

$\int_0^2 -\cos y^2(2y) + 2y\,dy = 4 - \sin 4$

29. $V = \int_{-4}^1 \int_{3x}^{4-x^2} (x + 4)\,dy\,dx = \int_{-4}^1 \left[(xy + 4y)\right]_{3x}^{4-x^2} dx = \int_{-4}^1 \left(-x^3 - 7x^2 - 8x + 16\right) dx = \frac{625}{12}$

31. $V = \int_0^2 \int_0^3 4 - y^2\,dx\,dy = \int_0^2 \left[4x - y^2x\right]_0^3 dy = \int_0^2 \left(12 - 3y^2\right) dy = 16$

33. $\int_1^3 \int_1^x \frac{1}{xy}\,dy\,dx = 0.603$, use the Calculus Toolkit

35. $\int_0^1 \int_0^1 \tan^{-1} xy\,dy\,dx = 0.233$, use the Calculus Toolkit

## 14.2 AREA, MOMENTS, AND CENTERS OF MASS

1. $\int_0^2 \int_2^{2-x} dy\,dx = \int_0^2 2 - x\,dx = 2$

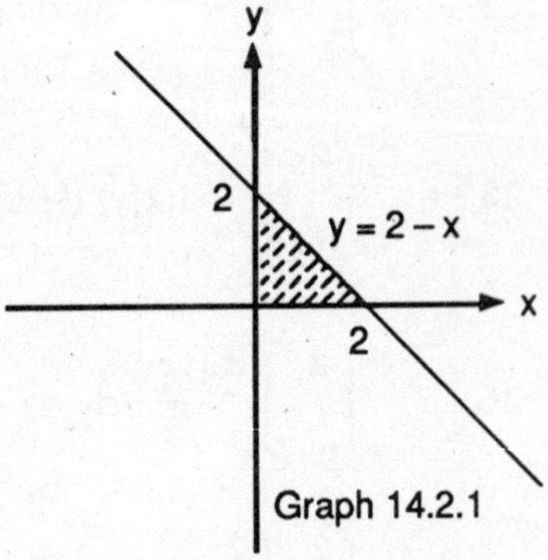

Graph 14.2.1

3. $\int_0^2 \int_{2x}^4 dy\,dx = \int_0^2 4 - 2x\,dx = 4$

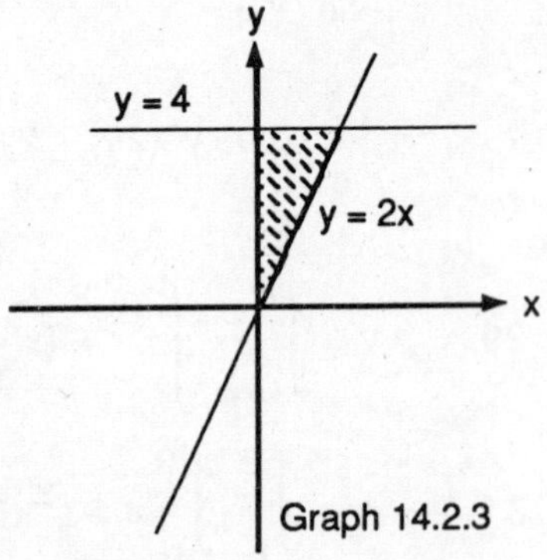

Graph 14.2.3

5. $\int_0^1 \int_{y^2}^{2y-y^2} dx\,dy = \int_0^1 2y - 2y^2\,dy = \frac{1}{3}$

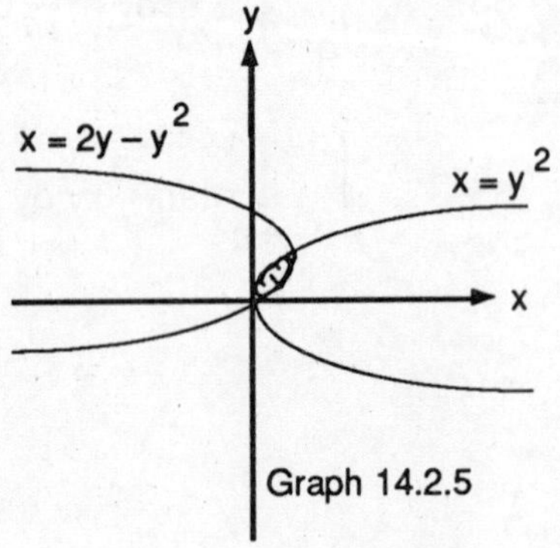

Graph 14.2.5

7. $2\int_0^1 \int_{-1}^{2\sqrt{1-x^2}} dy\,dx = 2\int_0^1 2\sqrt{1-x^2} + 1\,dx = \pi + 2$

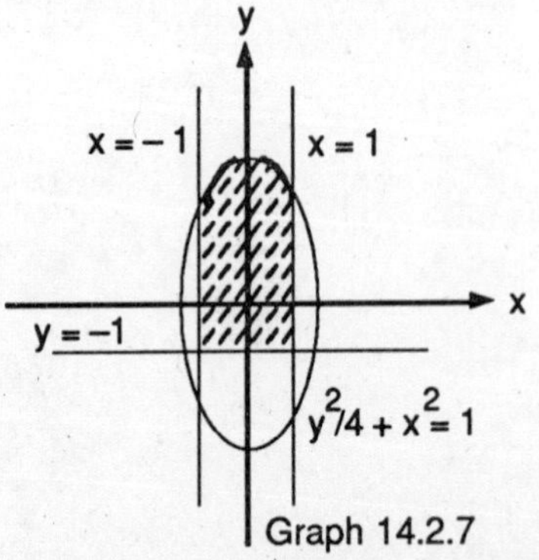

Graph 14.2.7

9. $\int_0^6 \int_{y^2/3}^{2y} dx\, dy = \int_0^6 \left(2y - y^2/3\right) dy = 12$

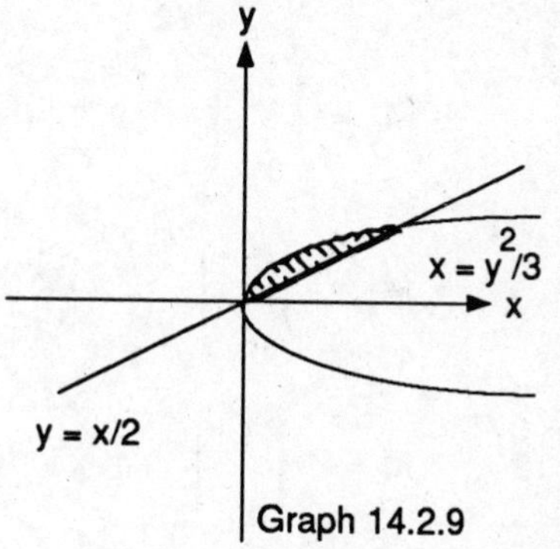

Graph 14.2.9

11. $\int_0^{\pi/4} \int_{\sin x}^{\cos x} dy\, dx = \int_0^{\pi/4} (\cos x - \sin x)\, dx = \sqrt{2} - 1$

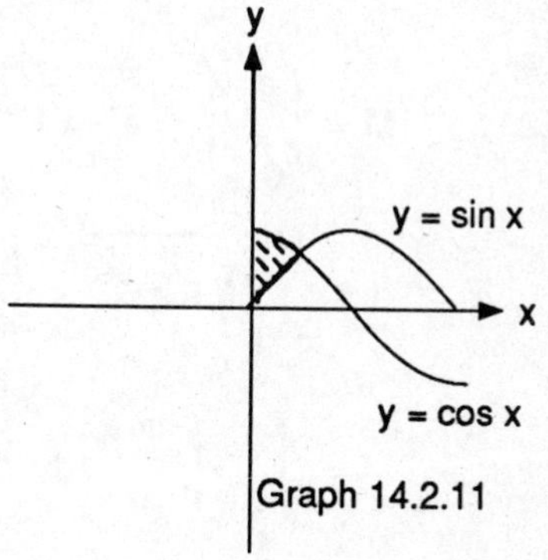

Graph 14.2.11

13. $\int_{-1}^{0} \int_{-2x}^{1-x} dy\, dx + \int_0^2 \int_{-x/2}^{1-x} dy\, dx = \int_{-1}^{0} (1 + x)\, dx + \int_0^2 (1 - x/2)\, dx = \frac{3}{2}$

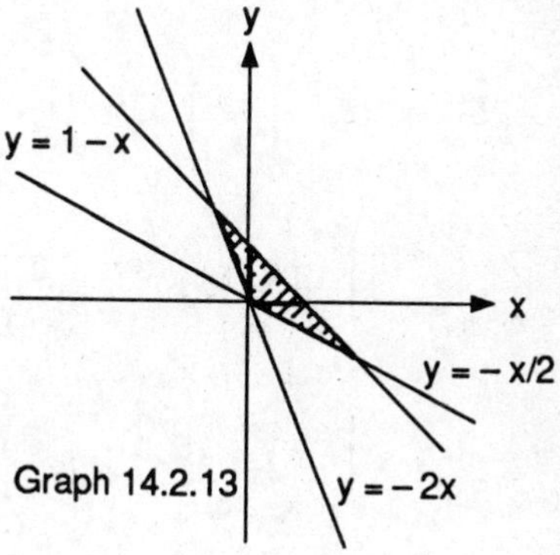

Graph 14.2.13

15. $M = \int_0^1 \int_x^{2-x^2} 3\, dy\, dx = 3\int_0^1 2 - x^2 - x\, dx = \frac{7}{2}$

$M_y = \int_0^1 \int_x^{2-x^2} 3x\, dy\, dx = 3\int_0^1 [xy]_x^{2-x^2} dx = 3\int_0^1 \left(2x - x^3 - x^2\right) dx = \frac{5}{4}$

$M_x = \int_0^1 \int_x^{2-x^2} 3y\, dy\, dx = \frac{3}{2}\int_0^1 \left[y^2\right]_x^{2-x^2} dx = \frac{3}{2}\int_0^1 4 - 5x^2 + x^4\, dx = \frac{19}{5} \Rightarrow \overline{x} = \frac{5}{14}, \overline{y} = \frac{38}{35}$

17. $M = \int_0^2 \int_{y^2/2}^{4-y} dx\, dy = \int_0^2 4 - y - \frac{y^2}{2}\, dy = \frac{14}{3}$

$M_y = \int_0^2 \int_{y^2/2}^{4-y} x\, dx\, dy = \frac{1}{2}\int_0^2 \left[x^2\right]_{y^2/2}^{4-y} dx = \frac{1}{2}\int_0^2 16 - 8y + y^2 - \frac{y^4}{4}\, dy = \frac{128}{15}$

$M_x = \int_0^2 \int_{y^2/2}^{4-y} y\, dx\, dy = \int_0^2 y\left(4 - y - \frac{y^2}{2}\right) dy = \frac{10}{3} \Rightarrow \overline{x} = \frac{64}{35},\ \overline{y} = \frac{5}{7}$

19. $M = 2\int_0^1 \int_0^{\sqrt{1-x^2}} dy\, dx = 2\int_0^1 \sqrt{1 - x^2}\, dx = \frac{\pi}{2}$

$M_x = 2\int_0^1 \int_0^{\sqrt{1-x^2}} y\, dy\, dx = \int_0^1 \left[y^2\right]_0^{\sqrt{1-x^2}} dx = \int_0^1 1 - x^2\, dx = \frac{2}{3} \Rightarrow \overline{x} = 0$, by symmetry; $\overline{y} = \frac{4}{3\pi}$

21. $M = \int_0^a \int_0^{\sqrt{a^2-x^2}} dy\, dx = \frac{\pi a^2}{4}$; $M_y = \int_0^a \int_0^{\sqrt{a^2-x^2}} x\, dy\, dx =$

$\int_0^a [xy]_0^{\sqrt{a^2-x^2}} dx = -\frac{1}{2}\int_0^a \sqrt{a^2 - y^2}(-2x)\, dx = \frac{a^3}{3} \Rightarrow \overline{x} = \overline{y} = \frac{4a}{3\pi}$, by symmetry

23. $M = \int_0^\pi \int_0^{\sin x} dy\, dx = \int_0^\pi \sin x\, dx = 2$; $M_x = \int_0^\pi \int_0^{\sin x} y\, dy\, dx = \frac{1}{2}\int_0^\pi \left[y^2\right]_0^{\sin x} dx =$

$\frac{1}{4}\int_0^\pi 1 - \cos 2x\, dx = \frac{\pi}{4} \Rightarrow \overline{x} = \frac{\pi}{2},\ \overline{y} = \frac{\pi}{8}$

25. $M = \int_0^2 \int_{-y}^{y-y^2} (x + y)\, dx\, dy = \int_0^2 \left[\frac{x^2}{2} + xy\right]_{-y}^{y-y^2} dy =$

$\int_0^2 \left(\frac{y^4}{2} - 2y^3 + 2y^2\right) dy = \frac{8}{15}$; $I_x = \int_0^2 \int_{-y}^{y-y^2} y^2(x + y)\, dx\, dy = \int_0^2 \left[\frac{x^2y^2}{2} + xy^3\right]_{-y}^{y-y^2} dy =$

$\int_0^2 \left(\frac{y^6}{2} - 2y^5 + 2y^4\right) dy = \frac{64}{105}$; $R_x = \sqrt{\frac{I_x}{M}} = \sqrt{\frac{8}{7}}$

27. $M = \int_0^1 \int_x^{2-x} (6x + 3y + 3)\, dy\, dx = \int_0^1 -12x^2 + 12\, dx = 8$; $M_y = \int_0^1 \int_x^{2-x} x(6x + 3y + 3)\, dy\, dx =$

$\int_0^1 12x - 12x^3\, dx = 3$; $M_x = \int_0^1 \int_x^{2-x} y(6x + 3y + 3)\, dy\, dx =$

$\int_0^1 14 - 6x - 6x^2 - 2x^3\, dx = \frac{17}{2} \Rightarrow \overline{x} = \frac{3}{8},\ \overline{y} = \frac{17}{16}$

$\overline{x} = \frac{8}{15},\ \overline{y} = \frac{8}{15}$; $I_x = \int_0^1 \int_{y^2}^{2y-y^2} y^2(y+1)\, dx\, dy = 2\int_0^1 y^3 - y^5\, dy = \frac{1}{6}$

29. $M = \int_0^1 \int_0^6 (x + y + 1)\,dx\,dy = \int_0^1 (6y + 24)\,dy = 27$

$M_x = \int_0^1 \int_0^6 y(x + y + 1)\,dx\,dy = \int_0^1 y(6y + 24)\,dy = 14$

$M_y = \int_0^1 \int_0^6 x(x + y + 1)\,dx\,dy = \int_0^1 (18y + 90)\,dy = 99 \Rightarrow \overline{x} = \frac{11}{3},\ \overline{y} = \frac{14}{27}$

$I_y = \int_0^1 \int_0^6 x^2(x + y + 1)\,dx\,dy = 216\int_0^1 \left(\frac{y}{3} + \frac{11}{6}\right) dy = 432 \Rightarrow R_y = \sqrt{\frac{I_y}{M}} = 4$

31. $M = \int_{-1}^1 \int_0^{x^2} (7y + 1)\,dy\,dx = \int_{-1}^1 \left(\frac{7x^4}{2} + x^2\right) dx = \frac{31}{15}$

$M_x = \int_{-1}^1 \int_0^{x^2} y(7y + 1)\,dy\,dx = \int_{-1}^1 \left(\frac{7x^6}{3} + \frac{x^4}{2}\right) dx = \frac{13}{15}$

$M_y = \int_{-1}^1 \int_0^{x^2} x(7y + 1)\,dy\,dx = \int_{-1}^1 \left(\frac{7x^5}{2} + x^3\right) dx = 0 \Rightarrow \overline{x} = 0,\ \overline{y} = \frac{13}{31}$

$I_y = \int_{-1}^1 \int_0^{x^2} x^2(7y + 1)\,dy\,dx = \int_{-1}^1 x^2\left(\frac{7x^4}{2} + x^2\right) dx = \frac{7}{5} \Rightarrow R_y = \sqrt{\frac{I_y}{M}} = \sqrt{\frac{21}{31}}$

33. $M = \int_0^1 \int_{-y}^{y} (1 + y)\,dx\,dy = \int_0^1 \left(2y^2 + 2y\right) dy = \frac{5}{3}$

$M_x = \int_0^1 \int_{-y}^{y} y(1 + y)\,dx\,dy = 2\int_0^1 \left(y^3 + y^2\right) dy = \frac{7}{6}$

$M_y = \int_0^1 \int_{-y}^{y} x(1 + y)\,dx\,dy = \int_0^1 0\,dy = 0 \Rightarrow \overline{x} = 0, \overline{y} = \frac{7}{10}$

$I_x = \int_0^1 \int_{-y}^{y} y^2(1 + y)\,dx\,dy = \int_0^1 y^2\left(2y^2 + 2y\right) dy = \frac{9}{10} \Rightarrow R_x = \sqrt{\frac{I_x}{M}} = \frac{3\sqrt{6}}{10}$

$I_y = \int_0^1 \int_{-y}^{y} x^2(1 + y)\,dx\,dy = \frac{1}{3}\int_0^1 y^2\left(2y^2 + 2y\right) dy = \frac{3}{10} \Rightarrow R_y = \sqrt{\frac{I_y}{M}} = \frac{3\sqrt{2}}{10}$

$I_o = I_x + I_y = \frac{6}{5}$ and $R_o = \sqrt{\frac{I_o}{M}} = \frac{3\sqrt{2}}{5}$

35. $f(a) = I_a = \int_0^4 \int_0^2 (y - a)^2\,dy\,dx = \int_0^4 \left(\frac{(2 - a)^3}{3} + \frac{a^3}{3}\right) dx = \frac{4}{3}\left[(2 - a)^3 + a^3\right] \Rightarrow f'(a) = 0 \Rightarrow$

$a = 1$, $f''(a) = 16 > 0 \Rightarrow f(1)$ is minimum

## 14.3 DOUBLE INTEGRALS IN POLAR FORM

1. $\int_{-1}^{1}\int_{0}^{\sqrt{1-x^2}} dy\,dx = \int_{0}^{\pi}\int_{0}^{1} r\,dr\,d\theta = \frac{1}{2}\int_{0}^{\pi} d\theta = \frac{\pi}{2}$

3. $\int_{0}^{1}\int_{0}^{\sqrt{1-y^2}} \left(x^2+y^2\right) dx\,dy = \int_{0}^{\pi/2}\int_{0}^{1} r^3\,dr\,d\theta = \frac{1}{4}\int_{0}^{\pi/2} d\theta = \frac{\pi}{8}$

5. $\int_{-a}^{a}\int_{-\sqrt{a^2-x^2}}^{\sqrt{a^2-x^2}} dy\,dx = \int_{0}^{2\pi}\int_{0}^{a} r\,dr\,d\theta = \frac{a^2}{2}\int_{0}^{2\pi} d\theta = \pi a^2$

7. $\int_{0}^{\pi/4}\int_{0}^{\sqrt{2}} \cos\theta\, r^2\,dr\,d\theta = \frac{2\sqrt{2}}{3}\int_{0}^{\pi/4} \cos\theta\,d\theta = \frac{2}{3}$

9. $\int_{0}^{3}\int_{0}^{\sqrt{3}x} \frac{1}{\sqrt{x^2+y^2}}\,dy\,dx = \int_{0}^{\pi/3}\int_{0}^{3\sec\theta} dr\,d\theta = \int_{9}^{\pi/3} 3\sec\theta\,d\theta = 3\ln\left(2+\sqrt{3}\right)$

11. $\int_{0}^{1}\int_{0}^{\sqrt{1-x^2}} 5\sqrt{x^2+y^2}\,dy\,dx = \int_{0}^{\pi/2}\int_{0}^{1} 5r^2\,dr\,d\theta = \frac{5}{3}\int_{0}^{\pi/2} d\theta = \frac{5\pi}{6}$

13. $\int_{0}^{\pi/2}\int_{0}^{2\sqrt{2-\sin 2\theta}} r\,dr\,d\theta = 2\int_{0}^{\pi/2}\left(2-\sin 2\theta\right) d\theta = 2(\pi-1)$

15. $2\int_{0}^{\pi/6}\int_{0}^{12\cos 3\theta} r\,dr\,d\theta = 144\int_{0}^{\pi/6} \cos^2 3\theta\,d\theta = 12\pi$

17. $A = \int_{0}^{\pi/2}\int_{0}^{1+\sin\theta} r\,dr\,d\theta = \frac{1}{2}\int_{0}^{\pi/2} \frac{3}{2} + 2\sin\theta - \frac{\cos 2\theta}{2}\,d\theta = \frac{3\pi+8}{8}$

19. $\int_{0}^{2\pi}\int_{0}^{\sqrt{3}/2} \frac{1}{1-r^2}\,r\,dr\,d\theta = \ln(2)\int_{0}^{2\pi} d\theta = \pi\ln 4$

21. $M_x = \int_{0}^{\pi}\int_{0}^{1-\cos\theta} 3r^2\sin\theta\,dr\,d\theta = \int_{0}^{\pi}\left(1-\cos\theta\right)^3\sin\theta\,d\theta = 4$

23. $M = 2\int_{0}^{\pi}\int_{0}^{1+\cos\theta} r\,dr\,d\theta = \int_{0}^{\pi}\left(1+\cos\theta\right)^2 d\theta = \frac{3\pi}{2}$

$M_y = \int_{0}^{2\pi}\int_{0}^{1+\cos\theta} \cos\theta\, r^2\,dr\,d\theta = \int_{0}^{2\pi} \frac{4\cos\theta}{3} + \frac{15}{24} + \cos 2\theta - \sin^2\theta\cos\theta + \frac{\cos 4\theta}{4}\,d\theta =$

$\frac{5\pi}{4} \Rightarrow \overline{x} = \frac{5}{6}$, $\overline{y} = 0$ by symmetry

25. $M = \int_{-\infty}^{0}\int_{0}^{\exp(x)} dy\,dx = \int_{-\infty}^{0} e^x\,dx = \lim_{t\to-\infty}\int_{t}^{0} e^x\,dx = 1$

$M_y = \int_{-\infty}^{0}\int_{0}^{\exp(x)} x\,dy\,dx = \int_{-\infty}^{0} x e^x\,dx = \lim_{t\to-\infty}\int_{t}^{0} x e^x\,dx = -1$

$M_x = \int_{-\infty}^{0}\int_{0}^{\exp(x)} y\,dy\,dx = \frac{1}{2}\int_{-\infty}^{0} e^{2x}\,dx = \frac{1}{2}\lim_{t\to-\infty}\int_{t}^{0} e^{2x}\,dx = \frac{1}{4} \Rightarrow \overline{x} = -1,\ \overline{y} = \frac{1}{4}$

## 14.4 TRIPLE INTEGRALS IN RECTANGULAR COORDINATES

1. $\int_0^1 \int_0^{1-z} \int_0^2 dx\,dy\,dz = 2\int_0^1 \int_0^{1-z} dy\,dz = 2\int_0^1 1 - z\,dz = 1$

3. $\int_0^1 \int_0^{2-2x} \int_0^{3-3x-3y/2} dz\,dy\,dx = \int_0^1 \int_0^{2-2x} 3 - 3x - \frac{3}{2}y\,dy\,dx = \int_0^1 3 - 6x + 3x^2\,dx = 1,$

$\int_0^2 \int_0^{x-y/2} \int_0^{3-3x-3y/2} dz\,dx\,dy,\quad \int_0^1 \int_0^{3-3x} \int_0^{2-2x-2z/3} dy\,dz\,dx,$

$\int_0^3 \int_0^{1-z/3} \int_0^{2-2x-2z/3} dy\,dx\,dz,\quad \int_0^2 \int_0^{3-3y/2} \int_0^{1-y/2-z/3} dx\,dz\,dy,$

$\int_0^3 \int_0^{2-2z/3} \int_0^{1-y/2-z/3} dx\,dy\,dz$

5. $\int_0^1 \int_0^1 \int_0^1 x^2 + y^2 + z^2\,dz\,dy\,dx = \int_0^1 \int_0^1 \left(x^2 + y^2 + \frac{1}{3}\right) dy\,dx = \int_0^1 x^2 + \frac{2}{3}\,dx = 1$

7. $\int_1^e \int_1^e \int_1^e \frac{1}{xyz}\,dx\,dy\,dz = \int_1^e \int_1^e \frac{\ln x}{yz}\,dy\,dz = \int_1^e \frac{1}{z}\,dz = 1$

9. $\int_0^1 \int_0^{\pi} \int_0^{\pi} y \sin z\,dx\,dy\,dz = \int_0^1 \int_0^{\pi} \pi y \sin z\,dy\,dz = \frac{\pi^3}{2}\int_0^1 \sin z\,dz = \frac{\pi^3}{2}(1 - \cos 1)$

11. $\int_0^3 \int_0^{\sqrt{9-x^2}} \int_0^{\sqrt{9-x^2}} dz\,dy\,dx = \int_0^3 \int_0^{\sqrt{9-x^2}} \sqrt{9 - x^2}\,dy\,dx = \int_0^3 \left(9 - x^2\right) dx = 18$

13. $\int_0^1 \int_0^{2-x} \int_0^{2-x-y} dz\,dy\,dx = \int_0^1 \int_0^{2-x} 2 - x - y\,dy\,dx = \int_0^1 \frac{x^2}{2} - 2x + 2\,dx = \frac{7}{6}$

15. a) $\int_{-1}^1 \int_0^{1-x^2} \int_{x^2}^{1-z} dy\,dz\,dx$ b) $\int_0^1 \int_{-\sqrt{1-z}}^{\sqrt{1-z}} \int_{x^2}^{1-z} dy\,dx\,dz$

c) $\int_0^1 \int_0^{1-z} \int_{-\sqrt{y}}^{\sqrt{y}} dx\,dy\,dz$ d) $\int_0^1 \int_0^{1-y} \int_{-\sqrt{y}}^{\sqrt{y}} dx\,dz\,dy$

e) $\int_0^1 \int_{-\sqrt{y}}^{\sqrt{y}} \int_0^{1-y} dz\,dx\,dy$

17. $V = \int_0^1 \int_{-1}^1 \int_0^{y^2} dz\,dy\,dx = \int_0^1 \int_{-1}^1 y^2\,dy\,dx = \frac{2}{3}\int_0^1 dx = \frac{2}{3}$

19. $V = \int_0^4 \int_0^{\sqrt{4-x}} \int_0^{2-y} dz\,dy\,dx = \int_0^4 \int_0^{\sqrt{4-x}} (2 - y)\,dy\,dx = \int_0^4 2\sqrt{4-x} - \left(\frac{4 - x}{2}\right) dx = \frac{20}{3}$

21. $V = \int_0^1 \int_0^{2-2x} \int_0^{3-3x-3y/2} dz\,dy\,dx = \int_0^1 \int_0^{2-2x} 3 - 3x - \frac{3}{2}y\,dy\,dx = \int_0^1 3 - 6x + 3x^2\,dx = 1$

23. $V = 8\int_0^1\int_0^{\sqrt{1-x^2}}\int_0^{\sqrt{1-x^2}} dz\,dy\,dx = 8\int_0^1\int_0^{\sqrt{1-x^2}}\sqrt{1-x^2}\,dy\,dx = 8\int_0^1 1-x^2\,dx = \frac{16}{3}$

25. $V = \int_0^4\int_0^{\left(\sqrt{16-y^2}\right)/2}\int_0^{4-y} dx\,dz\,dy = \int_0^4\int_0^{\left(\sqrt{16-y^2}\right)/2}(4-y)\,dz\,dy =$

$\int_0^4 \frac{\sqrt{16-y^2}}{2}(4-y)\,dy = 8\pi + \frac{32}{3}$

27. $\text{average} = \frac{1}{8}\int_0^2\int_0^2\int_0^2 x^2+9\,dz\,dy\,dx = \frac{1}{8}\int_0^2\int_0^2\left(2x^2+18\right)dy\,dx = \frac{1}{8}\int_0^2\left(4x^2+36\right)dx = \frac{31}{3}$

29. $\text{average} = \int_0^1\int_0^1\int_0^1 x^2+y^2+z^2\,dz\,dy\,dx = \int_0^1\int_0^1\left(x^2+y^2+\frac{1}{3}\right)dy\,dx = \int_0^1\left(x^2+\frac{2}{3}\right)dx = 1$

## 14.5 MASSES AND MOMENTS IN THREE DIMENSIONS

1. $I_x = \int_{-c/2}^{c/2}\int_{-b/2}^{b/2}\int_{-a/2}^{a/2}\left(y^2+z^2\right)dx\,dy\,dz = 4a\int_0^{c/2}\int_0^{b/2}\left(y^2+z^2\right)dy\,dz =$

$4a\int_0^{c/2}\left(\frac{b^3}{24}+\frac{z^2b}{2}\right)dx = \frac{abc}{12}\left(b^2+c^2\right) \Rightarrow I_x = \frac{M}{12}\left(b^2+c^2\right);\ R_x = \sqrt{\frac{b^2+c^2}{12}},$

$R_y = \sqrt{\frac{a^2+c^2}{12}},\ R_z = \sqrt{\frac{a^2+b^2}{12}}$

3. $I_x = \int_0^a\int_0^b\int_0^c\left(y^2+z^2\right)dz\,dy\,dx = \int_0^a\int_0^b cy^2+\frac{c^3}{3}\,dy\,dx = \int_0^a\frac{cb^3}{3}+\frac{c^3b}{3}\,dx = \frac{abc\left(b^2+c^2\right)}{3} =$

$\frac{M}{3}\left(b^2+c^2\right);\ I_y = \frac{M}{3}\left(a^2+c^2\right)$ and $I_z = \frac{M}{3}\left(a^2+b^2\right)$ by symmetry

5. $M = 4\int_0^1\int_0^1\int_{4y^2}^4 dz\,dy\,dx = 4\int_0^1\int_0^1 4-4y^2\,dy\,dx = 16\int_0^1\frac{2}{3}\,dx = \frac{32}{3}$

$M_{xy} = 4\int_0^1\int_0^1\int_{4y^2}^4 z\,dz\,dy\,dx = 2\int_0^1\int_0^1\left(16-16y^4\right)dy\,dx = \frac{128}{5}\int_0^1 dx = \frac{128}{5} \Rightarrow$

$\overline{z} = \frac{12}{5},\ \overline{x} = \overline{y} = 0$ by symmetry; $I_z = 4\int_0^1\int_0^1\int_{4y^2}^4\left(x^2+y^2\right)dz\,dy\,dx =$

$16\int_0^1\int_0^1 x^2-x^2y^2+y^2-y^4\,dy\,dx = 16\int_0^1\frac{2x^2}{3}+\frac{2}{15}\,dx = \frac{256}{45};$

$I_x = 4\int_0^1 \int_0^1 \int_{4y^2}^4 (y^2 + z^2)\, dz\, dy\, dx = 4\int_0^1 \int_0^1 \left(4y^2 + \frac{64}{3}\right) - \left(4y^4 + \frac{64y^6}{3}\right) dy\, dx =$

$4\int_0^1 \frac{1976}{105}\, dx = \frac{7904}{105}$; $I_y = 4\int_0^1 \int_0^1 \int_{4y^2}^4 (x^2 + z^2)\, dz\, dy\, dx =$

$4\int_0^1 \int_0^1 \left(4x^2 + \frac{64}{3}\right) - \left(4x^2y^2 + \frac{64y^6}{3}\right) dy\, dx = 4\int_0^1 \frac{8}{3}x^2 + \frac{128}{7}\, dx = \frac{4832}{63}$

7. a) $M = 4\int_0^2 \int_0^{\sqrt{4-x^2}} \int_{x^2+y^2}^4 dz\, dy\, dx = 4\int_0^{\pi/2} \int_0^2 \int_{r^2}^4 r\, dz\, dr\, d\theta =$

$4\int_0^{\pi/2} \int_0^2 4r - r^3\, dr\, d\theta = 4\int_0^{\pi/2} 4\, d\theta = 8\pi$; $M_{xy} = \int_0^{2\pi} \int_0^2 \int_{r^2}^4 r\, dz\, dr\, d\theta =$

$\int_0^{2\pi} \int_0^2 \frac{r}{2}\left(16 - r^4\right) dr\, d\theta = \frac{32\pi}{3}\int_0^{2\pi} d\theta = \frac{64\pi}{3} \Rightarrow \overline{z} = \frac{8}{3}$, $\overline{x} = \overline{y} = 0$ by symmetry

b) $M = 8\pi$; $4\pi = \int_0^{2\pi} \int_0^{\sqrt{c}} \int_{r^2}^c r\, dz\, dr\, d\theta = \int_0^{2\pi} \int_0^{\sqrt{c}} cr - r^3\, dr\, d\theta = \int_0^{2\pi} \left(\frac{c^2}{4}\right) d\theta = \frac{c^2\pi}{2} \Rightarrow$

$c^2 = 8 \Rightarrow c = 2\sqrt{2}$, since $c > 0$

9. $I_L = \int_{-2}^2 \int_{-2}^4 \int_{-1}^{1-y/2} \left((y-6)^2 + z^2\right) dz\, dy\, dx = \int_{-2}^2 \int_{-2}^4 \frac{(y-6)^2(4-y)}{2} + \frac{(2-y)^3}{24} + \frac{1}{3}\, dy\, dx =$

$4\int_{-2}^4 \frac{13t^3}{24} + 5t^2 + 16t + \frac{49}{3}\, dt = 1386$, where $t = 2 - y$; $M = 36$, $R_L = \sqrt{\frac{I_L}{M}} = \sqrt{\frac{77}{2}}$

11. $M = 8$, $I_L = \int_0^4 \int_0^2 \int_0^1 \left(z^2 + (y-2)^2\right) dz\, dy\, dx = \int_0^4 \int_0^2 y^2 - 4y + \frac{13}{3}\, dy\, dx = \frac{10}{3}\int_0^4 dx = \frac{40}{3} \Rightarrow$

$R_L = \sqrt{\frac{I_L}{M}} = \sqrt{\frac{5}{3}}$

13. $M = \int_0^2 \int_0^{2-x} \int_0^{2-x-y} 2x\, dz\, dy\, dx = \int_0^2 \int_0^{2-x} 4x - 2x^2 - 2xy\, dy\, dx = \int_0^2 x^3 - 4x^2 + 4x\, dx = \frac{4}{3}$

$M_{xy} = \int_0^2 \int_0^{2-x} \int_0^{2-x-y} 2xz\, dz\, dy\, dx = \int_0^2 \int_0^{2-x} x(2-x-y)^2\, dy\, dx = \int_0^2 \frac{x(2-x)^3}{3}\, dx = \frac{8}{15}$;

$M_{xz} = \frac{8}{15}$ by symmetry; $M_{yz} = \int_0^2 \int_0^{2-x} \int_0^{2-x-y} 2x^2\, dz\, dy\, dx = \int_0^2 \int_0^{2-x} 2x^2(2-x-y)\, dy\, dx =$

$\int_0^2 \left(2x - x^2\right)^2 dx = \frac{16}{15} \Rightarrow \overline{x} = \frac{4}{5}$, $\overline{y} = \overline{z} = \frac{2}{5}$

15. $M = \int_0^1\int_0^1\int_0^1 (x + y + z + 1)\,dz\,dy\,dx = \int_0^1\int_0^1 \left(x + y + \frac{3}{2}\right) dy\,dx = \int_0^1 (x + 2)\,dx = \frac{5}{2}$

$M_{xy} = \int_0^1\int_0^1\int_0^1 (x + y + z + 1)z\,dz\,dy\,dx = \frac{1}{2}\int_0^1\int_0^1 \left(x + y + \frac{5}{3}\right) dy\,dx =$

$\frac{1}{2}\int_0^1 \left(x + \frac{13}{6}\right) dx = \frac{4}{3} \Rightarrow M_{xy} = M_{yz} = M_{xz} = \frac{4}{3}$ by symmetry $\therefore\ \overline{x} = \overline{y} = \overline{z} = \frac{8}{15}$

$I_z = \int_0^1\int_0^1\int_0^1 (x + y + z + 1)\left(x^2 + y^2\right) dz\,dy\,dx = \int_0^1\int_0^1 \left(x + y + \frac{3}{2}\right)\left(x^2 + y^2\right) dy\,dx =$

$\int_0^1 x^3 + 2x^2 + \frac{1}{3}x + \frac{3}{4}\,dx = \frac{11}{6} \Rightarrow I_x = I_y = I_z = \frac{11}{6}$ by symmetry $\Rightarrow R_x = R_y = R_z = \sqrt{\frac{I_z}{M}} = \sqrt{\frac{11}{15}}$

## 14.6 TRIPLE INTEGRALS IN CYLINDRICAL AND SPHERICAL COORDINATES

1. $\int_0^{2\pi}\int_0^1\int_r^{\sqrt{2-r^2}} r\,dz\,dr\,d\theta = \int_0^{2\pi}\int_0^1 \left(\left(2 - r^2\right)^{1/2}r - r^2\right) dr\,d\theta = \int_0^{2\pi}\left(\frac{2^{3/2}}{3} - \frac{2}{3}\right) d\theta = \frac{4\pi\left(\sqrt{2} - 1\right)}{3}$

3. $\int_0^{2\pi}\int_0^{\theta/2\pi}\int_0^{3+24r^2} r\,dz\,dr\,d\theta = \int_0^{2\pi}\int_0^{\theta/2\pi}\left(3 + 24r^2\right) r\,dr\,d\theta = \frac{3}{2}\int_0^{2\pi} \frac{\theta^2}{4\pi^2} + \frac{4\theta^4}{16\pi^4}\,d\theta = \frac{17\pi}{5}$

5. $\int_0^{2\pi}\int_0^1\int_r^{\left(2-r^2\right)^{-1/2}} 3\,r\,dz\,dr\,d\theta = 3\int_0^{2\pi}\int_0^1 \left(2 - r^2\right)^{-1/2}r - r^2\,dr\,d\theta =$

$3\int_0^{2\pi}\left(\sqrt{2} - \frac{4}{3}\right) d\theta = \pi\left(6\sqrt{2} - 8\right)$

7. $\int_0^{\pi}\int_0^{\pi}\int_0^{2\sin\phi} \rho^2\sin\phi\,d\rho\,d\phi\,d\theta = \frac{8}{3}\int_0^{\pi}\int_0^{\pi} \sin^4\phi\,d\phi\,d\theta = \frac{2}{3}\int_0^{\pi} \frac{3\pi}{2}\,d\theta = \pi^2$

9. $\int_0^{2\pi}\int_0^{\pi}\int_0^{(1-\cos\phi)/2} \rho^2\sin\phi\,d\rho\,d\phi\,d\theta = \frac{1}{24}\int_0^{2\pi}\int_0^{\pi}\left(1 - \cos\phi\right)^3 \sin\phi\,d\phi\,d\theta =$

$\frac{1}{6}\int_0^{2\pi} d\theta = \frac{\pi}{3}$

11. $\int_0^{2\pi}\int_0^{\pi/3}\int_{\sec\phi}^{2} 3\rho^2\sin\phi\,d\rho\,d\phi\,d\theta = \int_0^{2\pi}\int_0^{\pi/3}\left(8 - \sec^3\phi\right)\sin\phi\,d\phi\,d\theta = \frac{5}{2}\int_0^{2\pi} d\theta = 5\pi$

13. a) $8\int_0^{\pi/2}\int_0^{\pi/2}\int_0^{2} \rho^2 \sin\phi \, d\rho \, d\phi \, d\theta$

b) $8\int_0^{\pi/2}\int_0^{2}\int_0^{\sqrt{4-r^2}} r \, dz \, dr \, d\theta$

c) $8\int_0^{2}\int_0^{\sqrt{4-x^2}}\int_0^{\sqrt{4-x^2-y^2}} dz \, dy \, dx$

15. $\int_{-\pi/2}^{\pi/2}\int_0^{\cos\theta}\int_0^{3r^2} f(r,\theta,z) \, r \, dz \, dr \, d\theta$

17. $V = 4\int_0^{\pi/2}\int_0^{1}\int_0^{r^2} r \, dz \, dr \, d\theta = 4\int_0^{\pi/2}\int_0^{1} r^3 \, dr \, d\theta = \int_0^{\pi/4} d\theta = \frac{\pi}{2}$

19. $V = 4\int_0^{\pi/2}\int_0^{2}\int_0^{4-r^2} r \, dz \, dr \, d\theta = 4\int_0^{\pi/2}\int_0^{2} \left(4r - r^3\right) dr \, d\theta = 16\int_0^{\pi/2} d\theta = 8\pi$

21. $V = 4\int_0^{\pi/2}\int_0^{1}\int_{4r^2}^{5-r^2} r \, dz \, dr \, d\theta = 4\int_0^{\pi/2}\int_0^{1} \left(5 - 5r^2\right) r \, dr \, d\theta = 5\int_0^{\pi/2} d\theta = \frac{5\pi}{2}$

23. $V = 8\int_0^{\pi/2}\int_0^{1}\int_0^{\sqrt{4-r^2}} r \, dz \, dr \, d\theta = 8\int_0^{\pi/2}\int_0^{1} \left(4 - r^2\right)^{1/2} r \, dr \, d\theta =$

$-\frac{8}{3}\int_0^{\pi/2} \left(3^{3/2} - 8\right) d\theta = \frac{4\pi\left(8 - 3\sqrt{3}\right)}{3}$

25. $\text{average} = \frac{1}{2\pi}\int_0^{2\pi}\int_0^{1}\int_{-1}^{1} r \, dz \, dr \, d\theta = \frac{1}{2\pi}\int_0^{2\pi}\int_0^{1} 2r^2 \, dr \, d\theta = \frac{1}{3\pi}\int_0^{2\pi} d\theta = \frac{2}{3}$

27. $M = 4\int_0^{\pi/2}\int_0^{1}\int_0^{r} r \, dz \, dr \, d\theta = 4\int_0^{\pi/2}\int_0^{1} r^2 \, dr \, d\theta = \frac{4}{3}\int_0^{\pi/2} d\theta = \frac{2\pi}{3}$

$M_{xy} = \int_0^{2\pi}\int_0^{1}\int_0^{r} r z \, dz \, dr \, d\theta = \frac{1}{2}\int_0^{2\pi}\int_0^{1} r^3 \, dr \, d\theta = \frac{1}{8}\int_0^{2\pi} d\theta = \frac{\pi}{4} \Rightarrow$

$\overline{z} = \frac{3}{8}, \ \overline{x} = \overline{y} = 0$ by symmetry

29. $M = 12\pi, \ I_z = \int_0^{2\pi}\int_1^{2}\int_0^{4} r^3 \, dz \, dr \, d\theta = 4\int_0^{2\pi}\int_1^{2} r^3 \, dr \, d\theta = 15\int_0^{2\pi} d\theta = 30\pi \Rightarrow$

$R_z = \sqrt{\frac{I_z}{M}} = \sqrt{\frac{5}{2}}$

$M_{xy} = = \int_0^{\pi/2}\int_0^{2}\int_0^{\sqrt{x^2+y^2}} r z \, dz \, dr \, d\theta = \frac{1}{2}\int_0^{\pi/2}\int_0^{2} r^2 \, dr \, d\theta = \frac{4}{3}\int_0^{\pi/2} d\theta = \frac{2\pi}{3} \Rightarrow \overline{z} = \frac{5}{6\sqrt{2}}$

31. a) $I_z = \int_0^{2\pi}\int_0^1\int_{-1}^1 r^2\,dz\,dr\,d\theta = 2\int_0^{2\pi}\int_0^1 r^3\,dr\,d\theta = \frac{1}{2}\int_0^{2\pi} d\theta = \pi$

b) $I_x = \int_0^{2\pi}\int_0^1\int_{-1}^1 \left(r^2\sin^2\theta + z^2\right)dz\,dr\,d\theta = \int_0^{2\pi}\int_0^1 \left(2r^2\sin^2\theta + \frac{2}{3}\right)r\,dr\,d\theta =$

$\int_0^{2\pi}\left(\frac{\sin^2\theta}{2} + \frac{1}{3}\right)d\theta = \frac{7\pi}{6}$

33. $I_z = \int_0^{2\pi}\int_0^a\int_{-\sqrt{a^2-r^2}}^{\sqrt{a^2-r^2}} r^3\,dz\,dr\,d\theta = -\int_0^{2\pi}\int_0^a r^2\left(a^2-r^2\right)^{1/2}(-2r)\,dr\,d\theta = \int_0^{2\pi}\frac{4a^5}{15}\,d\theta = \frac{8\pi a^5}{15}$

35. $\int_0^{2\pi}\int_{\pi/3}^{2\pi/3}\frac{a^3}{3}\sin\phi\,d\phi\,d\theta = \frac{a^3}{3}\int_0^{2\pi}d\theta = \frac{2\pi a^3}{3} \Rightarrow V = \frac{4}{3}\pi a^3 - \frac{2\pi a^3}{3} = \frac{2\pi a^3}{3}$

37. $V = \int_0^{2\pi}\int_0^{\pi/3}\int_{\sec\phi}^2 \rho^2\sin\phi\,d\rho\,d\phi\,d\theta = \frac{1}{3}\int_0^{2\pi}\int_0^{\pi/3} 8\sin\phi - \tan\phi\sec^2\phi\,d\phi\,d\theta =$

$\frac{5}{6}\int_0^{2\pi}d\theta = \frac{5\pi}{3}$

39. $\text{average} = \frac{3}{4\pi}\int_0^{2\pi}\int_0^{\pi}\int_0^1 \rho^3\sin\phi\,d\rho\,d\phi\,d\theta = \frac{3}{16\pi}\int_0^{2\pi}\int_0^{\pi}\sin\phi\,d\phi\,d\theta = \frac{3}{8\pi}\int_0^{2\pi}d\theta = \frac{3}{4}$

41. $M = \int_0^{2\pi}\int_0^{\pi/4}\int_0^a \rho^2\sin\phi\,d\rho\,d\phi\,d\theta = \frac{a^3}{3}\int_0^{2\pi}\int_0^{\pi/4}\sin\phi\,d\phi\,d\theta =$

$\frac{a^3}{3}\int_0^{2\pi}\frac{\sqrt{2}-1}{\sqrt{2}}\,d\theta = \frac{\pi a^3\left(2-\sqrt{2}\right)}{3}$

$M_{xy} = \int_0^{2\pi}\int_0^{\pi/4}\int_0^a \rho^3\sin\phi\cos\phi\,d\rho\,d\phi\,d\theta = \frac{a^4}{4}\int_0^{2\pi}\int_0^{\pi/4}\sin\phi\cos\phi\,d\phi\,d\theta =$

$\frac{a^4}{16}\int_0^{2\pi}d\theta = \frac{\pi a^4}{8} \Rightarrow \bar{z} = \frac{3\left(2+\sqrt{2}\right)a}{16}$, $\bar{x} = \bar{y} = 0$ by symmetry

## 14.7 SUBSTITUTIONS IN MULTIPLE INTEGRALS

1. $\int_0^4\int_{y/2}^{1+y/2}\frac{2x-y}{2}\,dx\,dy = \int_0^4\left[\frac{x^2}{2} - \frac{xy}{2}\right]_{y/2}^{1+y/2}dy = \frac{1}{2}\int_0^4 dy = 2$

3. a) $x = \frac{u+v}{3}$, $y = \frac{v-2u}{3}$, $J(u,v) = \begin{vmatrix} 1/3 & 1/3 \\ -2/3 & 1/3 \end{vmatrix} = \frac{1}{9} + \frac{2}{9} = \frac{1}{3}$

b) $\int_4^7\int_{-1}^2\frac{vu}{3}\,dv\,du = \frac{1}{2}\int_4^7 u\,du = \frac{33}{4}$

5. $J(u,v) = \begin{vmatrix} v^{-1} & -uv^{-2} \\ v & u \end{vmatrix} = v^{-1}u + v^{-1}u = \frac{2u}{v}$; $\int_1^3 \int_1^2 (v+u)\left(\frac{2u}{v}\right) dv\, du =$

$\int_1^3 2u + (\ln 4)\, u^2\, du = 8 + \frac{26 \ln 4}{3}$

7. $J(r,\theta) = \begin{vmatrix} a\cos\theta & -ar\sin\theta \\ b\sin\theta & br\cos\theta \end{vmatrix} = abr\cos^2\theta + abr\sin^2\theta = abr$

$I_o = \int_{-a}^{a} \int_{(-b/a)\sqrt{a^2-x^2}}^{(b/a)\sqrt{a^2-x^2}} \left(x^2+y^2\right) dy\, dx = \int_0^{2\pi} \int_0^1 r^3\left(a^2\cos^2\theta + b^2\sin^2\theta\right) abr\, dr\, d\theta =$

$\frac{ab}{4}\int_a^b a^2\cos^2\theta + b^2\sin^2\theta\, d\theta = \frac{\pi ab\left(a^2+b^2\right)}{4}$

9. $\begin{vmatrix} \sin\phi\cos\theta & \rho\cos\phi\cos\theta & -\rho\sin\phi\sin\theta \\ \sin\phi\sin\theta & \rho\cos\phi\sin\theta & \rho\sin\phi\cos\theta \\ \cos\phi & -\rho\sin\phi & 0 \end{vmatrix} = \cos\phi \begin{vmatrix} \rho\cos\phi\cos\theta & -\rho\sin\phi\sin\theta \\ \rho\cos\phi\sin\theta & \rho\sin\phi\cos\theta \end{vmatrix} +$

$\rho\sin\phi \begin{vmatrix} \sin\phi\cos\theta & -\rho\sin\phi\sin\theta \\ \sin\phi\sin\theta & \rho\sin\phi\cos\theta \end{vmatrix} = \rho^2\cos\phi\left(\sin\phi\cos\phi\cos^2\theta + \sin\phi\cos\phi\sin^2\theta\right) +$

$\rho^2\sin\phi\left(\sin^2\phi\cos^2\theta + \sin^2\phi\sin^2\theta\right) = \rho^2\sin\phi\cos^2\phi + \rho^2\sin^3\phi =$

$\rho^2\sin\phi\left(\cos^2\phi + \sin^2\phi\right) = \rho^2\sin\phi$

11. $J(u,v,w) = \begin{vmatrix} a & 0 & 0 \\ 0 & b & 0 \\ 0 & 0 & c \end{vmatrix} = abc$. The transformation takes the $\frac{x^2}{a^2} + \frac{y^2}{b^2} + \frac{z^2}{c^2} = 1$ region in the xyz–space into the $u^2 + v^2 + w^2 = 1$ region in the uvw–space which is a unit sphere with $V = \frac{4}{3}\pi$.

$\therefore\ V = \int\int_R\int dx\, dy\, dz = \int\int_G\int abc\, du\, dv\, dw = \frac{4\pi abc}{3}$

13. $J(u,v,w) = \begin{vmatrix} 1 & 0 & 0 \\ -v/u^2 & 1/u & 0 \\ 0 & 0 & 1/3 \end{vmatrix} = \frac{1}{3u}$; $\int\int_R\int x^2y + 3xyz\, dx\, dy\, dz =$

$\int\int_G\int u^2\left(\frac{v}{u}\right) + 3u\frac{v}{u}\frac{w}{3}\, J(u,v,w)\, du\, dv\, dw = \frac{1}{3}\int_0^3\int_0^2\int_1^2 v + \frac{vw}{u}\, du\, dv\, dw =$

$\frac{1}{3}\int_0^3\int_0^2 v + vw\ln 2\, dv\, dw = \frac{1}{3}\int_0^3 2 + (\ln 4)w\, dw = 2 + \ln 8$

## PRACTICE EXERCISES

1. $\int_1^{10}\int_0^{1/y} ye^{xy}\,dy\,dx = \int_1^{10}\left[e^{xy}\right]_0^{1/y} dx =$

$\int_1^{10}(e-1)\,dx = 9e-9$

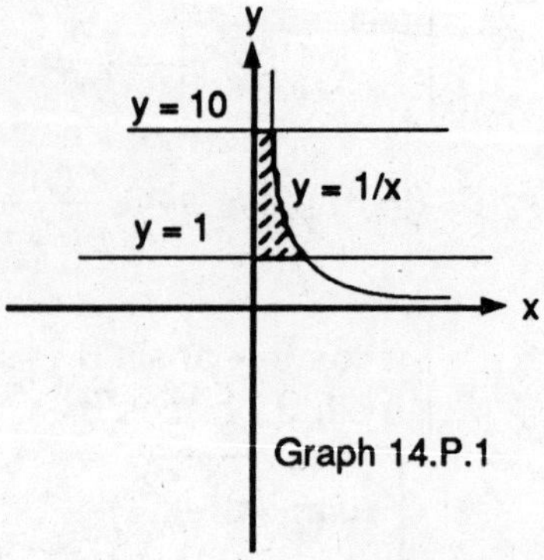

Graph 14.P.1

3. $\int_0^1\int_y^{\sqrt{y}} f(x,y)\,dx\,\,dy$

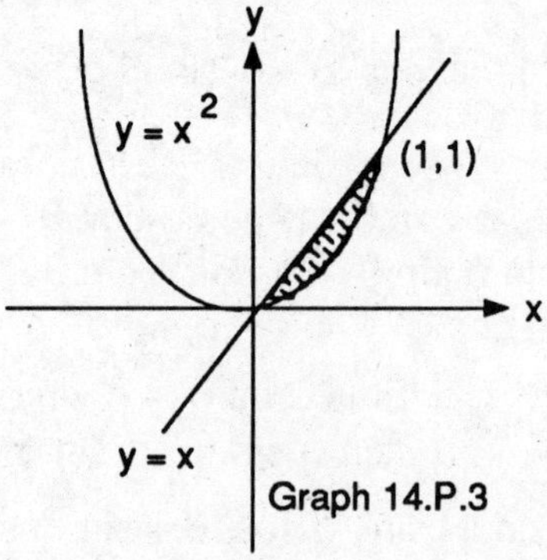

Graph 14.P.3

5. a) $\int_0^{3/2}\int_{-\sqrt{9-4y^2}}^{\sqrt{9-4y^2}} y\,dx\,dy = \int_0^{3/2}\left[\,yx\,\right]_{-\sqrt{9-4y^2}}^{\sqrt{9-4y^2}} dy =$

$\int_0^{3/2} 2y\sqrt{9-4y^2}\,dy = \frac{9}{2}$

b) $\int_{-3}^{3}\int_0^{(9-x^2)^{1/2}/2} y\,dy\,dx = \frac{1}{2}\int_{-3}^{3}\left[y^2\right]_0^{(9-x^2)^{1/2}/2} dx =$

$\frac{1}{8}\int_{-3}^{3}\left(9-x^2\right)dx = \frac{9}{2}$

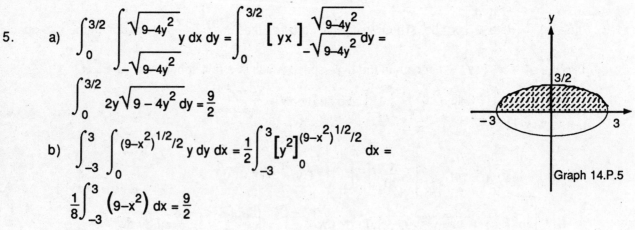

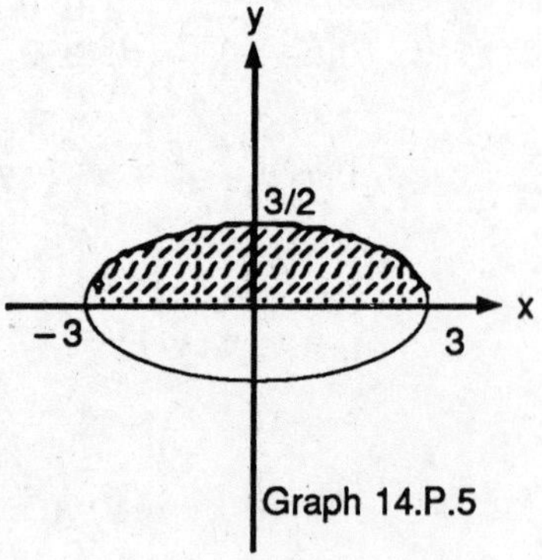

Graph 14.P.5

7. $\int_0^1\int_{2y}^{2} 4\cos x^2\,dx\,dy = \int_0^2\int_0^{x/2} 4\cos x^2\,dy\,dx = \int_0^2\left[\left(4\cos x^2\right)y\right]_0^{x/2} dx =$

$\int_0^2\left(\cos x^2\right)(2x)\,dx = \sin 4$

9. $\int_0^8\int_{\sqrt[3]{x}}^{2}\frac{1}{y^4+1}\,dy\,dx = \int_0^2\int_0^{y^3}\frac{1}{y^4+1}\,dx\,dy = \frac{1}{4}\int_0^2\frac{4y^3}{y^4+1}\,dy = \frac{\ln 17}{4}$

11. $V = \int_0^1 \int_x^{2-x} x^2 + y^2 \, dy \, dx = \int_0^1 \left[ x^2 y + \frac{x^3}{3} \right]_x^{2-x} dx =$

$\int_0^1 \left( -\frac{8}{3}x^3 + 4x^2 - 4x + \frac{8}{3} \right) dx = \frac{4}{3}$

13. $A = \int_{-2}^{0} \int_{2x+4}^{4-x^2} dy \, dx = \int_{-2}^{0} -x^2 - 2x \, dx = \frac{4}{3}$

15. average value $= \int_0^1 \int_0^1 xy \, dy \, dx = \int_0^1 \left[ \frac{xy^2}{2} \right]_0^1 dx = \int_0^1 \frac{x}{2} dx = \frac{1}{4}$

17. $M = \int_1^2 \int_{2/x}^{2} dy \, dx = \int_1^2 2 - \frac{2}{x} dx = 2 - \ln 4$

$M_y = \int_1^2 \int_{2/x}^{2} x \, dy \, dx = \int_1^2 x\left[ 2 - \frac{2}{x} \right] dx = 1$

$M_x = \int_1^2 \int_{2/x}^{2} y \, dy \, dx = \int_1^2 2 - \frac{2}{x^2} dx = 1 \Rightarrow \overline{x} = \frac{1}{2 - \ln 4}, \overline{y} = \frac{1}{2 - \ln 4}$

19. a) $I_o = \int_{-2}^{2} \int_{-1}^{1} \left( x^2 + y^2 \right) dy \, dx = \int_{-2}^{2} 2x^2 + \frac{2}{3} dx = \frac{40}{3}$

b) $I_x = \int_{-a}^{a} \int_{-b}^{b} y^2 \, dy \, dx = \int_{-a}^{a} \frac{2b^3}{3} dx = \frac{4ab^3}{3}, I_y = \int_{-b}^{b} \int_{-a}^{a} x^2 \, dx \, dy =$

$\int_{-b}^{b} \frac{2a^3}{3} dy = \frac{4a^3 b}{3} \Rightarrow I_o = I_x + I_y = \frac{4ab^3}{3} + \frac{4a^3 b}{3} = \frac{4ab\left(b^2 + a^2\right)}{3}$

21. $M = \int_{-1}^{1} \int_{-1}^{1} \left( x^2 + y^2 + \frac{1}{3} \right) dy \, dx = \int_{-1}^{1} 2x^2 + \frac{4}{3} dx = 4$

$M_x = \int_{-1}^{1} \int_{-1}^{1} y\left( x^2 + y^2 + \frac{1}{3} \right) dy \, dx = \int_{-1}^{1} 0 \, dx = 0$

$M_y = \int_{-1}^{1} \int_{-1}^{1} x\left( x^2 + y^2 + \frac{1}{3} \right) dy \, dx = \int_{-1}^{1} x\left( 2x^2 + \frac{4}{3} \right) dx = 0$

23. $\int_{-1}^{1} \int_{-\sqrt{1-x^2}}^{\sqrt{1-x^2}} \frac{2}{\left(1 + x^2 + y^2\right)^2} dy \, dx = \int_0^{2\pi} \int_0^1 \frac{2r}{\left(1 + r^2\right)^2} dr \, d\theta = \frac{1}{2} \int_0^{2\pi} d\theta = \pi$

25. $M = 2\int_0^{\pi/2} \int_1^{1+\cos\theta} r \, dr \, d\theta = \int_0^{\pi/2} 2\cos\theta + \frac{1 + \cos 2\theta}{2} d\theta = \frac{8 + \pi}{4}$

$M_y = \int_{-\pi/2}^{\pi/2} \int_1^{1+\cos\theta} \cos\theta \, r^2 \, dr \, d\theta = \int_{-\pi/2}^{\pi/2} \left( \cos^2\theta + \cos^3\theta + \frac{\cos^4\theta}{3} \right) d\theta = \frac{32 + 15\pi}{24} \Rightarrow$

$\overline{x} = \frac{15\pi + 32}{6\pi + 48}, \overline{y} = 0$ by symmetry

27. $M = \int_0^{\pi/2}\int_1^3 r\,dr\,d\theta = 4\int_0^{\pi/2} d\theta = 2\pi$

$M_y = \int_0^{\pi/2}\int_1^3 r^2\cos\theta\,dr\,d\theta = \frac{26}{3}\int_0^{\pi/2}\cos\theta\,d\theta = \frac{26}{3} \Rightarrow \overline{x} = \frac{13}{3\pi},\ \overline{y} = \frac{13}{3\pi}$ by symmetry

29. $\int_0^{\pi}\int_0^{\pi}\int_0^{\pi}\cos(x+y+z)\,dx\,dy\,dz = \int_0^{\pi}\int_0^{\pi}\sin(z+y+\pi) - \sin(z+y)\,dy\,dz =$

$\int_0^{\pi}\left(-\cos(z+2\pi) + \cos(z+\pi) + \cos(z) - \cos(z+\pi)\right)dz = 0$

31. $V = 2\int_0^{\pi/2}\int_{-\cos y}^{0}\int_0^{-2x} dz\,dx\,dy = -4\int_0^{\pi/2}\int_{-\cos y}^{0} x\,dx\,dy = 2\int_0^{\pi/2}\cos^2 y\,dy = \frac{\pi}{2}$

33. a) $4\int_0^{\sqrt{3}}\int_0^{\sqrt{3-x^2}}\int_0^{\sqrt{4-x^2-y^2}} dz\,dy\,dx$

b) $4\int_0^{\pi/2}\int_0^{\sqrt{3}}\int_0^{\sqrt{4-r^2}} r\,dz\,dr\,d\theta$

c) $4\int_0^{\pi/2}\int_0^{\pi/3}\int_{\sec\phi}^{2} \rho^2\sin\phi\,d\rho\,d\phi\,d\theta$

35. $V = \int_0^{\pi/2}\int_1^2\int_0^{r^2\sin\theta\cos\theta} r\,dz\,dr\,d\theta = \int_0^{\pi/2}\int_1^2 r^3\sin\theta\cos\theta\,dr\,d\theta =$

$\frac{15}{4}\int_0^{\pi/2}\sin\theta\cos\theta\,d\theta = \frac{15}{8}$

37. $V = 4\int_0^1\int_0^{\sqrt{1-x^2}}\int_{2x^2+2y^2}^{3-x^2-y^2} dz\,dy\,dx = 4\int_0^{\pi/2}\int_0^1\int_{2r^2}^{3-r^2} r\,dz\,dr\,d\theta =$

$4\int_0^{\pi/2}\int_0^1\left(3r - 3r^3\right)dr\,d\theta = 3\int_0^{\pi/2} d\theta = \frac{3\pi}{2}$

39. a) $M = 4\int_0^{\pi/2}\int_0^1\int_0^{r^2} z\,r\,dz\,dr\,d\theta = 2\int_0^{\pi/2}\int_0^1 r^5\,dr\,d\theta = \frac{1}{3}\int_0^{\pi/2} d\theta = \frac{\pi}{6}$

$M_{xy} = \int_0^{2\pi}\int_0^1\int_0^{r^2} z^2\,r\,dz\,dr\,d\theta = \frac{1}{3}\int_0^{2\pi}\int_0^1 r^7\,dr\,d\theta = \frac{1}{24}\int_0^{2\pi} d\theta = \frac{\pi}{12} \Rightarrow \overline{z} = \frac{1}{2}$,

$\overline{x} = \overline{y} = 0$ by symmetry; $I_z = \int_0^{2\pi}\int_0^1\int_0^{r^2} z\,r^3\,dz\,dr\,d\theta = \frac{1}{2}\int_0^{2\pi}\int_0^1 r^7\,dr\,d\theta =$

$\frac{1}{16}\int_0^{2\pi} d\theta = \frac{\pi}{8} \Rightarrow R_z = \sqrt{\frac{I_z}{M}} = \frac{\sqrt{3}}{2}$

b) $M = 4\int_0^{\pi/2}\int_0^1\int_0^{r^2} r^2\,dz\,dr\,d\theta = 4\int_0^{\pi/2}\int_0^1 r^4\,dr\,d\theta = \frac{4}{5}\int_0^{\pi/2} d\theta = \frac{2\pi}{5}$

$M_{xy} = \int_0^{2\pi}\int_0^1\int_0^{r^2} z\,r^2\,dz\,dr\,d\theta = \frac{1}{2}\int_0^{2\pi}\int_0^1 r^6\,dr\,d\theta = \frac{1}{14}\int_0^{2\pi} d\theta = \frac{\pi}{7} \Rightarrow \overline{z} = \frac{5}{14}$

$\overline{x} = \overline{y} = 0$ by symmetry; $I_z = \int_0^{2\pi}\int_0^1\int_0^{r^2} r^4\,dz\,dr\,d\theta = \int_0^{2\pi}\int_0^1 r^6\,dr\,d\theta =$

$\frac{1}{7}\int_0^{2\pi} d\theta = \frac{2\pi}{7} \Rightarrow R_z = \sqrt{\frac{I_z}{M}} = \sqrt{\frac{5}{7}}$

41. $V = 8\int_0^{\pi/2}\int_0^{\pi/2}\int_0^{2\sin\phi} \rho^2 \sin\phi\,d\rho\,d\phi\,d\theta = \frac{64}{3}\int_0^{\pi/2}\int_0^{\pi/2} \sin^4\phi\,d\phi\,d\theta = 4\pi\int_0^{\pi/2} d\theta = 2\pi^2$

43. $M = \frac{2}{3}\pi a^3$; $M_{xy} = \int_0^{2\pi}\int_0^{\pi/2}\int_0^a \rho^3 \sin\phi\cos\phi\,d\rho\,d\phi\,d\theta = \frac{a^4}{4}\int_0^{2\pi}\int_0^{\pi/2} \sin\phi\cos\phi\,d\phi\,d\theta =$

$\frac{a^4}{8}\int_0^{2\pi} d\theta = \frac{a^4\pi}{4} \Rightarrow \overline{z} = \frac{3a}{8}$, $\overline{x} = \overline{y} = 0$ by symmetry

# CHAPTER 15

# VECTOR ANALYSIS

## SECTION 15.1 LINE INTEGRALS

1. $\mathbf{r} = t\,\mathbf{i} + (1 - t)\,\mathbf{j} \Rightarrow x = t,\ y = 1 - t \Rightarrow y = 1 - x \Rightarrow$ c

3. $\mathbf{r} = (2\cos t)\,\mathbf{i} + (2\sin t)\,\mathbf{j} \Rightarrow x = 2\cos t,\ y = 2\sin t \Rightarrow x^2 + y^2 = 4 \Rightarrow$ g

5. $\mathbf{r} = t\,\mathbf{i} + t\,\mathbf{j} + t\,\mathbf{k} \Rightarrow x = t,\ y = t,\ z = t \Rightarrow$ d

7. $\mathbf{r} = (t^2 - 1)\,\mathbf{j} + 2t\,\mathbf{k} \Rightarrow y = t^2 - 1,\ z = 2t \Rightarrow y = \frac{z^2}{4} - 1 \Rightarrow$ f

9. $\mathbf{r} = t\,\mathbf{i} + (1 - t)\,\mathbf{j} \Rightarrow x = t,\ y = 1 - t,\ z = 0 \Rightarrow f(g(t),h(t),k(t)) = 1.\ \frac{dx}{dt} = 1,\ \frac{dy}{dt} = -1,\ \frac{dz}{dt} = 0 \Rightarrow$

$$\sqrt{\left(\frac{dx}{dt}\right)^2 + \left(\frac{dy}{dt}\right)^2 + \left(\frac{dz}{dt}\right)^2}\ dt = \sqrt{2}\ dt \Rightarrow \int_C f(x,y,z)\ ds = \int_0^1 \sqrt{2}\ dt = \sqrt{2}$$

11. $\mathbf{r} = 2t\,\mathbf{i} + t\,\mathbf{j} + (2 - 2t)\,\mathbf{k} \Rightarrow x = 2t,\ y = t,\ z = 2 - 2t \Rightarrow f(g(t),h(t),k(t)) = 2t^2 - t + 2.\ \frac{dx}{dt} = 2,\ \frac{dy}{dt} = 1,\ \frac{dz}{dt} = -2$

$$\Rightarrow \sqrt{\left(\frac{dx}{dt}\right)^2 + \left(\frac{dy}{dt}\right)^2 + \left(\frac{dz}{dt}\right)^2}\ dt = 3\ dt \Rightarrow \int_C f(x,y,z)\ ds = \int_0^1 (2t^2 - t + 2)3\ dt = \frac{13}{2}$$

13. $\mathbf{r} = \mathbf{i} + \mathbf{j} + t\,\mathbf{k} \Rightarrow x = 1,\ y = 1,\ z = t \Rightarrow f(g(t),h(t),k(t)) = 3t\sqrt{4 + t^2}.\ \frac{dx}{dt} = 0,\ \frac{dy}{dt} = 0,\ \frac{dz}{dt} = 1 \Rightarrow$

$$\sqrt{\left(\frac{dx}{dt}\right)^2 + \left(\frac{dy}{dt}\right)^2 + \left(\frac{dz}{dt}\right)^2}\ dt = 1\ dt = dt \Rightarrow \int_C f(x,y,z)\ ds = \int_{-1}^1 3t\sqrt{4 + t^2}\ dt = 0$$

15. $C_1: \mathbf{r} = t\,\mathbf{i} + t^2\,\mathbf{j} \Rightarrow x = t,\ y = t^2,\ z = 0 \Rightarrow f(g(t),h(t),k(t)) = t + \sqrt{t^2} = 2t$ since $0 \le t \le 1$. $\frac{dx}{dt} = 1,\ \frac{dy}{dt} = 2t,$

$$\frac{dz}{dt} = 0 \Rightarrow \sqrt{\left(\frac{dx}{dt}\right)^2 + \left(\frac{dy}{dt}\right)^2 + \left(\frac{dz}{dt}\right)^2}\ dt = \sqrt{1 + 4t^2}\ dt \Rightarrow \int_{C_1} f(x,y,z)\ ds = \int_0^1 2t\sqrt{1 + 4t^2}\ dt =$$

$\frac{1}{6}\left(5^{3/2}\right) - \frac{1}{6} = \frac{5}{6}\sqrt{5} - \frac{1}{6}$. $C_2: \mathbf{r} = \mathbf{i} + \mathbf{j} + t\,\mathbf{k} \Rightarrow x = 1,\ y = 1,\ z = t \Rightarrow f(g(t),h(t),k(t)) = 2 - t^2.\ \frac{dx}{dt} = 0,$

$$\frac{dy}{dt} = 0,\ \frac{dz}{dt} = 1 \Rightarrow \sqrt{\left(\frac{dx}{dt}\right)^2 + \left(\frac{dy}{dt}\right)^2 + \left(\frac{dz}{dt}\right)^2}\ dt = 1\ dt = dt \Rightarrow \int_{C_2} f(x,y,z)\ ds = \int_0^1 (2 - t^2)\ dt =$$

$$\frac{5}{3}.\ \therefore \int_C f(x,y,z)\ ds = \int_{C_1} f(x,y,z)\ ds + \int_{C_2} f(x,y,z)\ ds = \frac{5}{6}\sqrt{5} + \frac{3}{2}.$$

17. $\delta(x,y,z) = 2 - z$, $\mathbf{r} = (\cos t)\,\mathbf{j} + (\sin t)\,\mathbf{k}$, $0 \le t \le \pi$, ds = dt, x = 0, y = cos t, z = sin t, and $M = 2\pi - 2$ are all given or found in Example 3 in the text, page 946. $I_x = \int_C (y^2 + z^2)\delta\, ds$

19. Let $\delta$ be constant. Let x = a cos t, y = a sin t. Then $\frac{dx}{dt} = -a \sin t$, $\frac{dy}{dt} = a \cos t$, $0 \le t \le 2\pi$, $\frac{dz}{dt} = 0 \Rightarrow$

$\sqrt{\left(\frac{dx}{dt}\right)^2 + \left(\frac{dy}{dt}\right)^2 + \left(\frac{dz}{dt}\right)^2}\, dt = a\, dt$. $\therefore$ $I_z = \int_C (x^2 + y^2)\delta\, ds = \int_0^{2\pi} (a^2 \sin^2 t + a^2 \cos^2 t)a\delta\, dt =$

$\int_0^{2\pi} a^3\delta\, dt = 2\pi a^3\delta$. $M = \int_C \delta(x,y,z)\, ds = \int_0^{2\pi} \delta a\, dt = 2\pi\delta a$. $R_z = \sqrt{\frac{I_z}{M}} = \sqrt{\frac{2\pi a^3\delta}{2\pi a\delta}} = a$.

21. a) $\mathbf{r} = (\cos t)\,\mathbf{i} + (\sin t)\,\mathbf{j} + t\,\mathbf{k} \Rightarrow x = \cos t,\ y = \sin t,\ z = t \Rightarrow \frac{dx}{dt} = -\sin t$, $\frac{dy}{dt} = \cos t$, $\frac{dz}{dt} = 1 \Rightarrow$

$\sqrt{\left(\frac{dx}{dt}\right)^2 + \left(\frac{dy}{dt}\right)^2 + \left(\frac{dz}{dt}\right)^2}\, dt = \sqrt{2}\, dt$. $M = \int_C \delta(x,y,z)\, ds = \int_0^{2\pi} \delta\sqrt{2}\, dt = 2\pi\delta\sqrt{2}$.

$I_z = \int_C (x^2 + y^2)\delta\, ds = \int_0^{2\pi} (\cos^2 t + \sin^2 t)\delta\sqrt{2}\, dt = \int_0^{2\pi} \delta\sqrt{2}\, dt = 2\pi\delta\sqrt{2}$. $R_z = \sqrt{\frac{I_z}{M}} =$

$\sqrt{\frac{2\pi\delta\sqrt{2}}{2\pi\delta\sqrt{2}}} = 1$

b) $M = \int_C \delta(x,y,z)\, ds = \int_0^{4\pi} \delta\sqrt{2}\, dt = 4\pi\delta\sqrt{2}$. $I_z = \int_C (x^2 + y^2)\delta\, ds = \int_0^{4\pi} \delta\sqrt{2}\, dt = 4\pi\delta\sqrt{2}$.

$R_z = \sqrt{\frac{I_z}{M}} = \sqrt{\frac{4\pi\delta\sqrt{2}}{4\pi\delta\sqrt{2}}} = 1$

## SECTION 15.2 VECTOR FIELDS, WORK, CIRCULATION, AND FLUX

1. a) $\mathbf{F} = 3t\,\mathbf{i} + 2t\,\mathbf{j} + 4t\,\mathbf{k}$, $\frac{d\mathbf{r}}{dt} = \mathbf{i} + \mathbf{j} + \mathbf{k} \Rightarrow \mathbf{F}\cdot\frac{d\mathbf{r}}{dt} = 9t \Rightarrow W = \int_0^1 9t\,dt = \frac{9}{2}$

   b) $\mathbf{F} = 3t^2\,\mathbf{i} + 2t\,\mathbf{j} + 4t^4\,\mathbf{k}$, $\frac{d\mathbf{r}}{dt} = \mathbf{i} + 2t\,\mathbf{j} + 4t^3\,\mathbf{k} \Rightarrow \mathbf{F}\cdot\frac{d\mathbf{r}}{dt} = 7t^2 + 16t^7 \Rightarrow W = \int_0^1 (7t^2 + 16t^7)\,dt = \frac{13}{3}$

   c) $\mathbf{F}_1 = 3t\,\mathbf{i} + 2t\,\mathbf{j}$, $\frac{d\mathbf{r}_1}{dt} = \mathbf{i} + \mathbf{j} \Rightarrow \mathbf{F}_1\cdot\frac{d\mathbf{r}_1}{dt} = 5t \Rightarrow W_1 = \int_0^1 5t\,dt = \frac{5}{2}$. $\mathbf{F}_2 = 3\,\mathbf{i} + 2\,\mathbf{j} + 4t\,\mathbf{k}$, $\frac{d\mathbf{r}_2}{dt} = \mathbf{k} \Rightarrow$

   $\mathbf{F}_2\cdot\frac{d\mathbf{r}_2}{dt} = 4t \Rightarrow W_2 = \int_0^1 4t\,dt = 2$. $\therefore\ W = W_1 + W_2 = \frac{9}{2}$

3. a) $\mathbf{F} = \sqrt{t}\,\mathbf{i} - 2t\,\mathbf{j} + \sqrt{t}\,\mathbf{k}$, $\frac{d\mathbf{r}}{dt} = \mathbf{i} + \mathbf{j} + \mathbf{k} \Rightarrow \mathbf{F}\cdot\frac{d\mathbf{r}}{dt} = 2\sqrt{t} - 2t \Rightarrow W = \int_0^1 (2\sqrt{t} - 2t)\,dt = \frac{1}{3}$

   b) $\mathbf{F} = t^2\,\mathbf{i} - 2t\,\mathbf{j} + t\,\mathbf{k}$, $\frac{d\mathbf{r}}{dt} = \mathbf{i} + 2t\,\mathbf{j} + 4t^3\,\mathbf{k} \Rightarrow \mathbf{F}\cdot\frac{d\mathbf{r}}{dt} = 4t^4 - 3t^2 \Rightarrow W = \int_0^1 (4t^4 - 3t^2)\,dt = -\frac{1}{5}$

   c) $\mathbf{F}_1 = -2t\,\mathbf{j} + \sqrt{t}\,\mathbf{k}$, $\frac{d\mathbf{r}_1}{dt} = \mathbf{i} + \mathbf{j} \Rightarrow \mathbf{F}_1\cdot\frac{d\mathbf{r}_1}{dt} = -2t \Rightarrow W_1 = \int_0^1 -2t\,dt = -1$. $\mathbf{F}_2 = \sqrt{t}\,\mathbf{i} - 2\,\mathbf{j} + \mathbf{k}$,

   $\frac{d\mathbf{r}_2}{dt} = \mathbf{k} \Rightarrow \mathbf{F}_2\cdot\frac{d\mathbf{r}_2}{dt} = 1 \Rightarrow W_2 = \int_0^1 dt = 1$. $\therefore\ W = W_1 + W_2 = 0$

5. a) $\mathbf{F} = (3t^2 - 3t)\,\mathbf{i} + 3t\,\mathbf{j} + \mathbf{k}$, $\frac{d\mathbf{r}}{dt} = \mathbf{i} + \mathbf{j} + \mathbf{k} \Rightarrow \mathbf{F}\cdot\frac{d\mathbf{r}}{dt} = 3t^2 + 1 \Rightarrow W = \int_0^1 (3t^2 + 1)\,dt = 2$

   b) $\mathbf{F} = (3t^2 - 3t)\,\mathbf{i} + 3t^4\,\mathbf{j} + \mathbf{k}$, $\frac{d\mathbf{r}}{dt} = \mathbf{i} + 2t\,\mathbf{j} + 4t^3\,\mathbf{k} \Rightarrow \mathbf{F}\cdot\frac{d\mathbf{r}}{dt} = 6t^5 + 4t^3 + 3t^2 - 3t \Rightarrow$

   $W = \int_0^1 \left(6t^5 + 4t^3 + 3t^2 - 3t\right)dt = \frac{3}{2}$

   c) $\mathbf{F}_1 = (3t^2 - 3t)\,\mathbf{i} + \mathbf{k}$, $\frac{d\mathbf{r}_1}{dt} = \mathbf{i} + \mathbf{j} \Rightarrow \mathbf{F}_1\cdot\frac{d\mathbf{r}_1}{dt} = 3t^2 - 3t \Rightarrow W_1 = \int_0^1 (3t^2 - 3t)\,dt = -\frac{1}{2}$

   $\mathbf{F}_2 = 3t\,\mathbf{j} + \mathbf{k}$, $\frac{d\mathbf{r}_2}{dt} = \mathbf{k} \Rightarrow \mathbf{F}_2\cdot\frac{d\mathbf{r}_2}{dt} = 1 \Rightarrow W_2 = \int_0^1 dt = 1$. $\therefore\ W = W_1 + W_2 = \frac{1}{2}$

7. $\mathbf{F} = t^3\,\mathbf{i} + t^2\,\mathbf{j} - t^3\,\mathbf{k},\ \frac{d\mathbf{r}}{dt} = \mathbf{i} + 2t\,\mathbf{j} + \mathbf{k} \Rightarrow \mathbf{F}\cdot\frac{d\mathbf{r}}{dt} = 2t^3 \Rightarrow W = \int_0^1 2t^3\,dt = \frac{1}{2}$

9. $\mathbf{F} = t\,\mathbf{i} + (\sin t)\,\mathbf{j} + (\cos t)\,\mathbf{k},\ \frac{d\mathbf{r}}{dt} = (\cos t)\,\mathbf{i} - (\sin t)\,\mathbf{j} + \mathbf{k} \Rightarrow \mathbf{F}\cdot\frac{d\mathbf{r}}{dt} = t\cos t - \sin^2 t + \cos t \Rightarrow$

$W = \int_0^{2\pi} (t\cos t - \sin^2 t + \cos t)\,dt = -\pi$

11. $\mathbf{F} = -4t^3\,\mathbf{i} + 8t^2\,\mathbf{j} + 2\,\mathbf{k},\ \frac{d\mathbf{r}}{dt} = \mathbf{i} + 2t\,\mathbf{j} \Rightarrow \mathbf{F}\cdot\frac{d\mathbf{r}}{dt} = 12t^3 \Rightarrow \text{Flow} = \int_0^2 12t^3\,dt = 48$

13. $\mathbf{F} = (\cos t - \sin t)\,\mathbf{i} + (\cos t)\,\mathbf{k},\ \frac{d\mathbf{r}}{dt} = (-\sin t)\,\mathbf{i} + (\cos t)\,\mathbf{k} \Rightarrow \mathbf{F}\cdot\frac{d\mathbf{r}}{dt} = -\sin t\cos t + 1 \Rightarrow$

$\text{Flow} = \int_0^{\pi} (-\sin t\cos t + 1)\,dt = \pi$

15. a) $\mathbf{F} = (\cos t)\,\mathbf{i} + (\sin t)\,\mathbf{j},\ \frac{d\mathbf{r}}{dt} = (-\sin t)\,\mathbf{i} + (\cos t)\,\mathbf{j} \Rightarrow \mathbf{F}\cdot\frac{d\mathbf{r}}{dt} = 0 \Rightarrow \text{Circulation} = 0.$

$M = \cos t,\ N = \sin t,\ dx = -\sin t\,dt,\ dy = \cos t\,dt \Rightarrow \text{Flux} = \int_C M\,dy - N\,dx =$

$\int_0^{2\pi} (\cos^2 t + \sin^2 t)\,dt = \int_0^{2\pi} dt = 2\pi$

. b) $\mathbf{F} = (\cos t)\,\mathbf{i} + (4\sin t)\,\mathbf{j},\ \frac{d\mathbf{r}}{dt} = (-\sin t)\,\mathbf{i} + (4\cos t)\,\mathbf{j} \Rightarrow \mathbf{F}\cdot\frac{d\mathbf{r}}{dt} = 15\sin t\cos t \Rightarrow \text{Circ} =$

$\int_0^{2\pi} 15\sin t\cos t\,dt = 0.\ \ M = \cos t,\ N = \sin t,\ dx = -\sin t,\ dy = 4\cos t \Rightarrow \text{Flux} =$

$\int_C M\,dy - N\,dx = \int_0^{2\pi} (4\cos^2 t + 4\sin^2 t)\,dt = 8\pi$

17. $\mathbf{F}_1 = (a\cos t)\,\mathbf{i} + (a\sin t)\,\mathbf{j},\ \frac{d\mathbf{r}_1}{dt} = (-a\sin t)\,\mathbf{i} + (a\cos t)\,\mathbf{j} \Rightarrow \mathbf{F}_1\cdot\frac{d\mathbf{r}_1}{dt} = 0 \Rightarrow \text{Circ}_1 = 0.\ \ M_1 = a\cos t,\ N_1 =$

$a\sin t,\ dx = -a\sin t\,dt,\ dy = a\cos t\,dt \Rightarrow \text{Flux}_1 = \int_C M_1\,dy - N_1\,dx = \int_0^{\pi} (a^2\cos^2 t + a^2\sin^2 t)\,dt =$

$\int_0^{\pi} a^2\,dt = a^2\pi.$

17. (Continued)

$\mathbf{F}_2 = t\,\mathbf{i}, \frac{d\mathbf{r}_2}{dt} = \mathbf{i} \Rightarrow \mathbf{F}_2 \cdot \frac{d\mathbf{r}_2}{dt} = t \Rightarrow \text{Circ}_2 = \int_{-a}^{a} t\,dt = 0.$ $M_2 = t, N_2 = 0, dx = dt, dy = 0 \Rightarrow \text{Flux}_2 =$

$\int_C M_2 dy - N_2 dx = \int_{-a}^{a} 0\,dt = 0.$ $\therefore$ $\text{Circ} = \text{Circ}_1 + \text{Circ}_2 = 0, \text{Flux} = \text{Flux}_1 + \text{Flux}_2 = a^2\pi$

19. $\mathbf{F}_1 = (-a \sin t)\,\mathbf{i} + (a \cos t)\,\mathbf{j}, \frac{d\mathbf{r}_1}{dt} = (-a \sin t)\,\mathbf{i} + (a \cos t)\,\mathbf{j} \Rightarrow \mathbf{F}_1 \cdot \frac{d\mathbf{r}_1}{dt} = a^2 \sin^2 t + a^2 \cos^2 t = a^2 \Rightarrow$

$\text{Circ}_1 = \int_0^{\pi} a^2\,dt = a^2\pi.$

$\mathbf{F}_2 = t\,\mathbf{j}, \frac{d\mathbf{r}_2}{dt} = \mathbf{i} \Rightarrow \mathbf{F}_2 \cdot \frac{d\mathbf{r}_2}{dt} = 0 \Rightarrow \text{Circ}_2 = 0.$ $\therefore$ $\text{Circ} = \text{Circ}_1 + \text{Circ}_2 = a^2\pi$

$M_1 = -a \sin t, N_1 = a \cos t, dx = -a \sin t, dy = a \cos t \Rightarrow \text{Flux}_1 = \int_C M_1 dy - N_1 dx =$

$\int_0^{\pi} (-a^2 \sin t \cos t + a^2 \sin t \cos t)\,dt = 0.$ $M_2 = 0, N_2 = t, dx = dt, dy = 0 \Rightarrow \text{Flux}_2 =$

$\int_C M_2 dy - N_2 dx = \int_{-a}^{a} -t\,dt = 0.$ $\therefore$ $\text{Flux} = \text{Flux}_1 + \text{Flux}_2 = 0$

21. $\mathbf{F} = f(t)\,\mathbf{i}, \frac{d\mathbf{r}}{dt} = \mathbf{i} + f'(t)\,\mathbf{j} \Rightarrow \mathbf{F} \cdot \frac{d\mathbf{r}}{dt} = f(t) \Rightarrow \int_{t=a}^{t=b} \mathbf{F} \cdot d\mathbf{r} = \int_a^b f(t)\,dt.$ Since $t > 0$ for $a \le t \le b$, this integral yields the desired area.

## SECTION 15.3 GREEN'S THEOREM IN THE PLANE

1. Equation 15: $M = -y = -a \sin t, N = x = a \cos t, dx = -a \sin t\,dt, dy = a \cos t\,dt \Rightarrow \frac{\partial M}{\partial x} = 0, \frac{\partial M}{\partial y} = -1,$

$\frac{\partial N}{\partial x} = 1, \frac{\partial N}{\partial y} = 0 \Rightarrow \oint_C M dy - N dx = \int_0^{2\pi} ((-a \sin t)(a \cos t)\,dt - (a \cos t)(-a \sin t))dt =$

$0.$ $\iint_R \left(\frac{\partial M}{\partial x} + \frac{\partial N}{\partial y}\right) dx\,dy = \iint_R 0\,dx\,dy = 0$

Equation 16: $\oint_C M dx + N dy = \int_0^{2\pi} ((-a \sin t)(-a \sin t) + (a \cos t)(a \cos t))\,dt = 2\pi a^2$

1. (Continued)

$$\iint_R \left(\frac{\partial N}{\partial x} - \frac{\partial M}{\partial y}\right) dx\,dy = \int_{-a}^{a}\int_{-\sqrt{a^2+x^2}}^{\sqrt{a^2+x^2}} 2\,dx\,dy = \int_{-a}^{a} 4\sqrt{a^2 - x^2}\,dx = 2a^2\pi$$

3. $M = 2x = 2a\cos t$, $N = -3y = -3a\sin t$, $dx = -a\sin t$, $dy = a\cos t \Rightarrow \frac{\partial M}{\partial x} = 2$, $\frac{\partial M}{\partial y} = 0$, $\frac{\partial N}{\partial x} = 0$, $\frac{\partial N}{\partial y} = -3$

Equation 15: $\oint_C M\,dy - N\,dx = \int_0^{2\pi} (2a\cos t(a\cos t) + 3a\sin t(-a\sin t))dt =$

$\int_0^{2\pi} \left(2a^2\cos^2 t - 3a^2\sin^2 t\right) dt = -\pi a^2.$ $\iint_R \left(\frac{\partial M}{\partial x} + \frac{\partial N}{\partial y}\right) = \iint_R -1\,dx\,dy =$

$$\int_{-a}^{a}\int_{-\sqrt{a^2-x^2}}^{\sqrt{a^2-x^2}} -1\,dx\,dy = -\pi a^2$$

Equation 16: $\oint_C M\,dx + N\,dy = \int_0^{2\pi} (2a\cos t(-a\sin t) + (-3a\sin t)(a\cos t))\,dt =$

$\int_0^{2\pi} \left(-2a^2\sin t\cos t - 3a^2\sin t\cos t\right) dt = 0.$ $\iint_R 0\,dx\,dy = 0$

5. $M = x - y$, $N = y - x \Rightarrow \frac{\partial M}{\partial x} = 1$, $\frac{\partial M}{\partial y} = -1$, $\frac{\partial N}{\partial x} = -1$, $\frac{\partial N}{\partial y} = 1 \Rightarrow$ Flux $= \iint_R 2\,dx\,dy = \int_0^1\int_0^1 2\,dx\,dy$

$= 2$. Circ $= \iint_R (-1 - (-1))\,dx\,dy = 0$

7. $M = y^2 - x^2$, $N = x^2 + y^2 \Rightarrow \frac{\partial M}{\partial x} = -2x$, $\frac{\partial M}{\partial y} = 2y$, $\frac{\partial N}{\partial x} = 2x$, $\frac{\partial N}{\partial y} = 2y \Rightarrow$ Flux $= \iint_R (-2x + 2y)\,dx\,dy$

$= \int_0^3\int_0^x (-2x + 2y)dy\,dx = \int_0^3 (-2x^2 + x^2)\,dx = -9.$ Circ $= \iint_R (2x - 2y)\,dx\,dy =$

7. (Continued)

$$\int_0^3\int_0^x (2x - 2y)\, dy\, dx = \int_0^3 x^2\, dx = 9$$

9. $M = xy,\ N = y^2 \Rightarrow \frac{\partial M}{\partial x} = y,\ \frac{\partial M}{\partial y} = x,\ \frac{\partial N}{\partial x} = 0,\ \frac{\partial N}{\partial y} = 2y \Rightarrow \text{Flux} = \iint_R (y + 2y)\, dy\, dx = \int_0^1\int_{x^2}^{x} 3y\, dy\, dx$

$$= \int_0^1 \left(\frac{3x^2}{2} - \frac{3x^4}{2}\right) dx = \frac{1}{5}.\ \text{Circ} = \iint_R -x\, dy\, dx = \int_0^1\int_{x^2}^{x} -x\, dy\, dx = \int_0^1 (-x^2 + x^3)\, dx = -\frac{1}{12}$$

11. $M = y^2,\ N = x^2 \Rightarrow \frac{\partial M}{\partial y} = 2y,\ \frac{\partial N}{\partial x} = 2y \Rightarrow \oint_C y^2\, dx + x^2\, dy = \iint_R (2x - 2y) dy\, dx =$

$$\int_0^1\int_0^{-x+1} (2x - 2y)\, dy\, dx = \int_0^1 (-3x^2 + 4x - 1)\, dx = 0.$$

13. $M = 6y + x,\ N = y + 2x \Rightarrow \frac{\partial M}{\partial y} = 6,\ \frac{\partial N}{\partial x} = 2 \Rightarrow \oint_C (6y + x)dx + (y + 2x)dy = \iint_R (2 - 6)\, dy\, dx =$

$-4(\text{Area of the circle}) = -16\pi$

15. $M = 2xy^3,\ N = 4x^2y^2 \Rightarrow \frac{\partial M}{\partial y} = 6xy^2,\ \frac{\partial N}{\partial x} = 8xy^2 \Rightarrow \oint_C 2xy^3\, dx + 4x^2y^2\, dy = \iint_R (8xy^2 - 6xy^2)\, dx\, dy$

$$\int_0^1\int_0^{x^3} 2xy^2\, dy\, dx = \int_0^1 \frac{2}{3}x^{10}\, dx = \frac{2}{33}$$

17. a) $M = f(x),\ N = g(y) \Rightarrow \frac{\partial M}{\partial y} = 0,\ \frac{\partial N}{\partial x} = 0 \Rightarrow \oint_C f(x)\, dx + g(y)\, dy = \iint_R 0\, dy\, dx = 0$

b) $M = ky,\ N = hx \Rightarrow \frac{\partial M}{\partial y} = k,\ \frac{\partial N}{\partial x} = h \Rightarrow \oint_C ky\, dx + hx\, dy = \iint_R (h - k)\, dx\, dy =$

$(h - k)(\text{Area of the region})$

19. Area $= \frac{1}{2}\oint_C x\,dy - y\,dx$. $M = x = a\cos t$, $N = y = a\sin t \Rightarrow dx = -a\sin t\,dt$, $dy = a\cos t\,dt \Rightarrow$ Area $=$

$$\frac{1}{2}\int_0^{2\pi}\left(a^2\cos^2 t + a^2\sin^2 t\right)dt = \frac{1}{2}\int_0^{2\pi} a^2\,dt = \pi a^2$$

21. Area $= \frac{1}{2}\oint_C x\,dy - y\,dx$. $M = x = \cos^3 t$, $N = y = \sin^3 t \Rightarrow dx = -3\cos^2 t\sin t\,dt$, $dy = 3\sin^2 t\cos t\,dt$

$$\Rightarrow \text{Area} = \frac{1}{2}\int_0^{2\pi}(3\sin^2 t\cos^2 t(\cos^2 t + \sin^2 t))\,dt = \frac{1}{2}\int_0^{2\pi}(3\sin^2 t\cos^2 t)\,dt = \frac{3\pi}{8}$$

## SECTION 15.4 SURFACE AREA AND SURFACE INTEGRALS

1. $\mathbf{p} = \mathbf{k}$, $\nabla f = 2x\,\mathbf{i} + 2y\,\mathbf{j} - \mathbf{k} \Rightarrow |\nabla f| = \sqrt{(2x)^2 + (2y)^2 + (-1)^2} = \sqrt{4x^2 + 4y^2 + 1}$. $|\nabla f\cdot\mathbf{p}| = 1 \Rightarrow$

$$S = \iint_R \frac{|\nabla f|}{|\nabla f\cdot\mathbf{p}|}\,dA = \iint_R \sqrt{4x^2 + 4y^2 + 1}\,dx\,dy =$$

$$\iint_R \sqrt{4r^2\cos^2\theta + 4r^2\sin^2\theta + 1}\;r\,dr\,d\theta = \int_0^{2\pi}\int_0^{\sqrt{2}}\sqrt{4r^2 + 1}\;r\,dr\,d\theta = \frac{13}{3}\pi$$

3. $\mathbf{p} = \mathbf{k}$. $\nabla f = \mathbf{i} + 2\,\mathbf{j} + 2\,\mathbf{k} \Rightarrow |\nabla f| = 3$. $|\nabla f\cdot\mathbf{p}| = 2 \Rightarrow S = \iint_R \frac{|\nabla f|}{|\nabla f\cdot\mathbf{p}|}\,dA = \iint_R \frac{3}{2}\,dx\,dy =$

$$\int_{-1}^{1}\int_{y^2}^{2-y^2}\frac{3}{2}\,dx\,dy = \int_{-1}^{1}(3 - 3y^2)\,dy = 4$$

5. $\mathbf{p} = \mathbf{k}$. $\nabla f = 2x\,\mathbf{i} - 2\,\mathbf{j} - 2\,\mathbf{k} \Rightarrow |\nabla f| = \sqrt{(2x)^2 + (-2)^2 + (-2)^2} = \sqrt{4x^2 + 8}$. $|\nabla f\cdot\mathbf{p}| = 2 \Rightarrow S =$

$$\iint_R \frac{|\nabla f|}{|\nabla f\cdot\mathbf{p}|}\,dA = \iint_R \frac{\sqrt{4x^2 + 8}}{2}\,dx\,dy = \int_0^{2}\int_0^{3x}\sqrt{x^2 + 2}\,dy\,dx = \int_0^{2} 3x\sqrt{x^2 + 2}\,dx = 6\sqrt{6} - 2\sqrt{2}$$

7. $\mathbf{p} = \mathbf{k}$. $\nabla f = 2x\,\mathbf{i} + 2y\,\mathbf{j} + 2z\,\mathbf{k} \Rightarrow |\nabla f| = \sqrt{4x^2 + 4y^2 + 4z^2} = \sqrt{8} = 2\sqrt{2}$. $|\nabla f\cdot\mathbf{p}| = 2z \Rightarrow S =$

$$\iint_R \frac{|\nabla f|}{|\nabla f\cdot\mathbf{p}|}\,dA = \iint_R \frac{2\sqrt{2}}{2z}\,dA = \sqrt{2}\iint_R \frac{1}{z}\,dA =$$

7. (Continued)

$$\sqrt{2}\iint_R \frac{1}{\sqrt{2-(x^2+y^2)}}\,dA = \sqrt{2}\int_0^{2\pi}\int_0^1 \frac{r\,dr\,d\theta}{\sqrt{2-r^2}} = \sqrt{2}\int_0^{2\pi}(-1+\sqrt{2})\,d\theta = 2\pi(2-\sqrt{2})$$

9. $\mathbf{p} = \mathbf{k}$. $\nabla f = 2x\,\mathbf{i} + 2z\,\mathbf{j} \Rightarrow |\nabla f| = \sqrt{(2x)^2+(2z)^2} = 2$. $|\nabla f\cdot\mathbf{p}| = 2z$ for the upper surface, $z \ge 0 \Rightarrow$

$$S = \iint_R \frac{|\nabla f|}{|\nabla f\cdot\mathbf{p}|}\,dA = 2\iint_R \frac{2}{2z}\,dA = 2\iint_R \frac{1}{z}\,dA = 2\iint_R \frac{1}{\sqrt{1-x^2}}\,dy\,dx =$$

$$4\int_{-1/2}^{1/2}\int_0^{1/2} \frac{1}{\sqrt{1-x^2}}\,dy\,dx = 2\int_{-1/2}^{1/2}\frac{1}{\sqrt{1-x^2}}\,dx = \frac{2\pi}{3}$$

11. The bottom face of the cube is in the xy–plane $\Rightarrow z = 0 \Rightarrow g(x,y,0) = x + y$ and $f(x,y,z) = z = 0 \Rightarrow$

$\nabla f = \mathbf{k} \Rightarrow |\nabla f| = 1$. $\mathbf{p} = \mathbf{k} \Rightarrow |\nabla f\cdot\mathbf{p}| = 1 \Rightarrow d\sigma = dx\,dy \Rightarrow$ $\displaystyle\iint_{z=0}(x+y)\,dx\,dy = \int_0^1\int_0^1 (x+y)\,dx\,dy =$

1. Because of symmetry, you get 1 over the face of the cube in the xz–plane and 1 over the face of the cube in the yz–plane.
In the top of the cube, $g(x,y,z) = g(x,y,1) = x + y + 1$ and $f(x,y,z) = z = 1 \Rightarrow \nabla f = \mathbf{k} \Rightarrow |\nabla f| = 1$. $\mathbf{p} = \mathbf{k} \Rightarrow$

$|\nabla f\cdot\mathbf{p}| = 1 \Rightarrow d\sigma = dx\,dy \Rightarrow$ $\displaystyle\iint_{z=1}(x+y+1)\,dx\,dy = \int_0^1\int_0^1 (x+y+1)\,dx\,dy = 2.$ Because of

symmetry, the integral is 2 over each of the other two faces. $\therefore$ $\displaystyle\iint_{\text{cube}}(x+y+z)\,d\sigma = 9.$

13. On the faces in the coordinate planes, $g(x,y,z) = 0 \Rightarrow$ the integral over these faces is 0.
On the face, $x = a$, $f(x,y,z) = x = a$ and $g(x,y,z) = g(a,y,z) = ayz \Rightarrow \nabla f = \mathbf{i} \Rightarrow |\nabla f| = 1$. $\mathbf{p} = \mathbf{i} \Rightarrow |\nabla f\cdot\mathbf{p}| = 1$

$\Rightarrow d\sigma = dy\,dz \Rightarrow$ $\displaystyle\iint_{x=a} xyz\,d\sigma = \int_0^c\int_0^b ayz\,dy\,dz = \frac{ab^2c^2}{4}$

On the face, $y = b$, $f(x,y,z) = y = b$ and $g(x,y,z) = g(x,b,z) = bxz \Rightarrow \nabla f = \mathbf{j} \Rightarrow |\nabla f| = 1$. $\mathbf{p} = \mathbf{j} \Rightarrow |\nabla f\cdot\mathbf{p}| = 1$

$\Rightarrow d\sigma = dx\,dz \Rightarrow$ $\displaystyle\iint_{y=b} xyz\,dx\,dz = \int_0^c\int_0^a bxz\,dz\,dx = \frac{a^2bc^2}{4}$

On the face, $z = c$, $f(x,y,z) = z = c$ and $g(x,y,z) = g(x,y,c) = cxy \Rightarrow \nabla f = \mathbf{k} \Rightarrow |\nabla f| = 1$. $\mathbf{p} = \mathbf{k} \Rightarrow |\nabla f\cdot\mathbf{p}| = 1$

13. (Continued)

$$\Rightarrow d\sigma = dy\,dx \Rightarrow \iint_{z=c} xyz\,d\sigma = \int_0^b \int_0^a cxy\,dx\,dy = \frac{a^2b^2c}{4}$$

$$\therefore \iint_S g(x,y,z)\,d\sigma = \frac{abc(ab + ac + bc)}{4}$$

15. $\nabla f = 2\,\mathbf{i} + 2\,\mathbf{j} + \mathbf{k}$ and $g(x,y,z) = x + y + (2 - 2x - 2y) = 2 - x - y \Rightarrow |\nabla f| = 3$. $\mathbf{p} = \mathbf{k} \Rightarrow |\nabla f \cdot \mathbf{p}| = 1 \Rightarrow$

$$d\sigma = 3\,dy\,dx \Rightarrow \iint_S (x + y + z)\,d\sigma = 3\int_0^1 \int_0^{1-x} (2 - x - y)\,dy\,dx = 2$$

17. $\nabla G = 2x\,\mathbf{i} + 2y\,\mathbf{j} + 2z\,\mathbf{k} \Rightarrow |\nabla G| = \sqrt{4x^2 + 4y^2 + 4z^2} = 2a$. $\mathbf{n} = \dfrac{2x\,\mathbf{i} + 2y\,\mathbf{j} + 2z\,\mathbf{k}}{2\sqrt{x^2 + y^2 + z^2}} = \dfrac{x\,\mathbf{i} + y\,\mathbf{j} + z\,\mathbf{k}}{a} \Rightarrow$

$\mathbf{F} \cdot \mathbf{n} = \dfrac{z^2}{a}$. $|\nabla G \cdot \mathbf{k}| = 2z \Rightarrow d\sigma = \dfrac{2a}{2z}\,dA = \dfrac{a}{z}\,dA$. $\therefore$ Flux $= \displaystyle\iint_R \frac{z^2}{a}\left(\frac{a}{z}\right) dA = \iint_R z\,dA =$

$$\iint_R \sqrt{a^2 - (x^2 + y^2)}\,dx\,dy = \int_0^{\pi/2} \int_0^a \sqrt{a^2 - r^2}\;r\,dr\,d\theta = \frac{a^3\pi}{6}$$

19. $\mathbf{n} = \dfrac{x\,\mathbf{i} + y\,\mathbf{j} + z\,\mathbf{k}}{a}$, $d\sigma = \dfrac{a}{z}\,dA$ (See Exercise 17) and $\mathbf{F} \cdot \mathbf{n} = \dfrac{xy}{a} - \dfrac{xy}{a} + \dfrac{z}{a} = \dfrac{z}{a}$.

$$\therefore \text{Flux} = \iint_R \frac{z}{a}\left(\frac{a}{z}\right) dA = \iint_R 1\,dA = \frac{\pi a^2}{4}$$

21. $\mathbf{n} = \dfrac{x\,\mathbf{i} + y\,\mathbf{j} + z\,\mathbf{k}}{a}$, $d\sigma = \dfrac{a}{z}\,dA$ (See Exercise 17) and $\mathbf{F} \cdot \mathbf{n} = \dfrac{x^2}{a} + \dfrac{y^2}{a} + \dfrac{z^2}{a} = a \Rightarrow$ Flux $= \displaystyle\iint_R a\left(\frac{a}{z}\right) dA$

$$= \iint_R \frac{a^2}{z}\,dA = \iint_R \frac{a^2}{\sqrt{a^2 - (x^2 + y^2)}}\,dA = \int_0^{\pi/2} \int_0^a \frac{a^2}{\sqrt{a^2 - r^2}}\,r\,dr\,d\theta = \frac{a^3\pi}{2}$$

23. $\nabla G = 2y\,\mathbf{i} + \mathbf{k} \Rightarrow |\nabla G| = \sqrt{4y^2 + 1} \Rightarrow \mathbf{n} = \dfrac{2y\,\mathbf{i} + \mathbf{k}}{\sqrt{4y^2 + 1}} \Rightarrow \mathbf{F} \cdot \mathbf{n} = \dfrac{2xy - 3z}{\sqrt{4y^2 + 1}}$. $\mathbf{p} = \mathbf{k} \Rightarrow |\nabla G \cdot \mathbf{k}| = 1 \Rightarrow$

$$d\sigma = \sqrt{4y^2 + 1}\,dA \Rightarrow \text{Flux} = \iint_R \left(\frac{2xy - 3z}{\sqrt{4y^2 + 1}}\right)\sqrt{4y^2 + 1}\,dA = \iint_R (2xy - 3z)\,dA =$$

23. (Continued)

$$\iint_R (2xy - 3(4 - y^2))\, dA = \int_0^1 \int_{-2}^2 (2xy - 12 + 3y^2)\, dy\, dx = -32$$

25. $\nabla G = -e^x \mathbf{i} + \mathbf{j} \Rightarrow |\nabla G| = \sqrt{e^{2x} + 1}$. $\mathbf{p} = \mathbf{i} \Rightarrow |\nabla G \cdot \mathbf{i}| = e^x$. $\mathbf{n} = \dfrac{e^x \mathbf{i} - \mathbf{j}}{\sqrt{e^{2x} + 1}} \Rightarrow \mathbf{F} \cdot \mathbf{n} = \dfrac{-2e^x - 2y}{\sqrt{e^{2x} + 1}}$.

$$d\sigma = \frac{\sqrt{e^{2x} + 1}}{e^x} dA \Rightarrow \text{Flux} = \iint_R \frac{-2e^x - 2y}{\sqrt{e^{2x} + 1}} \left( \frac{\sqrt{e^{2x} + 1}}{e^x} \right) dA = \iint_R -4\, dA = \int_0^1 \int_1^2 -4\, dy\, dz$$

$= -4$

27. $\nabla F = 2x\,\mathbf{i} + 2y\,\mathbf{j} + 2z\,\mathbf{k} \Rightarrow |\nabla F| = \sqrt{4x^2 + 4y^2 + 4z^2} = 2$. $\mathbf{p} = \mathbf{k} \Rightarrow |\nabla F \cdot \mathbf{k}| = 2z$ since $z \geq 0 \Rightarrow d\sigma =$

$$\frac{2}{2z} dA = \frac{1}{z} dA. \therefore M = \iint_S \delta\, d\sigma = \frac{\pi}{2}\delta.\ M_{xy} = \iint_S z\delta\, d\sigma = \delta \iint_S z\left(\frac{1}{z}\right) dA =$$

$$\delta \int_0^1 \int_0^{\sqrt{1-x^2}} dy\, dx = \frac{\pi}{4}\delta.\ \therefore \bar{z} = \frac{\frac{\pi}{4}\delta}{\frac{\pi}{2}\delta} = \frac{1}{2}.$$ Because of symmetry, $\bar{x} = \bar{y} = \frac{1}{2}$. $\therefore$ Centroid $= \left(\frac{1}{2}, \frac{1}{2}, \frac{1}{2}\right)$

29. Because of symmetry, $\bar{x} = \bar{y} = 0$. $M = \iint_S \delta\, d\sigma = \delta \iint_S d\sigma = \delta(\text{Area of S}) = 3\pi\sqrt{2}\,\delta$.

$\nabla F = 2x\,\mathbf{i} + 2y\,\mathbf{j} - 2z\,\mathbf{k} \Rightarrow |\nabla F| = \sqrt{4x^2 + 4y^2 + 4z^2} = 2\sqrt{x^2 + y^2 + z^2}$. $\mathbf{p} = \mathbf{k} \Rightarrow |\nabla F \cdot \mathbf{k}| = 2z \Rightarrow$

$$d\sigma = \frac{2\sqrt{x^2 + y^2 + z^2}}{2z} dA = \frac{\sqrt{x^2 + y^2 + z^2}}{z} dA = \frac{\sqrt{x^2 + y^2 + (x^2 + y^2)}}{z} dA = \frac{\sqrt{2}\sqrt{x^2 + y^2}}{z} dA.$$

$$\therefore M_{xy} = \delta \iint_S z\left(\frac{\sqrt{2}\sqrt{x^2 + y^2}}{z}\right) dA = \delta \iint_S \sqrt{2}\sqrt{x^2 + y^2}\, dA = \delta \int_0^{2\pi} \int_1^2 \sqrt{2}\, r^2\, dr\, d\theta =$$

$$\frac{14\pi\sqrt{2}}{3}\delta.\ \bar{z} = \frac{\frac{14\pi\sqrt{2}}{3}\delta}{3\pi\sqrt{2}\,\delta} = \frac{14}{9}.\ \therefore (\bar{x}, \bar{y}, \bar{z}) = \left(0, 0, \frac{14}{9}\right).\ I_z = \iint_S (x^2 + y^2)\,\delta\, d\sigma =$$

$$\iint_S (x^2 + y^2)\left(\frac{\sqrt{2}\sqrt{x^2 + y^2}}{z}\right)\delta\, dA = \delta\sqrt{2} \iint_S (x^2 + y^2)\, dA = \delta\sqrt{2} \int_0^{2\pi} \int_1^2 r^3\, dr\, d\theta = \frac{15\pi\sqrt{2}}{2}\delta.$$

$$R_z = \sqrt{I_z/M} = \frac{\sqrt{10}}{2}$$

31. $f_x(x,y) = 2x,\ f_y(x,y) = 2y \Rightarrow \sqrt{f_x^2 + f_y^2 + 1} = \sqrt{4x^2 + 4y^2 + 1} \Rightarrow \text{Area} = \iint_R \sqrt{4x^2 + 4y^2 + 1}\ dx\ dy =$

$$\int_0^{2\pi}\int_0^{\sqrt{3}} \sqrt{4r^2 + 1}\ r\ dr\ d\theta = \frac{\pi}{6}\left(13\sqrt{13} - 1\right)$$

33. $f_z(y,z) = -2y,\ f_y(y,z) = -2z \Rightarrow. \sqrt{f_y^2 + f_z^2 + 1} = \sqrt{4y^2 + 4z^2 + 1} \Rightarrow \text{Area} = \iint_R \sqrt{4y^2 + 4z^2 + 1}\ dy\ d$

$$= \int_0^{2\pi}\int_0^{1} \sqrt{4r^2 + 1}\ r\ dr\ d\theta = \frac{\pi}{6}(5\sqrt{5} - 1)$$

35. $y = \frac{z^2}{2} \Rightarrow f_x(x,z) = 0,\ f_z(x,z) = z \Rightarrow \sqrt{f_x^2 + f_z^2 + 1} = \sqrt{z^2 + 1} \Rightarrow \text{Area} = \int_0^2\int_0^1 \sqrt{z^2 + 1}\ dx\ dz =$

$\sqrt{5} + \frac{1}{2}\ln\left(\sqrt{5} + 2\right)$ (Note: On integrating the second time with respect to z, use the substitution $z = \tan\theta$ which means the integration will go from 0 to $\tan^{-1}2$.)

## SECTION 15.5 THE DIVERGENCE THEOREM

1. $\frac{\partial}{\partial x}(y - x) = -1,\ \frac{\partial}{\partial y}(z - y) = -1,\ \frac{\partial}{\partial z}(y - x) = 0 \Rightarrow \nabla \cdot \mathbf{F} = -2 \Rightarrow \text{Flux} = \int_{-1}^{1}\int_{-1}^{1}\int_{-1}^{1} -2\ dx\ dy\ dz = -16$

3. $\frac{\partial}{\partial x}(x^2) = 2x,\ \frac{\partial}{\partial y}(y^2) = 2y,\ \frac{\partial}{\partial x}(z^2) = 2z \Rightarrow \nabla \cdot \mathbf{F} = 2x + 2y + 2z \Rightarrow$

$$\text{Flux} = \int_{-1}^{1}\int_{-1}^{1}\int_{-1}^{1} (2x + 2y + 2z)\ dx\ dy\ dz = 0$$

5. $\frac{\partial}{\partial x}(y) = 0, \frac{\partial}{\partial y}(xy) = x, \frac{\partial}{\partial z}(-z) = -1 \Rightarrow \nabla \cdot \mathbf{F} = x - 1 \Rightarrow \text{Flux} = \iiint_{\text{solid}} (x - 1)\, dz\, dy\, dx =$

$$\int_0^{2\pi}\int_0^{2}\int_0^{r^2} (r\cos\theta - 1)\, dz\, r\, dr\, d\theta = -8\pi$$

7. $\frac{\partial}{\partial x}(x^2) = 2x, \frac{\partial}{\partial y}(-2xy) = -2x, \frac{\partial}{\partial z}(3xz) = 3x \Rightarrow \text{Flux} = \iiint_D 3x\, dx\, dy\, dz =$

$$\int_0^{\pi/2}\int_0^{\pi/2}\int_0^{2} 3\rho\sin\phi\cos\theta(\rho^2\sin\phi)\, d\rho\, d\phi\, d\theta = 3\pi$$

9. $\frac{\partial}{\partial x}(2xz) = 2z, \frac{\partial}{\partial y}(-xy) = -x, \frac{\partial}{\partial z}(-z^2) = -2z \Rightarrow \nabla \cdot \mathbf{F} = -x \Rightarrow \text{Flux} = \iiint_D -x\, dV =$

$$\int_0^{2}\int_0^{\sqrt{16-4x^2}}\int_0^{4-y} -x\, dz\, dy\, dx = -\frac{40}{3}$$

11. Let $\rho = \sqrt{x^2 + y^2 + z^2}$. Then $\frac{\partial\rho}{\partial x} = \frac{x}{\rho}, \frac{\partial\rho}{\partial y} = \frac{y}{\rho}, \frac{\partial\rho}{\partial z} = \frac{z}{\rho} \Rightarrow \frac{\partial}{\partial x}(\rho x) = \frac{\partial\rho}{\partial x}x + \rho = \frac{x^2}{\rho} = \rho, \frac{\partial}{\partial y}(\rho y) =$ $\frac{\partial\rho}{\partial y}y + \rho = \frac{y^2}{\rho} + \rho, \frac{\partial}{\partial z}(\rho z) = \frac{\partial\rho}{\partial z}z + \rho = \frac{z^2}{\rho} + \rho \Rightarrow \nabla \cdot \mathbf{F} = \frac{x^2 + y^2 + z^2}{\rho} + 3\rho = 4\rho$ since $\rho = \sqrt{x^2 + y^2 + z^2}$

$\Rightarrow \text{Flux} = \iiint_D 4\rho\, dV = \iiint_D 4\sqrt{x^2 + y^2 + z^2}\, dx\, dy\, dz =$

$$\int_0^{2\pi}\int_0^{\pi}\int_1^{\sqrt{2}} (4\rho)\rho^2\sin\phi\, d\rho\, d\phi\, d\theta = 12\pi$$

13. $\frac{\partial}{\partial x}(x) = 1, \frac{\partial}{\partial y}(-2y) = -2, \frac{\partial}{\partial z}(z + 3) = 1 \Rightarrow \nabla \cdot \mathbf{F} = 0 \Rightarrow$ Flux = 0 over the solid. In the xy–plane, z = 0,

$\mathbf{n} = -\mathbf{k}$, and $\mathbf{F} = x\,\mathbf{i} - 2y\,\mathbf{j} + 3\,\mathbf{k} \Rightarrow \mathbf{F} \cdot \mathbf{n} = -3 \Rightarrow \text{Flux} = \iint_{z=0} -3\, d\sigma = -3(\text{Area of the square}) =$

$-3$. In the yz–plane, $x = 0$, $\mathbf{n} = -\mathbf{i}$, $\mathbf{F} = -2y\,\mathbf{j} + (z + 3)\,\mathbf{k} \Rightarrow \mathbf{F} \cdot \mathbf{n} = 0 \Rightarrow$ Flux = 0.
In the xz–plane, $y = 0$, $\mathbf{n} = -\mathbf{j}$, $\mathbf{F} = x\,\mathbf{i} + (z + 3)\,\mathbf{k} \Rightarrow \mathbf{F} \cdot \mathbf{n} = 0 \Rightarrow$ Flux = 0. $\therefore$ The total flux =
$-3 + 0 + 0 + 1 + (-3) + (\text{Flux of the top}) = 0 \Rightarrow$ Flux of the top = 5

15. a) $\frac{\partial}{\partial x}(x) = 1, \frac{\partial}{\partial y}(y) = 1, \frac{\partial}{\partial z}(z) = 1 \Rightarrow \nabla \cdot \mathbf{F} = 3 \Rightarrow \text{Flux} = \iiint_D 3\, dV = 3 \iiint_D dV =$

3(Volume of the solid)

b) If $\mathbf{F}$ is orthogonal to $\mathbf{n}$ at every point of S, then $\mathbf{F} \cdot \mathbf{n} = 0$ everywhere $\Rightarrow \text{Flux} = \iint_S \mathbf{F} \cdot \mathbf{n}\, d\sigma = 0.$

But the Flux is 3(Volume of the solid) $\neq 0$. $\therefore$ $\mathbf{F}$ is not orthogonal to $\mathbf{n}$ at every point.

## SECTION 15.6 STOKES' THEOREM

1. $\text{curl}\,\mathbf{F} = \nabla \times \mathbf{F} = 2\,\mathbf{k}$, $\mathbf{n} = \mathbf{k} \Rightarrow \text{curl}\,\mathbf{F} \cdot \mathbf{n} = 2 \Rightarrow d\sigma = dx\,dy \Rightarrow \oint_C \mathbf{F} \cdot d\mathbf{r} = \iint_R 2\, dA =$

2(Area of the ellipse) = $4\pi$

3. $\text{curl}\,\mathbf{F} = \nabla \times \mathbf{F} = -x\,\mathbf{i} - 2x\,\mathbf{j} + (z - 1)\,\mathbf{k}$, $\mathbf{n} = \frac{\mathbf{i} + \mathbf{j} + \mathbf{k}}{\sqrt{3}} \Rightarrow \text{curl}\,\mathbf{F} \cdot \mathbf{n} = \frac{1}{\sqrt{3}}(-3x + z - 1) \Rightarrow d\sigma = \frac{\sqrt{3}}{1}\, dA \Rightarrow$

$$\oint_C \mathbf{F} \cdot d\mathbf{r} = \iint_R \frac{1}{\sqrt{3}}(-3x + z - 1)\sqrt{3}\, dA = \int_0^1 \int_0^{1-x} (-3x + z - 1)\, dy\, dx =$$

$$\int_0^1 \int_0^{1-x} (-3x + (1 - x - y) - 1)\, dy\, dx = \int_0^1 \int_0^{1-x} (-4x - y)\, dy\, dx = -\frac{5}{6}$$

5. $\text{curl}\,\mathbf{F} = \nabla \times \mathbf{F} = (2y - 0)\,\mathbf{i} + (2z - 2x)\,\mathbf{j} + (2x - 2y)\,\mathbf{k} = 2y\,\mathbf{i} + (2z - 2x)\,\mathbf{j} + (2x - 2y)\,\mathbf{k}$,

$\mathbf{n} = \mathbf{k} \Rightarrow \text{curl}\,\mathbf{F} \cdot \mathbf{n} = 2x - 2y \Rightarrow d\sigma = dx\,dy \Rightarrow \oint_C \mathbf{F} \cdot d\mathbf{r} = \int_{-1}^{1} \int_{-1}^{1} (2x - 2y)\, dx\, dy = 0$

7. $x = 3\cos t$, $y = 2\sin t \Rightarrow \mathbf{F} = (2\sin t)\,\mathbf{i} + (9\cos^2 t)\,\mathbf{j} + (9\cos^2 t + 16\sin^4 t)\sin e^{\sqrt{6\sin t\cos t\,(0)}}$ and
$\mathbf{r} = (3\cos t)\,\mathbf{i} + (2\sin t)\,\mathbf{j} \Rightarrow d\mathbf{r} = (-3\sin t)\,dt\,\mathbf{i} + (2\cos t)\,dt\,\mathbf{j} \Rightarrow \mathbf{F} \cdot d\mathbf{r} = -6\sin^2 t\,dt + 18\cos^3 t\,dt \Rightarrow$

7. (Continued)

$$\iint_S \nabla \times \mathbf{F} \cdot \mathbf{n}\, d\sigma = \int_0^{2\pi} (-6 \sin^2 t + 18 \cos^3 t)\, dt = -6\pi$$

9. Let $S_1$ and $S_2$ be oriented surfaces that span C and that induce the same positive direction on C.

Then $$\iint_{S_1} \nabla \times \mathbf{F} \cdot \mathbf{n_1}\, d\sigma_1 = \int_C \mathbf{F} \cdot d\mathbf{r} = \iint_{S_2} \nabla \times \mathbf{F} \cdot \mathbf{n_2}\, d\sigma_2$$

11. a) $\mathbf{F} = M\,\mathbf{i} + N\,\mathbf{j} + P\,\mathbf{k} \Rightarrow \text{curl } \mathbf{F} = \left(\frac{\partial P}{\partial y} - \frac{\partial N}{\partial z}\right)\mathbf{i} + \left(\frac{\partial M}{\partial z} - \frac{\partial P}{\partial x}\right)\mathbf{j} + \left(\frac{\partial N}{\partial x} - \frac{\partial M}{\partial y}\right)\mathbf{k}.$

$\nabla \cdot \nabla \times \mathbf{F} = \text{div(curl } \mathbf{F}) = \frac{\partial}{\partial x}\left(\frac{\partial P}{\partial y} - \frac{\partial N}{\partial z}\right) + \frac{\partial}{\partial y}\left(\frac{\partial M}{\partial z} - \frac{\partial P}{\partial x}\right) + \frac{\partial}{\partial z}\left(\frac{\partial N}{\partial x} - \frac{\partial M}{\partial y}\right) = \frac{\partial^2 P}{\partial x \partial y} - \frac{\partial^2 N}{\partial x \partial z} +$
$\frac{\partial^2 M}{\partial y \partial z} - \frac{\partial^2 P}{\partial y \partial x} + \frac{\partial^2 N}{\partial z \partial x} - \frac{\partial^2 M}{\partial z \partial y} = 0$ if the partial derivatives are continuous.

b) $\iint_S \nabla \times \mathbf{F} \cdot \mathbf{n}\, d\sigma = \iiint_D \nabla \cdot \nabla \times \mathbf{F}\, dV = \iiint_D 0\, dV = 0$ if the divergence theorem applies.

13. $\mathbf{F} = \nabla f = -\frac{1}{2}(x^2 + y^2 + z^2)^{-3/2}(2x)\,\mathbf{i} - \frac{1}{2}(x^2 + y^2 + z^2)^{-3/2}(2y)\,\mathbf{j} - \frac{1}{2}(x^2 + y^2 + z^2)^{-3/2}(2z)\,\mathbf{k} =$
$-x(x^2 + y^2 + z^2)^{-3/2}\,\mathbf{i} - y(x^2 + y^2 + z^2)^{-3/2}\,\mathbf{j} - z(x^2 + y^2 + z^2)^{-3/2}\,\mathbf{k}$

a) $\mathbf{r} = (a \cos t)\,\mathbf{i} + (a \sin t)\,\mathbf{j},\ 0 \le t \le 2\pi \Rightarrow d\mathbf{r} = (-a \sin t)\, dt\,\mathbf{i} + (a \cos t)\, dt\,\mathbf{j} \Rightarrow$
$\mathbf{F} \cdot d\mathbf{r} = -x(x^2 + y^2 + z^2)^{-3/2}(-a \sin t)\, dt - y(x^2 + y^2 + z^2)^{-3/2}(a \cos t)\, dt =$
$-\frac{a \cos t}{a^3}(-a \sin t)\, dt - \frac{a \sin t}{a^3}(a \cos t)\, dt = 0 \Rightarrow \int_C \mathbf{F} \cdot d\mathbf{r} = 0$

b) $\oint_C \mathbf{F} \cdot d\mathbf{r} = \iint_S \text{curl } \mathbf{F} \cdot \mathbf{n}\, d\sigma = \iint_S \text{curl}(\nabla f) \cdot \mathbf{n}\, d\sigma = \iint_S \mathbf{0} \cdot \mathbf{n}\, d\sigma =$
$\iint_S 0\, d\sigma = 0$

## SECTION 15.7 PATH INDEPENDENCE, POTENTIAL FUNCTIONS, AND CONSERVATIVE FIELDS

1. $\frac{\partial P}{\partial y} = x = \frac{\partial N}{\partial z}, \frac{\partial M}{\partial z} = y = \frac{\partial P}{\partial x}, \frac{\partial N}{\partial x} = z = \frac{\partial M}{\partial y} \Rightarrow$ Conservative

3. $\frac{\partial P}{\partial y} = -1 \neq \frac{\partial N}{\partial z} \Rightarrow$ Not Conservative

5. $\frac{\partial N}{\partial x} = 0 \neq \frac{\partial M}{\partial y} \Rightarrow$ Not Conservative

7. $\frac{\partial f}{\partial x} = 2x \Rightarrow f(x,y,z) = x^2 + g(y,z)$. $\frac{\partial f}{\partial y} = \frac{\partial g}{\partial y} = 3y \Rightarrow g(y,z) = \frac{3y^2}{2} + h(z) \Rightarrow f(x,y,z) = x^2 + \frac{3y^2}{2} + h(z)$. $\frac{\partial f}{\partial z} = h'(z)$ $= 4z. \Rightarrow h(z) = 2z^2 + C \Rightarrow f(x,y,z) = x^2 + \frac{3y^2}{2} + 2z^2 + C$

9. $\frac{\partial f}{\partial x} = e^{y+2z} \Rightarrow f(x,y,z) = x\,e^{y+2z} + g(y,z)$. $\frac{\partial f}{\partial y} = x\,e^{y+2z} + \frac{\partial g}{\partial y} = x\,e^{y+2z} \Rightarrow \frac{\partial g}{\partial y} = 0$. Then $f(x,y,z) = x\,e^{y+2z} +$ $h(z)$. $\frac{\partial f}{\partial z} = 2x\,e^{y+2z} + h'(z) = 2x\,e^{y+2z} \Rightarrow h'(z) = 0 \Rightarrow h(z) = C$. $\therefore$ $f(x,y,z) = x\,e^{y+2z} + C$

11. Let $\mathbf{F}(x,y,z) = 2x\,\mathbf{i} + 2y\,\mathbf{j} + 2z\,\mathbf{k} \Rightarrow \frac{\partial P}{\partial y} = 0 = \frac{\partial N}{\partial z}, \frac{\partial M}{\partial z} = 0 = \frac{\partial P}{\partial x}, \frac{\partial N}{\partial x} = 0 = \frac{\partial M}{\partial y} \Rightarrow M\,dx + N\,dy + P\,dz$ is exact. $\frac{\partial f}{\partial x} = 2x \Rightarrow f(x,y,z) = x^2 + g(y,z)$. $\frac{\partial f}{\partial y} = \frac{\partial g}{\partial y} = 2y \Rightarrow g(y,z) = y^2 + h(z) \Rightarrow f(x,y,z) = x^2 + y^2 + h(z)$.

$\frac{\partial f}{\partial z} = h'(z) = 2z \Rightarrow h(z) = z^2 + C$. $\therefore$ $f(x,y,z) = x^2 + y^2 + z^2 + C \Rightarrow \int_{(0,0,0)}^{(2,3,-6)} 2x\,dx + 2y\,dy + 2z\,dz =$ $f(2,3,-6) - f(0,0,0) = 49$

13. Let $\mathbf{F}(x,y,z) = 2xy\,\mathbf{i} + (x^2 - z^2)\,\mathbf{j} - 2yz\,\mathbf{k} \Rightarrow \frac{\partial P}{\partial y} = -2z = \frac{\partial N}{\partial z}, \frac{\partial M}{\partial z} = 0 = \frac{\partial P}{\partial x}, \frac{\partial N}{\partial x} = 2x = \frac{\partial M}{\partial y} \Rightarrow M\,dx + N\,dy +$ $P\,dz$ is exact. $\frac{\partial f}{\partial x} = 2xy \Rightarrow f(x,y,z) = x^2y + g(y,z)$. $\frac{\partial f}{\partial y} = x^2 + \frac{\partial g}{\partial y} = x^2 - z^2 \Rightarrow \frac{\partial g}{\partial y} = -z^2 \Rightarrow g(y,z) = -yz^2 +$ $h(z) \Rightarrow f(x,y,z) = x^2y - yz^2 + h(z)$. $\frac{\partial f}{\partial z} = -2yz + h'(z) = -2yz \Rightarrow h'(z) = 0 \Rightarrow h(z) = C \Rightarrow f(x,y,z) =$ $x^2y - yz^2 + C \Rightarrow \int_{(0,0,0)}^{(1,2,3)} 2xy\,dx + (x^2 - z^2)\,dy - 2yz\,dz = f(1,2,3) - f(0,0,0) = -16$

15. Let $\mathbf{F}(x,y,z) = (\sin y \cos x)\,\mathbf{i} + (\cos y \sin x)\,\mathbf{j} + \mathbf{k} \Rightarrow \frac{\partial P}{\partial y} = 0 = \frac{\partial N}{\partial z}, \frac{\partial M}{\partial z} = 0 = \frac{\partial P}{\partial x}, \frac{\partial N}{\partial x} = \cos y \cos x = \frac{\partial M}{\partial y} \Rightarrow$ $M\,dx + N\,dy + P\,dz$ is exact. $\frac{\partial f}{\partial x} = \sin y \cos x \Rightarrow f(x,y,z) = \sin y \sin x + g(y,z)$. $\frac{\partial f}{\partial y} = \cos y \sin x + \frac{\partial g}{\partial y} =$ $\cos y \sin x \Rightarrow \frac{\partial g}{\partial y} = 0 \Rightarrow g(y,z) = h(z) \Rightarrow f(x,y,z) = \sin y \sin x + h(z)$. $\frac{\partial f}{\partial z} = h'(z) = 1 \Rightarrow h(z) = z + C \Rightarrow$ $f(x,y,z) = \sin y \sin x + z + C \Rightarrow \int_{(1,0,0)}^{(0,1,1)} \sin y \cos x\,dx + \cos y \sin x\,dy + dz = f(0,1,1) - f(1,0,0) = 1$

17. Let $\mathbf{F}(x,y,z) = (2\cos y)\,\mathbf{i} + \left(\frac{1}{y} - 2x\sin y\right)\mathbf{j} + \frac{1}{z}\mathbf{k} \Rightarrow \frac{\partial P}{\partial y} = 0 = \frac{\partial N}{\partial z},\ \frac{\partial M}{\partial z} = 0 = \frac{\partial P}{\partial x},\ \frac{\partial N}{\partial x} = -2\sin y = \frac{\partial M}{\partial y} \Rightarrow$ $M\,dx + N\,dy + P\,dz$ is exact. $\frac{\partial f}{\partial x} = 2\cos y \Rightarrow f(x,y,z) = 2x\cos y + g(y,z).\ \frac{\partial f}{\partial y} = -2x\sin y + \frac{\partial g}{\partial y} = \frac{1}{y} - 2x\sin y$ $\Rightarrow \frac{\partial g}{\partial y} = \frac{1}{y} \Rightarrow g(y,z) = \ln y + h(z) \Rightarrow f(x,y,z) = 2x\cos y + \ln y + h(z).\ \frac{\partial f}{\partial z} = h'(z) = \frac{1}{z} \Rightarrow h(z) = \ln z + C \Rightarrow$

$$f(x,y,z) = 2x\cos y + \ln y + \ln z + C \Rightarrow \int_{(0,2,1)}^{(1,\pi/2,2)} 2\cos y\,dx + \left(\frac{1}{y} - 2x\sin y\right)dy + \frac{1}{z}dz =$$

$$f\left(1,\frac{\pi}{2},2\right) - f(0,2,1) = \ln\frac{\pi}{2}$$

19. Let $\mathbf{F}(x,y,z) = \frac{1}{y}\mathbf{i} + \left(\frac{1}{z} - \frac{x}{y^2}\right)\mathbf{j} - \frac{y}{z^2}\mathbf{k} \Rightarrow \frac{\partial P}{\partial y} = -\frac{1}{z^2} = \frac{\partial N}{\partial z},\ \frac{\partial M}{\partial z} = 0 = \frac{\partial P}{\partial x},\ \frac{\partial N}{\partial x} = -\frac{1}{y^2} = \frac{\partial M}{\partial y} \Rightarrow$ $M\,dx + N\,dy + P\,dz$ is exact. $\frac{\partial f}{\partial x} = \frac{1}{y} \Rightarrow f(x,y,z) = \frac{x}{y} + g(y,z).\ \frac{\partial f}{\partial y} = -\frac{x}{y^2} + \frac{\partial g}{\partial y} = \frac{1}{z} - \frac{x}{y^2} \Rightarrow \frac{\partial g}{\partial y} = \frac{1}{z} \Rightarrow$ $g(y,z) = \frac{y}{z} + h(z) \Rightarrow f(x,y,z) = \frac{x}{y} + \frac{y}{z} + h(z).\ \frac{\partial f}{\partial z} = -\frac{y}{z^2} + h'(z) = -\frac{y}{z^2} \Rightarrow h'(z) = 0 \Rightarrow h(z) = C \Rightarrow$

$$f(x,y,z) = \frac{x}{y} + \frac{y}{z} + C \Rightarrow \int_{(1,1,1)}^{(2,2,2)} \frac{1}{y}dx + \left(\frac{1}{z} - \frac{x}{y^2}\right)dy - \frac{y}{z^2}dz = f(2,2,2) - f(1,1,1) = 0$$

21. Let $x - 1 = t,\ y - 1 = 2t,\ z - 1 = -2t,\ 0 \le t \le 1 \Rightarrow dx = dt,\ dy = 2\,dt,\ dz = -2\,dt \Rightarrow$

$$\int_{(1,1,1)}^{(2,3,-1)} y\,dx + x\,dy + 4\,dz = \int_0^1 (2t+1)dt + (t+1)2\,dt + 4(-2\,dt) = \int_0^1 (4t-5)\,dt = -3$$

23. $\frac{\partial P}{\partial y} = 0 = \frac{\partial N}{\partial z},\ \frac{\partial M}{\partial z} = 2z = \frac{\partial P}{\partial x},\ \frac{\partial N}{\partial x} = 0 = \frac{\partial M}{\partial y} \Rightarrow M\,dx + N\,dy + P\,dz$ is exact $\Rightarrow \mathbf{F}$ is conservative $\Rightarrow$ path independence.

25. $\frac{\partial P}{\partial y} = 0,\ \frac{\partial N}{\partial z} = 0,\ \frac{\partial M}{\partial z} = 0,\ \frac{\partial P}{\partial x} = 0,\ \frac{\partial N}{\partial x} = \frac{y^2 - x^2}{(x^2+y^2)^2},\ \frac{\partial M}{\partial y} = \frac{y^2 - x^2}{(x^2+y^2)^2} \Rightarrow \text{curl}\,\mathbf{F} = \left(\frac{y^2 - x^2}{(x^2+y^2)^2} - \frac{y^2 - x^2}{(x^2+y^2)^2}\right)\mathbf{k}$ $= \mathbf{0}.\ x^2 + y^2 = 1 \Rightarrow \mathbf{r} = (a\cos t)\,\mathbf{i} + (a\sin t)\,\mathbf{j} \Rightarrow d\mathbf{r} = (-a\sin t)\,\mathbf{i} + (a\cos t)\,\mathbf{j} \Rightarrow \mathbf{F} = \frac{-a\sin t}{a^2}\mathbf{i} + \frac{a\cos t}{a^2}\mathbf{j} +$

$$z\,\mathbf{k} \Rightarrow \mathbf{F}\cdot d\mathbf{r} = \frac{a^2\sin^2 t}{a^2} + \frac{a^2\cos^2 t}{a^2} = 1 \Rightarrow \int_C \mathbf{F}\cdot d\mathbf{r} = \int_0^{2\pi} 1\,dt = 2\pi$$

## PRACTICE EXERCISES

1. Path 1: $\mathbf{r} = t\mathbf{i} + t\mathbf{j} + t\mathbf{k} \Rightarrow x = t,\ y = t,\ z = t,\ 0 \le t \le 1 \Rightarrow f(g(t),h(t),h(t)) = 3 - 3t^2$ and $\frac{dx}{dt} = 1, \frac{dy}{dt} = 1,$

$\frac{dz}{dt} = 1 \Rightarrow \sqrt{\left(\frac{dx}{dt}\right)^2 + \left(\frac{dy}{dt}\right)^2 + \left(\frac{dz}{dt}\right)^2}\, dt = \sqrt{3}\, dt \Rightarrow \int_C f(x,y,z)\, ds = \int_0^1 \sqrt{3}\left(3 - 3t^2\right) dt = 2\sqrt{3}$

Path 2: $\mathbf{r}_1 = t\mathbf{i} + t\mathbf{j},\ 0 \le t \le 1 \Rightarrow x = t,\ y = t,\ z = 0 \Rightarrow f(g(t),h(t),h(t)) = 2t - 3t^2 + 3$ and $\frac{dx}{dt} = 1, \frac{dy}{dt} = 1,$

$\frac{dz}{dt} = 0 \Rightarrow \sqrt{\left(\frac{dx}{dt}\right)^2 + \left(\frac{dy}{dt}\right)^2 + \left(\frac{dz}{dt}\right)^2}\, dt = \sqrt{2}\, dt \Rightarrow \int_{C_1} f(x,y,z)\, ds = \int_0^1 \sqrt{2}\left(2t - 3t^2 + 3\right) dt =$

$3\sqrt{2}$. $\mathbf{r}_2 = \mathbf{i} + \mathbf{j} + t\mathbf{k} \Rightarrow x = 1,\ y = 1,\ z = t \Rightarrow f(g(t),h(t),h(t)) = 2 - 2t$ and $\frac{dx}{dt} = 0, \frac{dy}{dt} = 0, \frac{dz}{dt} = 1 \Rightarrow$

$\sqrt{\left(\frac{dx}{dt}\right)^2 + \left(\frac{dy}{dt}\right)^2 + \left(\frac{dz}{dt}\right)^2}\, dt = dt \Rightarrow \int_{C_2} f(x,y,z)\, ds = \int_0^1 (2 - 2t)\, dt = 1.\ \therefore \int_C f(x,y,z)\, ds =$

$\int_{C_1} f(x,y,z)\, ds + \int_{C_2} f(x,y,z)\, ds = 3\sqrt{2} + 1$

3. $\mathbf{r} = (a \cos t)\mathbf{j} + (a \sin t)\mathbf{k} \Rightarrow x = 0,\ y = a \cos t,\ z = a \sin t \Rightarrow f(g(t),h(t),h(t)) = \sqrt{a^2 \sin^2 t} = a\,|\sin t|$ and

$\frac{dx}{dt} = 0, \frac{dy}{dt} = -a \sin t, \frac{dz}{dt} = a \cos t \Rightarrow \sqrt{\left(\frac{dx}{dt}\right)^2 + \left(\frac{dy}{dt}\right)^2 + \left(\frac{dz}{dt}\right)^2}\, dt = a\, dt \Rightarrow \int_C f(x,y,z)\, ds =$

$\int_0^{2\pi} a^2\,|\sin t|\, dt = \int_0^{\pi} a^2 \sin t\, dt + \int_{\pi}^{2\pi} -a^2 \sin t\, dt = 4a^2$

5. a) $\mathbf{r} = \sqrt{2}\,t\,\mathbf{i} + \sqrt{2}\,\mathbf{j} + (4 - t^2)\,\mathbf{k},\ 0 \le t \le 1 \Rightarrow x = \sqrt{2}\,t,\ y = \sqrt{2}\,t,\ z = 4 - t^2 \Rightarrow \frac{dx}{dt} = \sqrt{2}, \frac{dy}{dt} = \sqrt{2}, \frac{dz}{dt} = -2t$

$\Rightarrow \sqrt{\left(\frac{dx}{dt}\right)^2 + \left(\frac{dy}{dt}\right)^2 + \left(\frac{dz}{dt}\right)^2}\, dt = \sqrt{4 + t^2}\, dt \Rightarrow M = \int_C \delta(x,y,z)\, ds = \int_0^1 3t\sqrt{4 + 4t^2}\, dt =$

$4\sqrt{2} - 2$

b) $M = \int_C \delta(x,y,z)\, ds = \int_0^1 \sqrt{4 + 4t^2}\, dt = \sqrt{2} + \ln(1 + \sqrt{2})$

7. $\mathbf{r} = t\mathbf{i} + \frac{2\sqrt{2}}{3} t^{3/2}\mathbf{j} + \frac{t^2}{2}\mathbf{k},\ 0 \le t \le 2 \Rightarrow x = t,\ y = \frac{2\sqrt{2}}{3} t^{3/2},\ z = \frac{t^2}{2} \Rightarrow \frac{dx}{dt} = t, \frac{dy}{dt} = \sqrt{2}\, t^{1/2}, \frac{dz}{dt} = t \Rightarrow$

$\sqrt{\left(\frac{dx}{dt}\right)^2 + \left(\frac{dy}{dt}\right)^2 + \left(\frac{dz}{dt}\right)^2}\, dt = \sqrt{(t + 1)^2}\, dt = |t + 1|\, dt = (t + 1)dt$ on the domain given. Then $M_{yz} =$

7. (Continued)

$\int_C x\delta\, ds = \int_0^2 t\left(\frac{1}{t+1}\right)(t+1)\, dt = \int_0^2 t\, dt = 2.$ $M = \int_C \delta\, ds = \int_0^2 \frac{1}{t+1}(t+1)\, dt = \int_0^2 dt = 2$

$M_{xz} = \int_C y\delta\, ds = \int_0^2 \frac{2\sqrt{2}}{3} t^{3/2}\left(\frac{1}{t+1}\right)(t+1)\, dt = \int_0^2 \frac{2\sqrt{2}}{3} t^{3/2}\, dt = \frac{32}{15}.$ $M_{xy} = \int_C z\delta\, ds =$

$\int_0^2 \frac{t^2}{2}\left(\frac{1}{t+1}\right)(t+1)\, dt = \int_0^2 \frac{t^2}{2}\, dt = \frac{4}{3}.$ $\therefore\ \bar{x} = M_{yz}/M = \frac{2}{2} = 1,\ \bar{y} = M_{xz}/M = \frac{32/15}{2} = \frac{16}{15},\ \bar{z} = M_{xy}/M =$

$\frac{4/3}{2} = \frac{2}{3}.$ $I_x = \int_C (y^2 + z^2)\delta\, ds = \int_0^2 \left(\frac{8}{9}t^3 + \frac{t^4}{4}\right)dt = \frac{232}{45}.$ $I_y = \int_C (x^2 + z^2)\delta\, ds =$

$\int_0^2 \left(t^2 + \frac{t^4}{4}\right)dt = \frac{64}{15}$ $I_z = \int_C (y^2 + x^2)\delta\, ds = \int_0^2 \left(t^2 + \frac{8}{9}t^3\right) dt = \frac{56}{9}.$ $R_x = \sqrt{I_x/M}$

$= \sqrt{\frac{232/45}{2}} = \frac{2}{3}\sqrt{\frac{29}{5}}.$ $R_y = \sqrt{I_y/M} = \sqrt{\frac{64/15}{2}} = 4\sqrt{\frac{2}{15}}.$ $R_z = \sqrt{I_z/M} = \sqrt{\frac{56/9}{2}} = \frac{2}{3}\sqrt{7}$

9. a) $x^2 + y^2 = 1 \Rightarrow \mathbf{r} = (\cos t)\,\mathbf{i} + (\sin t)\,\mathbf{j},\ 0 \le t \le \pi \Rightarrow x = \cos t,\ y = \sin t \Rightarrow \mathbf{F} = (\cos t + \sin t)\,\mathbf{i} - \mathbf{j}$ and $\frac{d\mathbf{r}}{dt} = (-\sin t)\,\mathbf{i} + (\cos t)\,\mathbf{j} \Rightarrow \mathbf{F}\cdot\frac{d\mathbf{r}}{dt} = -\sin t\cos t - \sin^2 t - \cos t \Rightarrow$ Flow $=$

$\int_0^\pi (-\sin t\cos t - \sin^2 t - \cos t)\, dt = -\frac{1}{2}\pi$

b) $\mathbf{r} = -t\,\mathbf{i},\ -1 \le t \le 1 \Rightarrow x = -t,\ y = 0 \Rightarrow \mathbf{F} = -t\,\mathbf{i} - t^2\,\mathbf{j}$ and $\frac{d\mathbf{r}}{dt} = -\mathbf{i} \Rightarrow \mathbf{F}\cdot\frac{d\mathbf{r}}{dt} = t \Rightarrow$ Flow $= \int_{-1}^1 t\, dt =$

0

c) $\mathbf{r}_1 = (1-t)\,\mathbf{i} - t\,\mathbf{j},\ 0 \le t \le 1 \Rightarrow \mathbf{F}_1 = (1-2t)\,\mathbf{i} - (1 - 2t - 2t^2)\,\mathbf{j}$ and $\frac{d\mathbf{r}_1}{dt} = -\mathbf{i} - \mathbf{j} \Rightarrow \mathbf{F}_1\cdot\frac{d\mathbf{r}_1}{dt} = 2t^2 \Rightarrow$

$\text{Flow}_1 = \int_0^1 2t^2\, dt = \frac{2}{3}.$ $\mathbf{r}_2 = -t\,\mathbf{i} + (t-1)\,\mathbf{j},\ 0 \le t \le 1 \Rightarrow \mathbf{F}_2 = -\mathbf{i} - (2t^2 - 2t + 1)\,\mathbf{j}$ and $\frac{d\mathbf{r}_2}{dt} = -\mathbf{i} +$

$\mathbf{j} \Rightarrow \mathbf{F}_2\cdot\frac{d\mathbf{r}_2}{dt} = -2t^3 + 4t^2 - 2t + 1 \Rightarrow \text{Flow}_2 = \int_0^1 (-2t^3 + 4t^2 - 2t + 1)\, dt = \frac{1}{3}.$ $\therefore$ Flow $= \text{Flow}_1 +$

$\text{Flow}_2 = 1$

11. $M = 2xy + x,\ N = xy - y \Rightarrow \frac{\partial M}{\partial x} = 2y + 1,\ \frac{\partial M}{\partial y} = 2x,\ \frac{\partial N}{\partial x} = y,\ \frac{\partial N}{\partial y} = x - 1 \Rightarrow$ Flux =

$$\iint_R (2y + 1 + x - 1)\,dy\,dx = \int_0^1\int_0^1 (2y + x)\,dy\,dx = \frac{3}{2}.\ \text{Circ} = \iint_R (y - 2x)\,dy\,dx =$$

$$\int_0^1\int_0^1 (y - 2x)\,dy\,dx = -\frac{1}{2}$$

13. Let $M = 4x^3y,\ N = x^4 \Rightarrow \frac{\partial M}{\partial y} = 4x^3,\ \frac{\partial N}{\partial x} = 4x^3 \Rightarrow \oint_C 4x^3y\,dx + x^4\,dy = \iint_R (4x^3 - 4x^3)\,dx\,dy = 0$

15. Let $M = 8x \sin y,\ N = -8y \cos x \Rightarrow \frac{\partial M}{\partial y} = 8x \cos y,\ \frac{\partial N}{\partial x} = 8y \sin x \Rightarrow \int_C 8x \sin y\,dx - 8y \cos x\,dy =$

$$\iint_R (8y \sin x - 8x \cos y)\,dy\,dx = \int_0^{\pi/2}\int_0^{\pi/2} (8y \sin x - 8x \cos y)\,dy\,dx = 0$$

17. Let $z = 1 - x - y \Rightarrow f_x(x,y) = -1,\ f_y(x,y) = -1 \Rightarrow \sqrt{f_x^2 + f_y^2 + 1} = \sqrt{3} \Rightarrow \text{Area} = \iint_R \sqrt{3}\,dx\,dy =$

$\sqrt{3}$(Area of the circlular region in the xy–plane) $= \pi\sqrt{3}$

19. $\nabla f = 2x\,\mathbf{i} + 2y\,\mathbf{j} + 2z\,\mathbf{k},\ \mathbf{p} = \mathbf{k} \Rightarrow |\nabla f| = \sqrt{4x^2 + 4y^2 + 4z^2} = 2\sqrt{x^2 + y^2 + z^2} = 2$ and $|\nabla f \cdot \mathbf{p}| = |2z| = 2z$

since $z \ge 0 \Rightarrow S = \iint_R \frac{2}{2z}\,dA = \iint_R \frac{1}{z}\,dA = \iint_R \frac{1}{\sqrt{1 - x^2 - y^2}}\,dx\,dy =$

$$\int_0^{2\pi}\int_0^{1/2} \frac{1}{\sqrt{1 - r^2}}\,r\,dr\,d\theta = \int_0^{2\pi} \left(1 - \frac{\sqrt{3}}{2}\right) d\theta = 2\pi - \pi\sqrt{3}$$

21. a) $\nabla f = 2y\,\mathbf{j} - \mathbf{k}$, $\mathbf{p} = \mathbf{k} \Rightarrow |\nabla f| = \sqrt{4y^2+1}$ and $|\nabla f \cdot \mathbf{p}| = 1 \Rightarrow d\sigma = \sqrt{4y^2+1}\ dx\,dy \Rightarrow$

$$\iint_S g(x,y,z)\,d\sigma = \iint_S \frac{z}{\sqrt{4y^2+1}}\sqrt{4y^2+1}\,dx\,dy = \iint_S y(y^2-1)\,d\sigma =$$

$$\int_{-1}^{1}\int_{0}^{3} (y^3 - y)\,dx\,dy = 0$$

b) $$\iint_S g(x,y,z)\,d\sigma = \iint_S \frac{z}{\sqrt{4y^2+1}}\sqrt{4y^2+1}\,dx\,dy = \int_{-1}^{1}\int_{0}^{3} (y^2-1)\,dx\,dy = -4$$

23. On the face, $z = 1$: $G(x,y,z) = G(x,y,1) = z \Rightarrow \nabla G = \mathbf{k} \Rightarrow |\nabla G| = 1$. $\mathbf{n} = \mathbf{k} \Rightarrow \mathbf{F} \cdot \mathbf{n} = 2xz = 2x$ since $z = 1$

$d\sigma = dA \Rightarrow \text{Flux} = \iint_R 2x\,dx\,dy = \int_0^1\int_0^1 2x\,dx\,dy = 1$. On the face, $z = 0$: $G(x,y,z) = G(x,y,0) =$

$z \Rightarrow \nabla G = \mathbf{k} \Rightarrow |\nabla G| = 1$. $\mathbf{n} = -\mathbf{k} \Rightarrow \mathbf{F} \cdot \mathbf{n} = -2xz = 0$ since $z = 0 \Rightarrow \text{Flux} = \iint_R 0\,dx\,dy = 0$

On the face, $x = 1$: $G(x,y,z) = G(1,y,z) = x \Rightarrow \nabla G = \mathbf{i} \Rightarrow |\nabla G| = 1$. $\mathbf{n} = \mathbf{i} \Rightarrow \mathbf{F} \cdot \mathbf{n} = 2xy = 2y$ since $x = 1$

$\text{Flux} = \int_0^1\int_0^1 2y\,dy\,dz = 1$ On the face, $x = 0$: $G(x,y,z) = G(0,y,z) = x \Rightarrow \nabla G = \mathbf{i} \Rightarrow |\nabla G| = 1$. $\mathbf{n} = -\mathbf{i}$

$\Rightarrow \mathbf{F} \cdot \mathbf{n} = -2xy = 0$ since $x = 0 \Rightarrow \text{Flux} = 0$. On the face, $y = 1$: $G(x,y,z) = G(x,1,z) = y \Rightarrow \nabla G = \mathbf{j} \Rightarrow |\nabla G|$

$= 1$. $\mathbf{n} = \mathbf{j} \Rightarrow \mathbf{F} \cdot \mathbf{n} = 2yz = 2z$ since $y = 1 \Rightarrow \text{Flux} = \int_0^1\int_0^1 2z\,dz\,dx = 1$. On the face, $y = 0$: $G(x,y,z) =$

$G(z,0,z) = y \Rightarrow \nabla G = \mathbf{j} \Rightarrow |\nabla G| = 1$. $\mathbf{n} = -\mathbf{j} \Rightarrow \mathbf{F} \cdot \mathbf{n} = -2yz = 0$ since $y = 0 \Rightarrow \text{Flux} = 0$.
$\therefore$ Total Flux = 3

25. Because of symmetry $\bar{x} = \bar{y} = 0$. Let $F(x,y,z) = x^2 + y^2 + z^2 = 25 \Rightarrow \nabla F = 2x\,\mathbf{i} + 2y\,\mathbf{j} + 2z\,\mathbf{k} \Rightarrow$

$|\nabla F| = \sqrt{4x^2 + 4y^2 + 4z^2} = 10$, $\mathbf{p} = \mathbf{k} \Rightarrow |\nabla F \cdot \mathbf{p}| = 2z$ since $z \ge 0 \Rightarrow M = \iint_R \delta(x,y,z)\,d\sigma =$

$\iint_R z\left(\frac{10}{2z}\right) dA = \iint_R 5\,dA = 5(\text{Area of the circular region}) = 80\pi$. $M_{xy} = \iint_R z\delta\,d\sigma =$

25. (Continued)

$$\iint_R z\,dA = \iint_R 5\sqrt{25 - x^2 - y^2}\,dx\,dy = \int_0^{2\pi}\int_0^4 5\sqrt{25 - r^2}\,r\,dr\,d\theta = \int_0^{2\pi} \frac{490}{3}\,d\theta = \frac{980}{3}\pi$$

$$\therefore \bar{z} = \frac{\frac{980}{3}\pi}{80\pi} = \frac{49}{12}.\ \text{Thus } (\bar{x},\bar{y},\bar{z}) = \left(0,0,\frac{49}{12}\right).\ I_z = \iint_R (x^2 + y^2)\delta\,d\sigma = \iint_R 5(x^2 + y^2)\,dx\,dy =$$

$$\int_0^{2\pi}\int_0^4 5r^3\,dr\,d\theta = \int_0^{2\pi} 320\,d\theta = 640\pi.\ R_z = \sqrt{I_z/M} = \sqrt{\frac{640\pi}{80\pi}} = 2\sqrt{2}$$

27. $\frac{\partial}{\partial x}(2xy) = 2y,\ \frac{\partial}{\partial y}(2yz) = 2z,\ \frac{\partial}{\partial z}(2xz) = 2x \Rightarrow \nabla \cdot \mathbf{F} = 2x + 2y + 2z \Rightarrow \text{Flux} = \iiint_D (2x + 2y + 2z)\,dV$

$$\int_0^1\int_0^1\int_0^1 (2x + 2y + 2z)\,dx\,dy\,dz = \int_0^1\int_0^1 (1 + 2y + 2z)\,dy\,dz = \int_0^1 (2 + 2z)\,dz = 3$$

29. $\frac{\partial}{\partial x}(-2x) = -2,\ \frac{\partial}{\partial y}(-3y) = -3,\ \frac{\partial}{\partial z}(z) = 1 \Rightarrow \nabla \cdot \mathbf{F} = -4 \Rightarrow \text{Flux} = \iiint_D -4\,dV =$

$$-4\int_0^{2\pi}\int_0^1\int_{r^2}^{\sqrt{2-r^2}} dz\,r\,dr\,d\theta = -4\int_0^{2\pi}\int_0^1 (r\sqrt{2 - r^2} - r^3)\,dr\,d\theta = -4\int_0^{2\pi}\left(-\frac{7}{12} + \frac{2}{3}\sqrt{2}\right)d\theta =$$

$$\frac{2}{3}\pi\left(7 - 8\sqrt{2}\right)$$

31. $\nabla f = 2\,\mathbf{i} + 6\,\mathbf{j} - 3\,\mathbf{k} \Rightarrow \nabla \times \mathbf{F} = -2y\,\mathbf{k}.\ \mathbf{n} = \frac{2\,\mathbf{i} + 6\,\mathbf{j} - 3\,\mathbf{k}}{\sqrt{4 + 36 + 9}} = \frac{2\,\mathbf{i} + 6\,\mathbf{j} - 3\,\mathbf{k}}{7} \Rightarrow \nabla \times \mathbf{F} \cdot \mathbf{n} = \frac{6}{7}y.\ \mathbf{p} = \mathbf{k} \Rightarrow$

$$|\nabla f \cdot \mathbf{p}| = 3 \Rightarrow d\sigma = \frac{7}{3}\,dA \Rightarrow \oint_C \mathbf{F} \cdot d\mathbf{r} = \iint_R \frac{6}{7}y\,d\sigma = \iint_R \frac{6}{7}y\left(\frac{7}{3}\,dA\right) = \iint_R 2y\,dx\,dy =$$

$$\int_0^{2\pi}\int_0^1 2r\sin\theta\ r\,dr\,d\theta = \int_0^{2\pi} \frac{2}{3}\sin\theta\,d\theta = 0$$

33. $\frac{\partial P}{\partial y} = 0 = \frac{\partial N}{\partial z}, \frac{\partial M}{\partial z} = 0 = \frac{\partial P}{\partial x}, \frac{\partial N}{\partial x} = 0 = \frac{\partial M}{\partial y} \Rightarrow$ Conservative

35. $\frac{\partial P}{\partial y} = 0 \neq ye^z = \frac{\partial N}{\partial xz} \Rightarrow$ Not Conservative

37. $\frac{\partial f}{\partial x} = 2 \Rightarrow f(x,y,z) = 2x + g(y,z)$. $\frac{\partial f}{\partial y} = \frac{\partial g}{\partial y} = 2y + z \Rightarrow g(y,z) = y^2 + zy + h(z) \Rightarrow f(x,y,z) = 2x + y^2 + zy + h(z)$.
$\frac{\partial f}{\partial z} = y + h'(z) = y + 1 \Rightarrow h'(z) = 1 \Rightarrow h(z) = z + C$. $\therefore$ $f(x,y,z) = 2x + y^2 + zy + z + C$

39. $\frac{\partial P}{\partial y} = -\frac{1}{2}(x+y+z)^{-3/2} = \frac{\partial N}{\partial z}, \frac{\partial M}{\partial z} = -\frac{1}{2}(x+y+z)^{-3/2} = \frac{\partial P}{\partial x}, \frac{\partial N}{\partial x} = -\frac{1}{2}(x+y+z)^{-3/2} = \frac{\partial M}{\partial y} \Rightarrow M\,dx +$ $N\,dy + P\,dz$ is exact. $\frac{\partial f}{\partial x} = \frac{1}{\sqrt{x+y+z}} \Rightarrow f(x,y,z) = 2\sqrt{x+y+z} + g(y,z)$. $\frac{\partial f}{\partial y} = \frac{1}{\sqrt{x+y+z}} + \frac{\partial g}{\partial y} = \frac{1}{\sqrt{x+y+z}}$ $\Rightarrow \frac{\partial g}{\partial y} = 0 \Rightarrow g(y,z) = h(z) \Rightarrow f(x,y,z) = 2\sqrt{x+y+z} + h(z)$. $\frac{\partial f}{\partial z} = \frac{1}{\sqrt{x+y+z}} + h'(z) = \frac{1}{\sqrt{x+y+z}} \Rightarrow h'(z) = 0$ $\Rightarrow h(z) = C \Rightarrow f(x,y,z) = 2\sqrt{x+y+z} + C \Rightarrow$ $\int_{(-1,1,1)}^{(4,-3,0)} \frac{dx + dy + dz}{\sqrt{x+y+z}} = f(4,-3,0) - f(-1,1,1) = 0$

41. Over Path 1: $\mathbf{r} = t\,\mathbf{i} + t\,\mathbf{j} + t\,\mathbf{k} \Rightarrow x = t, y = t, z = t$ and $d\mathbf{r} = (\mathbf{i} + \mathbf{j} + \mathbf{k})\,dt \Rightarrow \mathbf{F} = 2t^2\,\mathbf{i} + \mathbf{j} + t^2\,\mathbf{k} \Rightarrow$
$\mathbf{F} \cdot d\mathbf{r} = (3t^2 + 1)\,dt \Rightarrow \text{Work} = \int_0^1 (3t^2 + 1)\,dt = 2$
Over Path 2: $\mathbf{r}_1 = t\,\mathbf{i} + t\,\mathbf{j}$, $0 \le t \le 1 \Rightarrow x = t, y = t, z = 0$ and $d\mathbf{r}_1 = (\mathbf{i} + \mathbf{j})\,dt \Rightarrow \mathbf{F}_1 = 2t^2\,\mathbf{i} + \mathbf{j} + t^2\,\mathbf{k} \Rightarrow$
$\mathbf{F}_1 \cdot d\mathbf{r}_1 = (2t^2 + 1)\,dt \Rightarrow \text{Work}_1 = \int_0^1 (2t^2 + 1)\,dt = \frac{5}{3}$. $\mathbf{r}_2 = \mathbf{i} + \mathbf{j} + t\,\mathbf{k}$, $0 \le t \le 1 \Rightarrow x = 1, y = 1, z = t$ and
$d\mathbf{r}_2 = \mathbf{k}\,dt \Rightarrow \mathbf{F}_2 = 2\,\mathbf{i} + \mathbf{j} + \mathbf{k} \Rightarrow \mathbf{F}_2 \cdot d\mathbf{r}_2 = dt \Rightarrow \text{Work}_2 = \int_0^1 dt = 1$. $\therefore$ $\text{Work} = \text{Work}_1 + \text{Work}_2 = \frac{8}{3}$

43. $dx = (-2\sin t + 2\sin 2t)\,dt$, $dy = (2\cos t - 2\cos 2t)\,dt$. Area $= \frac{1}{2}\oint_C x\,dy - y\,dx =$

$$\frac{1}{2}\int_0^{2\pi} \big((2\cos t - \cos 2t)(2\cos t - 2\cos 2t) - (2\sin t - \sin 2t)(-2\sin t + 2\sin 2t)\big)\,dt =$$

$$\frac{1}{2}\int_0^{2\pi} (6 - 6\cos t)\,dt = 6\pi$$

45. $dx = \cos 2t\,dt$, $dy = \cos t\,dt$. Area $= \frac{1}{2}\oint_C x\,dy - y\,dx = \frac{1}{2}\int_0^{\pi} \left(\frac{1}{2}\sin 2t\cos t - \sin t\cos 2t\right)dt =$

$$\frac{1}{2}\int_0^{\pi} (-\sin t\cos^2 t + \sin t)\,dt = \frac{2}{3}$$

# CHAPTER 16

# PREVIEW OF DIFFERENTIAL EQUATIONS

## SECTION 16.1 SEPARABLE FIRST ORDER EQUATIONS

1. First Order, Non–Linear

3. Fourth Order, Linear

5. a) $y = x^2 \Rightarrow y' = 2x,\ y'' = 2 \Rightarrow y'' - y' = 2x - 2x = 0$

   b) $y = 1 \Rightarrow y' = 0,\ y'' = 0 \Rightarrow xy'' - y' = 0$

   c) $y = C_1x^2 + C_2 \Rightarrow y' = 2C_1x,\ y'' = 2C_1 \Rightarrow xy'' - y' = x(2C_1) - 2C_1x = 0$

7. a) $y = e^{-x} \Rightarrow y' = -e^{-x} \Rightarrow 2y' + 3y = 2(-e^{-x}) + 3e^{-x} = e^{-x}$

   b) $y = e^{-x} + e^{-3x/2} \Rightarrow y' = -e^{-x} - \frac{3}{2}e^{-3x/2} \Rightarrow 2y' + 3y = 2\left(-e^{-x} - \frac{3}{2}e^{-3x/2}\right) + 3\left(e^{-x} + e^{-3x/2}\right) = e^{-x}$

   c) $y = e^{-x} + Ce^{-3x/2} \Rightarrow y' = -e^{-x} - \frac{3}{2}Ce^{-3x/2} \Rightarrow$
   $2y' + 3y = 2\left(-e^{-x} - \frac{3}{2}Ce^{-3x/2}\right) + 3\left(e^{-x} + Ce^{-3x/2}\right) = e^{-x}$

9. $y'' = -32 \Rightarrow y' = -32x + C_1$. $y'(5) = 0 \Rightarrow -32(5) + C_1 = 0 \Rightarrow C_1 = 160 \Rightarrow y' = -32x + 160 \Rightarrow y = -16x^2 + 160x + C_2$. $y(5) = 400 \Rightarrow -16(5)^2 + 160(5) + C_2 = 400 \Rightarrow C_2 = 0 \Rightarrow y = 160x - 16x^2$

11. $y = 3\cos 2t - \sin 2t \Rightarrow y(0) = 3\cos 0 - \sin 0 = 3$ and $y' = -6\sin 2t - 2\cos 2t \Rightarrow y'(0) = -6\sin 0 - 2\cos 0 = -2$ and $y'' = -12\cos 2t + 4\sin 2t \Rightarrow y'' + 4y = -12\cos 2t + 4\sin 2t + 4(3\cos 2t - \sin 2t) = 0$

13. a) $\frac{dx}{dt} = 1000 + 0.10\,x \Rightarrow dx = (1000 + 0.10\,x)\,dt \Rightarrow \frac{dx}{1000 + 0.10\,x} = dt \Rightarrow \int \frac{dx}{1000 + 0.10\,x} = \int dt$
$\Rightarrow 10\ln(1000 + 0.10\,x) = t + C \Rightarrow 1000 + 0.10\,x = e^{0.10(t + C)} = e^{0.10t + 0.10C} = e^{0.10t}\,e^{0.10C} = C_1e^{0.10t} \Rightarrow x = 10C_1\,e^{0.10t} - 10000 \Rightarrow x = C_2\,e^{0.10t} - 10000$. $x(0) = 1000 \Rightarrow 1000 = C_2e^{0.10(0)} - 10000 \Rightarrow C_2 = 11000 \Rightarrow x = 11000\,e^{0.10t} - 10000$

    b) $100\,000 = 11000\,e^{0.10t} - 10000 \Rightarrow 10 = e^{0.10t} \Rightarrow \ln 10 = 0.10\,t \Rightarrow t = 10\ln 10 \approx 23.03$

15. $(x^2 + y^2)\,dx + xy\,dy = 0 \Rightarrow \left(1 + \frac{y^2}{x^2}\right)dx + \frac{xy}{x^2}dy = 0 \Rightarrow \left(1 + \left(\frac{y}{x}\right)^2\right)dx + \frac{y}{x}dy = 0 \Rightarrow \Rightarrow \frac{y}{x}dy = -\left(1 + \left(\frac{y}{x}\right)^2\right) \Rightarrow \frac{dy}{dx} = \frac{-\left(1 + \left(\frac{y}{x}\right)^2\right)}{y/x} \Rightarrow$ Homogeneous. $v = \frac{y}{x} \Rightarrow F(v) = \frac{-(1 + v^2)}{v} \Rightarrow \frac{dx}{x} + \frac{dv}{v + \frac{1 + v^2}{v}} =$

15. (Continued)

$0 \Rightarrow \frac{dx}{x} + \frac{v\,dv}{1+2v^2} = 0 \Rightarrow \ln|x| + \frac{1}{4}\ln(1+2v^2) = C \Rightarrow |x|\left(1+2v^2\right)^{1/4} = e^C \Rightarrow x^4(1+2v^2) = C_1$ where $C_1 = (e^C)^4 \Rightarrow x^4\left(1+2\left(\frac{y^2}{x^2}\right)\right) = C_1 \Rightarrow x^4 + 2x^2y^2 = C_1$

17. $\left(x\,e^{y/x} + y\right)dx - x\,dy = 0 \Rightarrow \left(e^{y/x} + \frac{y}{x}\right)dx - dy = 0 \Rightarrow \frac{dy}{dx} = e^{y/x} + \frac{y}{x} \Rightarrow$ Homogeneous. $v = \frac{y}{x} \Rightarrow$ $F(v) = e^v + v \Rightarrow \frac{dx}{x} + \frac{dv}{v-(e^v+v)} = 0 \Rightarrow \frac{dx}{x} + \frac{dv}{-e^v} = 0 \Rightarrow \ln|x| + e^{-v} = C \Rightarrow \ln|x| + e^{-y/x} = C$

19. $\frac{dy}{dx} = \frac{y}{x} + \cos\left(\frac{y-x}{x}\right) \Rightarrow \frac{dy}{dx} = \frac{y}{x} + \cos\left(\frac{y}{x} - 1\right) \Rightarrow$ Homogeneous. $v = \frac{y}{x} \Rightarrow F(v) = v + \cos(v-1) \Rightarrow$ $\frac{dx}{x} + \frac{dv}{v-(v+\cos(v-1))} = 0 \Rightarrow \frac{dx}{x} - \sec(v-1)\,dv = 0 \Rightarrow \ln|x| + \ln|\sec(v-1) + \tan(v-1)| = C \Rightarrow$ $\frac{x}{\sec(v-1)+\tan(v-1)} = \pm e^C \Rightarrow \frac{x}{\sec\left(\frac{y-x}{x}\right) + \tan\left(\frac{y-x}{x}\right)} = C_1 \Rightarrow x = C_1\left(\sec\left(\frac{y-x}{x}\right) + \tan\left(\frac{y-x}{x}\right)\right)$

$y(2) = 2 \Rightarrow 2 = C_1\left(\sec\left(\frac{2-2}{2}\right) + \tan\left(\frac{2-2}{2}\right)\right) \Rightarrow C_1 = 2 \Rightarrow x = 2\left(\sec\left(\frac{y-x}{x}\right) + \tan\left(\frac{y-x}{x}\right)\right)$

21. $x\,dx + 2y\,dy = 0 \Rightarrow \frac{x^2}{2} + y^2 = C$. $2y\,dx - x\,dy = 0 \Rightarrow \frac{2\,dx}{x} - \frac{dy}{y} = 0 \Rightarrow 2\ln|x| - \ln|y| = C \Rightarrow \frac{x^2}{|y|} = e^C \Rightarrow$ $\frac{x^2}{y} = \pm e^C = C_1 \Rightarrow x^2 = C_1 y$

23. $y\,dx + x\,dy = 0 \Rightarrow \frac{dx}{x} + \frac{dy}{y} = 0 \Rightarrow \ln|x| + \ln|y| = C \Rightarrow |xy| = e^C \Rightarrow xy = \pm e^C = C_1 \Rightarrow xy = C_1$, a family of hyperbolas with transverse axis the line $y = x$ or $y = -x$. $x\,dx - y\,dy = 0 \Rightarrow \frac{x^2}{2} - \frac{y^2}{2} = C \Rightarrow x^2 - y^2 = C_1$, a family of hyperbolas with transverse axis the x or y axis.

## SECTION 16.2 EXACT DIFFERENTIAL EQUATIONS

1. $\frac{\partial M}{\partial y} = 0,\ \frac{\partial N}{\partial x} = 0 \Rightarrow$ Exact

3. $\frac{\partial M}{\partial y} = -\frac{1}{y^2},\ \frac{\partial N}{\partial x} = -\frac{1}{y^2} \Rightarrow$ Exact

5. $\frac{\partial M}{\partial y} = e^y,\ \frac{\partial N}{\partial x} = e^y \Rightarrow$ Exact

7. $\frac{\partial M}{\partial y} = 2,\ \frac{\partial N}{\partial x} = -2 \Rightarrow$ Not Exact

9. $\frac{\partial M}{\partial y} = 1 = \frac{\partial N}{\partial x} \Rightarrow$ Exact. $\frac{\partial f}{\partial x} = x + y \Rightarrow f(x,y) = \frac{x^2}{2} + xy + k(y)$. $\frac{\partial f}{\partial y} = x + k'(y) = x + y^2 \Rightarrow k(y) = \frac{y^3}{3} + C \Rightarrow$ $f(x,y) = \frac{x^2}{2} + xy + \frac{y^3}{3} + C = C_1 \Rightarrow 3x^2 + 6xy + 2y^3 = C_2$

11. $\frac{\partial M}{\partial y} = 2x + 2y = \frac{\partial N}{\partial x} \Rightarrow$ Exact. $\frac{\partial f}{\partial x} = 2xy + y^2 \Rightarrow f(x,y) = x^2y + xy^2 + k(y)$. $\frac{\partial f}{\partial y} = x^2 + 2xy + k'(y) =$ $x^2 + 2xy - y \Rightarrow k'(y) = -y \Rightarrow k(y) = -\frac{y^2}{2} + C \Rightarrow f(x,y) = x^2y + xy^2 - \frac{y^2}{2} + C = C_1 \Rightarrow 2x^2y + 2xy^2 - y^2 = C_2$

13. $\frac{\partial M}{\partial y} = \frac{1}{y} + \frac{1}{x} = \frac{\partial N}{\partial x} \Rightarrow$ Exact. $\frac{\partial f}{\partial x} = e^x + \ln y + \frac{y}{x} \Rightarrow f(x,y) = e^x + x \ln y + y \ln x + k(y)$. $\frac{\partial f}{\partial y} = \frac{x}{y} + \ln x + k'(y) = \frac{x}{y} + \ln x + \sin y \Rightarrow k'(y) = \sin y \Rightarrow k(y) = -\cos y + C \Rightarrow f(x,y) = e^x + x \ln y + y \ln x - \cos y + C = C_1 \Rightarrow$
$e^x + x \ln y + y \ln x - \cos y = C_2$

15. $\rho = \frac{1}{y^2} \Rightarrow \frac{1}{y^2}\left(xy^2 + y\right) dx - \frac{1}{y^2}(x\, dy) = 0 \Rightarrow \left(x + \frac{1}{y}\right) dx - \frac{x}{y^2} dy = 0 \Rightarrow \frac{\partial M}{\partial y} = -\frac{1}{y^2} = \frac{\partial N}{\partial x} \Rightarrow$ Exact.
$\frac{\partial f}{\partial x} = x + \frac{1}{y} \Rightarrow f(x,y) = \frac{x^2}{2} + \frac{x}{y} + k(y)$. $\frac{\partial f}{\partial y} = -\frac{x}{y^2} + k'(y) = -\frac{x}{y^2} \Rightarrow k'(y) = 0 \Rightarrow k(y) = C \Rightarrow f(x,y) = \frac{x^2}{2} + \frac{x}{y} + C =$
$C_1 \Rightarrow x^2 + \frac{2x}{y} = C_2$

17. a) $\rho = \frac{1}{xy} \Rightarrow \frac{1}{xy}(y\, dx) + \frac{1}{xy}(x\, dy) = 0 \Rightarrow \frac{1}{x} dx + \frac{1}{y} dy = 0 \Rightarrow \frac{\partial M}{\partial y} = 0 = \frac{\partial N}{\partial x} \Rightarrow$ Exact. $\frac{\partial f}{\partial x} = \frac{1}{x} \Rightarrow f(x,y) =$
$\ln|x| + k(y)$. $\frac{\partial f}{\partial y} = k'(y) = \frac{1}{y} \Rightarrow k(y) = \ln|y| + C \Rightarrow f(x,y) = \ln|x| + \ln|y| + C = C_1 \Rightarrow \ln|xy| = C_2 \Rightarrow$
$xy = \pm e^{C_2} \Rightarrow xy = C_3$

b) $\rho = \frac{1}{(xy)^2} \Rightarrow \frac{1}{(xy)^2}(y\, dx) + \frac{1}{(xy)^2}(x\, dy) = 0 \Rightarrow \frac{1}{x^2 y} dx + \frac{1}{xy^2} dy = 0 \Rightarrow \frac{\partial M}{\partial y} = -\frac{1}{x^2 y^2} = \frac{\partial N}{\partial x} \Rightarrow$ Exact.
$\frac{\partial f}{\partial x} = \frac{1}{x^2 y} \Rightarrow f(x,y) = -\frac{1}{xy} + k(y)$. $\frac{\partial f}{\partial y} = \frac{1}{xy^2} + k'(y) = \frac{1}{xy^2} \Rightarrow k'(y) = 0 \Rightarrow k(y) = C \Rightarrow f(x,y) = -\frac{1}{xy} + C = C_1$
$\Rightarrow xy = C_2$

19. $\frac{\partial M}{\partial y} = 2y = \frac{\partial N}{\partial x} \Rightarrow$ Exact. $\frac{\partial f}{\partial x} = x + y^2 \Rightarrow f(x,y) = \frac{x^2}{2} + xy^2 + k(y)$. $\frac{\partial f}{\partial y} = 2xy + k'(y) = 2xy + 1 \Rightarrow k'(y) = 1$
$\Rightarrow k(y) = y + C \Rightarrow f(x,y) = \frac{x^2}{2} + xy^2 + y + C = C_1 \Rightarrow x^2 + 2xy^2 + 2y = C_2$. $y(0) = 2 \Rightarrow C_2 = 4 \Rightarrow$
$x^2 + 2xy^2 + 2y = 4$

21. $\frac{\partial M}{\partial y} = -1 = \frac{\partial N}{\partial x} \Rightarrow$ Exact. $\frac{\partial f}{\partial x} = \frac{1}{x} - y \Rightarrow f(x,y) = \ln|x| - xy + k(y)$. $\frac{\partial f}{\partial y} = -x + k'(y) = \frac{1}{y} - x \Rightarrow k'(y) = \frac{1}{y} \Rightarrow$
$k(y) = \ln|y| + C \Rightarrow f(x,y) = \ln|x| - xy + \ln|y| + C = C_1 \Rightarrow \ln|xy| = xy + C_2 \Rightarrow xy = \pm e^{xy + C_2} = \pm e^{C_2} e^{xy} \Rightarrow$
$xy = C_3 e^{xy}$. $y(1) = 1 \Rightarrow C_3 = \frac{1}{e} \Rightarrow xy = \frac{1}{e} e^{xy} \Rightarrow xy = e^{xy-1}$

23. $\frac{\partial M}{\partial y} = 2y$, $\frac{\partial N}{\partial x} = ay \Rightarrow a = 2$. $\therefore$ $(3x^2 + y^2)\, dx + 2xy\, dy = 0$ is exact. $\frac{\partial f}{\partial x} = 3x^2 + y^2 \Rightarrow f(x,y) = x^3 + xy^2 +$
$k(y)$. $\frac{\partial f}{\partial y} = 2xy + k'(y) = 2xy \Rightarrow k'(y) = 0 \Rightarrow k(y) = C \Rightarrow f(x,y) = x^3 + xy^2 + C = C_1 \Rightarrow x^3 + xy^2 = C_2$.

## SECTION 16.3 LINEAR FIRST ORDER EQUATIONS

1. $\frac{dy}{dx} + 2y = e^{-x} \Rightarrow P(x) = 2,\ Q(x) = e^{-x} \Rightarrow \int P(x)\, dx = 2x \Rightarrow \rho(x) = e^{2x} \Rightarrow y = \frac{1}{e^{2x}} \int e^{2x}(e^{-x})\, dx \Rightarrow$
$y = \frac{1}{e^{2x}} \int e^x\, dx = \frac{1}{e^{2x}}\left(e^x + C\right) = e^{-x} + C\, e^{-2x}$

3. $\frac{dy}{dx} + \frac{3}{x} y = \frac{\sin x}{x^3} \Rightarrow P(x) = \frac{3}{x},\ Q(x) = \frac{\sin x}{x^3} \Rightarrow \int P(x)\, dx = 3 \ln|x| = \ln|x|^3 \Rightarrow \rho(x) = e^{\ln|x|^3} = |x|^3 = x^3$ if

3. (Continued)

$x > 0 \Rightarrow y = \frac{1}{x^3} \int x^3 \left(\frac{\sin x}{x^3}\right) dx = \frac{1}{x^3} \int \sin x \, dx \Rightarrow y = \frac{1}{x^3}(-\cos x + C) = -\frac{\cos x}{x^3} + \frac{C}{x^3}$

5. $\frac{dy}{dx} + \frac{4y}{x-1} = \frac{x+1}{(x-1)^3} \Rightarrow P(x) = \frac{4}{x-1},\ Q(x) = \frac{x+1}{(x-1)^3} \Rightarrow \int P(x)\,dx = 4\ln|x-1| = \ln(x-1)^4 \Rightarrow$

$\rho(x) = e^{\ln(x-1)^4} = (x-1)^4 \Rightarrow y = \frac{1}{(x-1)^4} \int (x-1)^4 \frac{x+1}{(x-1)^3}\,dx = \frac{1}{(x-1)^4} \int (x^2-1)\,dx =$

$\frac{1}{(x-1)^4}\left(\frac{x^3}{3} - x + C\right) \Rightarrow y = \frac{x^3}{3(x-1)^4} - \frac{x}{(x-1)^4} + \frac{C}{(x-1)^4}$

7. $\frac{dy}{dx} + (\cot x)\,y = \sec x \Rightarrow P(x) = \cot x,\ Q(x) = \sec x \Rightarrow \int P(x)\,dx = \ln|\sin x| \Rightarrow \rho(x) = e^{\ln|\sin x|} = \sin x$ if

$\sin x > 0 \Rightarrow y = \frac{1}{\sin x} \int \sin x(\sec x)\,dx = \frac{1}{\sin x} \int \tan x\,dx = \frac{1}{\sin x}(\ln|\sec x| + C) = \csc x(\ln|\sec x| + C)$

9. $\frac{dy}{dx} + 2y = x \Rightarrow P(x) = 2,\ Q(x) = x \Rightarrow \int P(x)\,dx = 2x \Rightarrow \rho(x) = e^{2x} \Rightarrow y = \frac{1}{e^{2x}} \int e^{2x}(x\,dx) =$

$\frac{1}{e^{2x}}\left(\frac{1}{2}x\,e^{2x} - \frac{1}{4}e^{2x} + C\right) \Rightarrow y = \frac{1}{2}x - \frac{1}{4} + C\,e^{-2x}.\ y(0) = 1 \Rightarrow -\frac{1}{4} + C = 1 \Rightarrow C = \frac{5}{4} \Rightarrow$

$y = \frac{1}{2}x - \frac{1}{4} + \frac{5}{4}e^{-2x}$

11. $\frac{dy}{dx} + \frac{y}{x} = \frac{\sin x}{x} \Rightarrow P(x) = \frac{1}{x},\ Q(x) = \frac{\sin x}{x} \Rightarrow \int P(x)\,dx = \ln|x| \Rightarrow \rho(x) = e^{\ln|x|} = |x| \Rightarrow$

$y = \frac{1}{|x|} \int |x| \frac{\sin x}{x}\,dx = \frac{1}{x} \int x \frac{\sin x}{x}\,dx$ for $x \neq 0 \Rightarrow y = \frac{1}{x} \int \sin x\,dx = \frac{1}{x}(-\cos x + C) \Rightarrow$

$y = -\frac{1}{x}\cos x + \frac{C}{x}.\ y\left(\frac{\pi}{2}\right) = 1 \Rightarrow C = \frac{\pi}{2} \Rightarrow y = -\frac{1}{x}\cos x + \frac{\pi}{2x}$

13. $\frac{dy}{dx} - ky = 0 \Rightarrow P(x) = -k,\ Q(x) = 0 \Rightarrow \int P(x)\,dx = -kx \Rightarrow \rho(x) = e^{-kx} \Rightarrow$

$y = \frac{1}{e^{-kx}} \int e^{-kx}(0)\,dx = Ce^{kx}.\ y(0) = y_0 \Rightarrow C = y_0 \Rightarrow y = y_0\,e^{kx}$

15. $\frac{dp}{dx} + p = 0 \Rightarrow \frac{dp}{dx} = -p \Rightarrow dp = -p\,dx \Rightarrow \frac{1}{p}\,dp = -dx \Rightarrow \ln|p| = -x + C \Rightarrow p = C_1e^{-x} \Rightarrow \frac{dy}{dx} = C_1\,e^{-x} \Rightarrow$

$y = -C_1e^{-x} + C_2 = C_3e^{-x} + C_2$

17. $x\frac{dp}{dx} + p = 0 \Rightarrow \frac{dp}{dx} + \frac{1}{x}p = 0 \Rightarrow P(x) = \frac{1}{x},\ Q(x) = 0 \Rightarrow \int P(x)\,dx = \ln|x| \Rightarrow \rho(x) = e^{\ln|x|} = |x| \Rightarrow$

$p = \frac{1}{|x|} \int |x|\,(0)\,dx = \frac{C}{|x|} = \frac{C}{x}$ if $x > 0 \Rightarrow \frac{dy}{dx} = \frac{C}{x} \Rightarrow y = C\ln x + C_1$

19. $x\frac{dp}{dx} + p = x^2 \Rightarrow \frac{dp}{dx} + \frac{1}{x}p = x \Rightarrow P(x) = \frac{1}{x},\ Q(x) = x \Rightarrow \int P(x)\,dx = \ln x$ if $x > 0 \Rightarrow \rho(x) = e^{\ln x} = x \Rightarrow$

$p = \frac{1}{x} \int x\,(x)\,dx = \frac{1}{x} \int x^2\,dx = \frac{1}{x}\left(\frac{x^3}{3} + C\right) = \frac{x^2}{3} + \frac{C}{x} \Rightarrow \frac{dy}{dx} = \frac{x^2}{3} + \frac{C}{x}.\ y'(1) = 1 \Rightarrow C = \frac{2}{3} \Rightarrow$

19. (Continued)

$\frac{dy}{dx} = \frac{x^2}{3} + \frac{2}{3x} \Rightarrow y = \frac{x^3}{9} + \frac{2}{3}\ln x + C_1$. $y(1) = 0 \Rightarrow C_1 = -\frac{1}{9} \Rightarrow y = \frac{x^3}{9} + \frac{2}{3}\ln x - \frac{1}{9}$

21. Steady State $= \frac{V}{R} \Rightarrow$ we want $i = \frac{1}{2}\left(\frac{V}{R}\right) \Rightarrow \frac{1}{2}\left(\frac{V}{R}\right) = \frac{V}{R}\left(1 - e^{-Rt/L}\right) \Rightarrow \frac{1}{2} = 1 - e^{-Rt/L} \Rightarrow -\frac{1}{2} = -e^{-Rt/L}$

$\Rightarrow \ln\frac{1}{2} = -\frac{Rt}{L} \Rightarrow -\frac{L}{R}\ln\frac{1}{2} = t \Rightarrow t = \frac{L}{R}\ln 2$

23. $t = \frac{3L}{R} \Rightarrow i = \frac{V}{R}\left(1 - e^{(-R/L)(3L/R)}\right) = \frac{V}{R}\left(1 - e^{-3}\right) \approx 0.9502\,\frac{V}{R}$ or about 95% of the steady state value.

## SECTION 16.4 SECOND ORDER LINEAR HOMOGENEOUS EQUATIONS

1. $r^2 + 2r = 0 \Rightarrow r_1 = 0, r_2 = -2 \Rightarrow y = C_1e^{0x} + C_2e^{-2x} = C_1 + C_2e^{-2x}$

3. $r^2 + 6r + 5 = 0 \Rightarrow (r+5)(r+1) = 0 \Rightarrow r_1 = -5, r_2 = -1 \Rightarrow y = C_1e^{-5x} + C_2{}^{-x}$

5. $r^2 - 4r + 4 = 0 \Rightarrow (r-2)^2 = 0 \Rightarrow r_1 = r_2 = 2 \Rightarrow y = \left(C_1x + C_2\right)e^{2x}$

7. $r^2 - 10r + 25 = 0 \Rightarrow (r-5)^2 = 0 \Rightarrow r_1 = r_2 = 5 \Rightarrow y = \left(C_1x + C_2\right)e^{5x}$

9. $r^2 + r + 1 = 0 \Rightarrow r = \frac{-1 \pm i\sqrt{3}}{2} \Rightarrow y = e^{-x/2}\left(C_1 \cos\frac{\sqrt{3}}{2}x + C_2 \sin\frac{\sqrt{3}}{2}x\right)$

11. $r^2 - 2r + 4 = 0 \Rightarrow r = 1 \pm i\sqrt{3} \Rightarrow y = e^x\left(C_1 \cos\sqrt{3}\,x + C_2 \sin\sqrt{3}\,x\right)$

13. $r^2 - 1 = 0 \Rightarrow r_1 = 1, r_2 = -1 \Rightarrow y = C_1e^x + C_2e^{-x}$. $y(0) = 1 \Rightarrow C_1 + C_2 = 1$. $y' = C_1e^x - C_2e^{-x}$. $y'(0) = -2$

$\Rightarrow -2 = C_1 - C_2$. $\therefore\ C_1 = -\frac{1}{2}, C_2 = \frac{3}{2} \Rightarrow y = -\frac{1}{2}e^x + \frac{3}{2}e^{-x}$

15. $r^2 - 4 = 0 \Rightarrow r_1 = 2, r_2 = -2 \Rightarrow y = C_1e^{2x} + C_2e^{-2x}$. $y(0) = 0 \Rightarrow C_1 + C_2$. $y' = 2C_1e^{2x} - 2C_2e^{-2x}$.

$y'(0) = 3 \Rightarrow 3 = 2C_1 - 2C_2$. $\therefore\ C_1 = \frac{3}{4}, C_2 = -\frac{3}{4} \Rightarrow y = \frac{3}{4}e^{2x} - \frac{3}{4}e^{-2x}$

17. $r^2 + 2r + 1 = 0 \Rightarrow (r+1)^2 = 0 \Rightarrow r_1 = r_2 = -1 \Rightarrow y = \left(C_1x + C_2\right)e^{-x}$. $y(0) = 0 \Rightarrow C_2 = 0$. $y' =$

$C_1\left(e^{-x} - x\,e^{-x}\right)$. $y'(0) = 1 \Rightarrow C_1 = 1 \Rightarrow y = x\,e^{-x}$

19. $4r^2 + 12r + 9 = 0 \Rightarrow (2r+3)^2 = 0 \Rightarrow r_1 = r_2 = -\frac{3}{2} \Rightarrow y = \left(C_1x + C_2\right)e^{-3x/2}$. $y(0) = 0 \Rightarrow C_2 = 0 \Rightarrow$

$y = C_1x\,e^{-3x/2} \Rightarrow y' = C_1\left(e^{-3x/2} - \frac{3}{2}x\,e^{-3x/2}\right)$. $y'(0) = -1 \Rightarrow C_1 = -1 \Rightarrow y = -x\,e^{-3x/2}$

21. $r^2 + 4 = 0 \Rightarrow r = \pm 2i \Rightarrow y = C_1 \cos 2x + C_2 \sin 2x$. $y(0) = 0 \Rightarrow C_1 = 0 \Rightarrow y = C_2 \sin 2x$. $y' = 2C_2 \cos 2x$.

$y'(0) = 2 \Rightarrow C_2 = 1 \Rightarrow y = \sin 2x$

23. $r^2 - 2r + 3 = 0 \Rightarrow r = 1 \pm i\sqrt{2} \Rightarrow y = e^x\left(C_1 \cos\sqrt{2}\,x + C_2 \sin\sqrt{2}\,x\right)$. $y(0) = 2 \Rightarrow C_1 = 2 \Rightarrow y =$

$e^x\left(2 \cos\sqrt{2}\,x + C_2 \sin\sqrt{2}\,x\right)$. $y' = e^x\left(2 \cos\sqrt{2}\,x + C_2 \sin\sqrt{2}\,x\right) +$

23. (Continued)

$e^x\left(-2\sqrt{2}\sin\sqrt{2}\,x + \sqrt{2}\,C_2\cos\sqrt{2}\,x\right)$. $y'(0) = 1 \Rightarrow 1 = 2 + \sqrt{2}\,C_2 \Rightarrow C_2 = -\frac{1}{\sqrt{2}} \Rightarrow$

$y = e^x\left(2\cos\sqrt{2}\,x - \frac{1}{\sqrt{2}}\sin\sqrt{2}\,x\right)$

## SECTION 16.5 SECOND ORDER NONHOMOGENEOUS LINEAR EQUATIONS

1. $\frac{d^2y}{dx^2} + \frac{dy}{dx} = 0 \Rightarrow r^2 + r = 0 \Rightarrow r_1 = 0,\ r_2 = -1 \Rightarrow y_h = C_1 + C_2e^{-x} \Rightarrow u_1 = 1,\ u_2 = e^{-x} \Rightarrow D = \begin{vmatrix} 1 & e^{-x} \\ 0 & -e^{-x} \end{vmatrix}$ $= -e^{-x} \Rightarrow v'_1 = -\frac{u_2\,F(x)}{D} = x \Rightarrow v_1 = \int x\,dx = \frac{x^2}{2} + C_1$. $v'_2 = \frac{u_1\,F(x)}{D} = -x\,e^x \Rightarrow v_2 = \int x\,e^x\,dx =$ $-x\,e^x + e^x + C_2 \Rightarrow y = \left(\frac{x^2}{2} + C_1\right) + \left(-x\,e^x + e^x + C_2\right)e^{-x} = \frac{x^2}{2} - x + C_3 + C_2\,e^{-x}$

3. $\frac{d^2y}{dx^2} + y = 0 \Rightarrow r^2 + 1 = 0 \Rightarrow r = \pm i \Rightarrow y_h = C_1\cos x + C_2\sin x \Rightarrow u_1 = \cos x,\ u_2 = \sin x \Rightarrow D =$ $\begin{vmatrix} \cos x & \sin x \\ -\sin x & \cos x \end{vmatrix} = 1 \Rightarrow v'_1 = -\frac{u_2\,F(x)}{D} = -\sin^2 x \Rightarrow v_1 = \int -\sin^2 x\,dx = -\frac{1}{2}x + \frac{\sin 2x}{4} + C_1$. $v'_2 = \frac{u_1\,F(x)}{D} = \cos x\sin x \Rightarrow v_2 = \int \sin x\cos x\,dx = \frac{\sin^2 x}{2} + C_2 \Rightarrow y = \cos x\left(-\frac{1}{2}x + \frac{\sin 2x}{4} + C_1\right) +$ $\sin x\left(\frac{\sin^2 x}{2} + C_2\right) = -\frac{1}{2}x\cos x + C_1\cos x + C_3\sin x$

5. $\frac{d^2y}{dx^2} + 2\frac{dy}{dx} + y = 0 \Rightarrow r^2 + 2r + 1 = 0 \Rightarrow r_1 = r_2 = -1 \Rightarrow y_h = C_1x\,e^{-x} + C_2\,e^{-x} \Rightarrow u_1 = x\,e^{-x},\ u_2 = e^{-x}$ $\Rightarrow D = \begin{vmatrix} x\,e^{-x} & e^{-x} \\ e^{-x} - x\,e^{-x} & -e^{-x} \end{vmatrix} = -e^{-2x} \Rightarrow v'_1 = -\frac{u_2\,F(x)}{D} = 1 \Rightarrow v_1 = \int 1\,dx = x + C_1$. $v'_2 = \frac{u_1\,F(x)}{D} =$ $-x \Rightarrow v_2 = \int -x\,dx = -\frac{x^2}{2} + C_2 \Rightarrow y = \left(x + C_1\right)xe^{-x} + \left(-\frac{x^2}{2} + C_2\right)e^{-x} = C_1x\,e^{-x} + C_2e^{-x} + \frac{1}{2}x^2\,e^{-x}$

7. $\frac{d^2y}{dx^2} - y = 0 \Rightarrow r^2 - 1 = 0 \Rightarrow r_1 = 1,\ r_2 = -1 \Rightarrow y_h = C_1\,e^x + C_2\,e^{-x} \Rightarrow u_1 = e^x,\ u_2 = e^{-x} \Rightarrow$ $D = \begin{vmatrix} e^x & e^{-x} \\ e^x & -e^{-x} \end{vmatrix} = -2 \Rightarrow v'_1 = -\frac{u_2\,F(x)}{D} = \frac{1}{2} \Rightarrow v_1 = \frac{1}{2}x + C_1$. $v'_2 = \frac{u_1\,F(x)}{D} = -\frac{1}{2}e^{2x} \Rightarrow v_2 = -\frac{1}{4}e^{2x} +$ $C_2 \Rightarrow y = \left(\frac{1}{2}x + C_1\right)e^x + \left(-\frac{1}{4}e^{2x} + C_2\right)e^{-x} = \frac{1}{2}x\,e^x + C_3\,e^x + C_2\,e^{-x}$

9. $\frac{d^2y}{dx^2} + 4\frac{dy}{dx} + 5y = 0 \Rightarrow r^2 + 4r + 5 = 0 \Rightarrow r = -2 \pm i \Rightarrow y_h = e^{-2x}\left(C_1 \cos x + C_2 \sin x\right) \Rightarrow$

$u_1 = e^{-2x}\cos x,\ u_2 = e^{-2x}\sin x \Rightarrow D = \begin{vmatrix} e^{-2x}\cos x & e^{-2x}\sin x \\ -2e^{-2x}\cos x - e^{-2x}\sin x & -2e^{-2x}\sin x + e^{-2x}\cos x \end{vmatrix} =$

$e^{-4x} \Rightarrow v'_1 = -\frac{u_2 F(x)}{D} = -10\, e^{2x}\sin x \Rightarrow v_1 = \int -10\, e^{2x}\sin x\, dx = -2\, e^{2x}(2\sin x - \cos x) + C_1.$

$v'_2 = \frac{u_1 F(x)}{D} = 10\, e^{2x}\cos x \Rightarrow v_2 = \int 10\, e^{2x}\cos x\, dx = 10\left(\frac{e^{2x}}{5}(2\cos x + \sin x)\right) + C_2 \Rightarrow$

$y = \left(-2\, e^{2x}(2\sin x - \cos x) + C_1\right)e^{-2x}\cos x + \left(2\, e^{2x}(2\cos x + \sin x) + C_2\right)e^{-2x}\sin x =$

$2 + C_1\, e^{-2x}\cos x + C_2\, e^{-2x}\sin x$

11. $\frac{d^2y}{dx^2} + y = 0 \Rightarrow r^2 + 1 = 0 \Rightarrow r = \pm i \Rightarrow y_h = C_1 \cos x + C_2 \sin x \Rightarrow u_1 = \cos x,\ u_2 = \sin x \Rightarrow$

$D = \begin{vmatrix} \cos x & \sin x \\ -\sin x & \cos x \end{vmatrix} = 1 \Rightarrow v'_1 = -\frac{u_2 F(x)}{D} = -\tan x \Rightarrow v_1 = \int -\tan x\, dx = -\ln(\sec x) + C_1.$

$v'_2 = \frac{u_1 F(x)}{D} = 1 \Rightarrow v_2 = \int dx = x + C_2 \Rightarrow y = \left(-\ln(\sec x) + C_1\right)\cos x + \left(x + C_2\right)\sin x = \cos x \ln(\cos x)$

$+ x\sin x + C_1 \cos x + C_2 \sin x$

13. $\frac{d^2y}{dx^2} - 3\frac{dy}{dx} - 10y = 0 \Rightarrow r^2 - 3r - 10 = 0 \Rightarrow r_1 = 5,\ r_2 = -2 \Rightarrow y_h = C_1\, e^{5x} + C_2\, e^{-2x}.\ y_p = C \Rightarrow$

$\frac{d^2y}{dx^2} = 0,\ \frac{dy}{dx} = 0 \Rightarrow -10C = -3 \Rightarrow C = \frac{3}{10} \Rightarrow y_p = \frac{3}{10} \Rightarrow y = \frac{3}{10} + C_1\, e^{5x} + C_2\, e^{-2x}$

15. $\frac{d^2y}{dx^2} - \frac{dy}{dx} = 0 \Rightarrow r^2 - r = 0 \Rightarrow r_1 = 0,\ r_2 = 1 \Rightarrow y_h = C_1 + C_2\, e^x.\ y_p = B\cos x + C\sin x \Rightarrow \frac{dy}{dx} = -B\sin x +$

$C\cos x,\ \frac{d^2y}{dx^2} = -B\cos x - C\sin x \Rightarrow -B\cos x - C\sin x + B\sin x - C\cos x = \sin x \Rightarrow B = \frac{1}{2},\ C = -\frac{1}{2} \Rightarrow$

$y_p = \frac{1}{2}\cos x - \frac{1}{2}\sin x \Rightarrow y = \frac{1}{2}\cos x - \frac{1}{2}\sin x + C_1 + C_2\, e^x$

17. $\frac{d^2y}{dx^2} + y = 0 \Rightarrow r^2 + 1 = 0 \Rightarrow r = \pm i \Rightarrow y_h = C_1 \cos x + C_2 \sin x.\ y_p = B\cos 3x + C\sin 3x \Rightarrow y'_p =$

$-3B\sin 3x + 3C\cos 3x \Rightarrow y''_p = -9B\cos 3x - 9C\sin 3x \Rightarrow -9B\cos 3x - 9C\sin 3x + B\cos 3x +$

$C\sin 3x = \cos 3x \Rightarrow B = -\frac{1}{8},\ C = 0 \Rightarrow y_p = -\frac{1}{8}\cos 3x \Rightarrow y = -\frac{1}{8}\cos 3x + C_1\cos x + C_2\sin x$

19. $\frac{d^2y}{dx^2} - \frac{dy}{dx} - 2y = 0 \Rightarrow r^2 - r - 2 = 0 \Rightarrow r_1 = 2,\ r_2 = -1 \Rightarrow y_h = C_1\, e^{2x} + C_2\, e^{-x}.\ y_p = B\cos x + C\sin x \Rightarrow$

$y'_p = -B\sin x + C\cos x \Rightarrow y''_p = -B\cos x - C\sin x \Rightarrow -B\cos x - C\sin x - (-B\sin x + C\cos x) -$

$2(B\cos x + C\sin x) = 20\cos x \Rightarrow B = -6,\ C = -2 \Rightarrow y_p = -6\cos x - 2\sin x \Rightarrow y = -6\cos x - 2\sin x +$

$C_1\, e^{2x} + C_2\, e^{-x}$

21. $\frac{d^2y}{dx^2} - y = 0 \Rightarrow r^2 - 1 = 0 \Rightarrow r_1 = 1, r_2 = -1 \Rightarrow y_h = C_1\,e^x + C_2\,e^{-x}$. $y_p = Ax\,e^x + Dx^2 + Ex + F \Rightarrow$ $y'_p = A\,e^x + Ax\,e^x + 2Dx + E \Rightarrow y''_p = 2A\,e^x + Ax\,e^x + 2D \Rightarrow 2A\,e^x + Ax\,e^x + 2D -$ $\left(Ax\,e^x + Dx^2 + Ex + F\right) = e^x + x^2 \Rightarrow A = \frac{1}{2}, D = -1, E = 0, F = -2 \Rightarrow y_p = \frac{1}{2}x\,e^x - x^2 - 2 \Rightarrow$ $y = \frac{1}{2}x\,e^x - x^2 - 2 + C_1\,e^x + C_2\,e^{-x}$

23. $\frac{d^2y}{dx^2} - \frac{dy}{dx} - 6y = 0 \Rightarrow r^2 - r - 6 = 0 \Rightarrow r_1 = 3, r_2 = -2 \Rightarrow y_h = C_1\,e^{3x} + C_2\,e^{-2x}$. $y_p = A\,e^{-x} + B\cos x +$ $D\sin x \Rightarrow y'_p = -A\,e^{-x} - B\sin x + C\cos x \Rightarrow y''_p = A\,e^{-x} - B\cos x - C\sin x \Rightarrow A\,e^{-x} - B\cos x -$ $C\sin x - \left(-A\,e^{-x} - B\sin x + C\cos x\right) - 6\left(A\,e^{-x} + B\cos x + C\sin x\right) = e^{-x} - 7\cos x \Rightarrow A = -\frac{1}{4}$, $B = \frac{49}{50}, C = \frac{7}{50} \Rightarrow y_p = -\frac{1}{4}e^{-x} + \frac{49}{50}\cos x + \frac{7}{50}\sin x \Rightarrow y = -\frac{1}{4}e^{-x} + \frac{49}{50}\cos x + \frac{7}{50}\sin x + C_1\,e^{3x} +$ $C_2\,e^{-2x}$

25. $\frac{d^2y}{dx^2} + 5\frac{dy}{dx} = 0 \Rightarrow r^2 + 5r = 0 \Rightarrow r_1 = 0, r_2 = -5 \Rightarrow y_h = C_1 + C_2\,e^{-5x}$. $y_p = Dx^3 + Ex^2 + Fx \Rightarrow$ $y'_p = 3Dx^2 + 2Ex + F \Rightarrow y''_p = 6Dx + 2E \Rightarrow 6Dx + 2E + 5\left(3Dx^2 + 2Ex + F\right) = 15x^2 \Rightarrow D = 1, E = -\frac{3}{5}$, $F = \frac{6}{25} \Rightarrow y_p = x^3 - \frac{3}{5}x^2 + \frac{6}{25}x \Rightarrow y = x^3 - \frac{3}{5}x^2 + \frac{6}{25}x + C_1 + C_2\,e^{-5x}$

27. $\frac{d^2y}{dx^2} - 3\frac{dy}{dx} = 0 \Rightarrow r^2 - 3r = 0 \Rightarrow r_1 = 0, r_2 = 3 \Rightarrow y_h = C_1 + C_2\,e^{3x}$. $y_p = Ax\,e^{3x} + Dx^2 + Ex \Rightarrow$ $y'_p = A\,e^{3x} + 3Ax\,e^{3x} + 2Dx + E \Rightarrow y''_p = 6A\,e^{3x} + 9Ax\,e^{3x} + 2D \Rightarrow 6A\,e^{3x} + 9Ax\,e^{3x} + 2D -$ $3\left(A\,e^{3x} + 3Ax\,e^{3x} + 2Dx + E\right) = e^{3x} - 12x \Rightarrow A = \frac{1}{3}, D = 2, E = \frac{4}{3} \Rightarrow y_p = \frac{1}{3}x\,e^{3x} + 2x^2 + \frac{4}{3}x \Rightarrow$ $y = \frac{1}{3}x\,e^{3x} + 2x^2 + \frac{4}{3}x + C_1 + C_2\,e^{3x}$

29. $\frac{d^2y}{dx^2} - 5\frac{dy}{dx} = 0 \Rightarrow r^2 - 5r = 0 \Rightarrow r_1 = 0, r_2 = 5 \Rightarrow y_h = C_1 + C_2\,e^{5x}$. $y_p = Ax^2\,e^{5x} + Bx\,e^{5x} \Rightarrow$ $y'_p = (2A + 5B)x\,e^{5x} + 5Ax^2\,e^{5x} + B\,e^{5x} \Rightarrow y''_p = (2A + 10B)e^{5x} + (20A + 25B)x\,e^{5x} + 25Ax^2\,e^{5x} \Rightarrow$ $(2A + 10B)e^{5x} + (20A + 25B)x\,e^{5x} + 25Ax^2\,e^{5x} - 5\left((2A + 5B)e^{5x} + 5Ax^2\,e^{5x} + B\,e^{5x}\right) = x\,e^{5x} \Rightarrow$ $A = \frac{1}{10}, B = -\frac{1}{25} \Rightarrow y_p = \frac{1}{10}x^2\,e^{5x} - \frac{1}{25}x\,e^{5x} \Rightarrow y = \frac{1}{10}x^2\,e^{5x} - \frac{1}{25}x\,e^{5x} + C_1 + C_2\,e^{5x}$

31. $\frac{d^2y}{dx^2} + y = 0 \Rightarrow r^2 + 1 = 0 \Rightarrow r = \pm i \Rightarrow y_h = C_1\cos x + C_2\sin x$. $y_p = Ax\cos x + Bx\sin x \Rightarrow y'_p =$ $A\cos x - Ax\sin x + B\sin x + Bx\cos x \Rightarrow y''_p = -2A\sin x - Ax\cos x + 2B\cos x - Bx\sin x \Rightarrow$ $-2A\sin x - Ax\cos x + 2B\cos x - Bx\sin x + Ax\cos x + Bx\sin x = 2\cos x + \sin x$ $\Rightarrow A = -\frac{1}{2}, B = 1 \Rightarrow y_p = -\frac{1}{2}x\cos x + x\sin x \Rightarrow y = -\frac{1}{2}x\cos x + x\sin x + C_1\cos x + C_2\sin x$

33. a) $\frac{d^2y}{dx^2} - \frac{dy}{dx} = 0 \Rightarrow r^2 - r = 0 \Rightarrow r_1 = 0,\ r_2 = 1 \Rightarrow y_h = C_1 + C_2 e^x \Rightarrow u_1 = 1,\ u_2 = e^x \Rightarrow$

$D = \begin{vmatrix} 1 & e^x \\ 0 & e^x \end{vmatrix} = e^x \Rightarrow v'_1 = -\frac{u_2 F(x)}{D} = -e^x - e^{-x} \Rightarrow v_1 = \int -e^x - e^{-x}\,dx = -e^x + e^{-x} + C_1.$

$v'_2 = \frac{u_1 F(x)}{D} = 1 + e^{-2x} \Rightarrow v_2 = \int \left(1 + e^{-2x}\right) dx = x - \frac{1}{2} e^{-2x} + C_2 \Rightarrow$

$y = \left(-e^x + e^{-x} + C_1\right) + \left(x - \frac{1}{2} e^{-2x} + C_2\right) e^x = \frac{1}{2} e^{-x} + x\,e^x + C_1 + C_3 e^x$

b) $y_h = C_1 + C_2 e^x$. $y_p = Ax\,e^x + B\,e^{-x} \Rightarrow y'_p = A\,e^x + Ax\,e^x - B\,e^{-x} \Rightarrow y''_p = 2A\,e^x + Ax\,e^x + B\,e^{-x} \Rightarrow$

$2A\,e^x + Ax\,e^x + B\,e^{-x} - \left(A\,e^x + Ax\,e^x - B\,e^{-x}\right) = e^x + e^{-x} \Rightarrow A = 1,\ B = \frac{1}{2} \Rightarrow y_p = x\,e^x + \frac{1}{2} e^{-x} \Rightarrow$

$y = x\,e^x + \frac{1}{2} e^{-x} + C_1 + C_2 e^x$

35. a) $\frac{d^2y}{dx^2} - 4\frac{dy}{dx} - 5y = 0 \Rightarrow r^2 - 4r - 5 = 0 \Rightarrow r_1 = 5,\ r_2 = -1 \Rightarrow y_h = C_1 e^{5x} + C_2 e^{-x} \Rightarrow u_1 = e^{5x},$

$u_2 = e^{-x} \Rightarrow D = \begin{vmatrix} e^{5x} & e^{-x} \\ 5\,e^{5x} & -e^{-x} \end{vmatrix} = -6\,e^{4x} \Rightarrow v'_1 = -\frac{u_2 F(x)}{D} = \frac{1}{6} e^{-4x} + \frac{2}{3} e^{-5x} \Rightarrow$

$v_1 = \int \left(\frac{1}{6} e^{-4x} + \frac{2}{3} e^{-5x}\right) dx = -\frac{1}{24} e^{-4x} - \frac{2}{15} e^{-5x} + C_1.$ $v'_2 = \frac{u_1 F(x)}{D} = -\frac{1}{6} e^{2x} - \frac{2}{3} e^x \Rightarrow$

$v_2 = \int \left(-\frac{1}{6} e^{2x} - \frac{2}{3} e^x\right) dx = -\frac{1}{12} e^{2x} - \frac{2}{3} e^x + C_2 \Rightarrow y = \left(-\frac{1}{24} e^{-4x} - \frac{2}{15} e^{-5x} + C_1\right) e^{5x} +$

$\left(-\frac{1}{12} e^{2x} - \frac{2}{3} e^x + C_2\right) e^{-x} = -\frac{1}{8} e^x - \frac{4}{5} + C_1 e^{5x} + C_2 e^{-x}$

b) $y_h = C_1 e^{5x} + C_2 e^{-x}$. $y_p = A\,e^x + B\,x + C \Rightarrow y'_p = A\,e^x + B \Rightarrow y''_p = A\,e^x \Rightarrow A\,e^x - 4\left(A\,e^x + B\right) -$

$5\left(A\,e^x + Bx + C\right) = e^x + 4 \Rightarrow A = -\frac{1}{8},\ B = 0,\ C = -\frac{4}{5} \Rightarrow y_p = -\frac{1}{8} e^x - \frac{4}{5} \Rightarrow\ y = -\frac{1}{8} e^x - \frac{4}{5} + C_1 e^{5x} +$

$C_2 e^{-x}$

37. $\frac{d^2y}{dx^2} + y = 0 \Rightarrow r^2 + 1 = 0 \Rightarrow r = \pm i \Rightarrow y_h = C_1 \cos x + C_2 \sin x \Rightarrow u_1 = \cos x,\ u_2 = \sin x \Rightarrow$

$D = \begin{vmatrix} \cos x & \sin x \\ -\sin x & \cos x \end{vmatrix} = 1 \Rightarrow v'_1 = -\frac{u_2 F(x)}{D} = -\cos x \Rightarrow v_1 = \int -\cos x\,dx = -\sin x + C_1.$

$v'_2 = \frac{u_1 F(x)}{D} = \csc x - \sin x \Rightarrow v_2 = \int (\csc x - \sin x)\,dx = -\ln|\csc x + \cot x| + \cos x + C_2 \Rightarrow$

$y = \left(-\sin x + C_1\right)\cos x + \left(-\ln|\csc x + \cot x| + \cos x + C_2\right)\sin x = C_1 \cos x + C_2 \sin x -$

$\sin x(\ln|\csc x + \cot x|)$

39. $\frac{d^2y}{dx^2} - 8\frac{dy}{dx} = 0 \Rightarrow r^2 - 8r = 0 \Rightarrow r_1 = 0,\ r_2 = 8 \Rightarrow y_h = C_1 + C_2 e^{8x}$ $y_p = Ax\,e^{8x} \Rightarrow y'_p = A\,e^{8x} + 8Ax\,e^{8x}$

$\Rightarrow y''_p = 16A\,e^{8x} + 64Ax\,e^{8x} \Rightarrow 16A\,e^{8x} + 64Ax\,e^{8x} - 8\left(A\,e^{8x} + 8Ax\,e^{8x}\right) = e^{8x} \Rightarrow A = \frac{1}{8} \Rightarrow$

$y_p = \frac{1}{8} x\,e^{8x} \Rightarrow y = \frac{1}{8} x\,e^{8x} + C_1 + C_2 e^{8x}$

41. $\frac{d^2y}{dx^2} - \frac{dy}{dx} = 0 \Rightarrow r^2 - r = 0 \Rightarrow r_1 = 0, r_2 = 1 \Rightarrow y_h = C_1 + C_2\,e^x.\ y_p = Dx^4 + Ex^3 + Fx^2 + Gx \Rightarrow$
$y'_p = 4Dx^3 + 3Ex^2 + 2Fx + G \Rightarrow y''_p = 12Dx^2 + 6Ex + 2F \Rightarrow 12Dx^2 + 6Ex + 2F -$
$\left(4Dx^3 + 3Ex^2 + 2Fx + G\right) = x^3 \Rightarrow D = -\frac{1}{4}, E = -1, F = -3, G = -6 \Rightarrow y_p = -\frac{1}{4}x^4 - x^3 - 3x^2 - 6x \Rightarrow$
$y = -\frac{1}{4}x^4 - x^3 - 3x^2 - 6x + C_1 + C_2\,e^x$

43. $\frac{d^2y}{dx^2} + 2\frac{dy}{dx} = 0 \Rightarrow r^2 + 2r = 0 \Rightarrow r_1 = 0, r_2 = -2 \Rightarrow y_h = C_1 + C_2\,e^{-2x}.\ y_p = Ax^3 + Bx^2 + Cx + E\,e^x \Rightarrow$
$y'_p = 3Ax^2 + 2Bx + C + E\,e^x \Rightarrow y''_p = 6Ax + 2B + E\,e^x \Rightarrow 6Ax + 2B + E\,e^x + 2\left(3Ax^2 + 2Bx + C + E\,e^x\right)$
$= x^2 - e^x \Rightarrow A = \frac{1}{6}, B = -\frac{1}{4}, C = \frac{1}{4}, E = -\frac{1}{3} \Rightarrow y_p = \frac{1}{6}x^3 - \frac{1}{4}x^2 + \frac{1}{4}x - \frac{1}{3}e^x \Rightarrow y = \frac{1}{6}x^3 - \frac{1}{4}x^2 + \frac{1}{4}x -$
$\frac{1}{3}e^x + C_1 + C_2\,e^{-2x}$

45. $\frac{d^2y}{dx^2} + y = 0 \Rightarrow r^2 + 1 = 0 \Rightarrow r = \pm i \Rightarrow y_h = C_1 \cos x + C_2 \sin x \Rightarrow u_1 = \cos x, u_2 = \sin x \Rightarrow$
$D = \begin{vmatrix} \cos x & \sin x \\ -\sin x & \cos x \end{vmatrix} = 1 \Rightarrow v'_1 = -\frac{u_2\,F(x)}{D} = -\tan^2 x \Rightarrow v_1 = \int -\tan^2 x\,dx = -\tan x + x + C_1.$
$v'_2 = \frac{u_1\,F(x)}{D} = \tan x \Rightarrow v_2 = \int \tan x\,dx = \ln|\sec x| + C_2 \Rightarrow y = \left(-\tan x + x + C_1\right)\cos x +$
$\left(\ln|\sec x| + C_2\right)\sin x \Rightarrow y = x\cos x + \sin x \ln(\sec x) + C_1 \cos x + C_3 \sin x$

47. $\frac{dy}{dx} - 3y = 0 \Rightarrow r - 3 = 0 \Rightarrow r = 3 \Rightarrow y_h = C_1\,e^{3x}.\ y_p = A\,e^x \Rightarrow y'_p = A\,e^x \Rightarrow A\,e^x - 3A\,e^x = e^x \Rightarrow A =$
$-\frac{1}{2} \Rightarrow y_p = -\frac{1}{2}e^x \Rightarrow y = C_1\,e^{3x} - \frac{1}{2}e^x$

49. $\frac{dy}{dx} - 3y = 0 \Rightarrow r - 3 = 0 \Rightarrow r = 3 \Rightarrow y_h = C_1\,e^{3x}.\ y_p = Ax\,e^{3x} \Rightarrow y'_p = A\,e^{3x} + 3Ax\,e^{3x} \Rightarrow A\,e^{3x} +$
$3Ax\,e^{3x} - 3Ax\,e^{3x} = 5\,e^{3x} \Rightarrow A = 5 \Rightarrow y_p = 5x\,e^{3x} \Rightarrow y = C_1e^{3x} + 5x\,e^{3x}$

51. $\frac{d^2y}{dx^2} + y = 0 \Rightarrow r^2 + 1 = 0 \Rightarrow r = \pm i \Rightarrow y_h = C_1 \cos x + C_2 \sin x \Rightarrow u_1 = \cos x, u_2 = \sin x \Rightarrow$
$D = \begin{vmatrix} \cos x & \sin x \\ -\sin x & \cos x \end{vmatrix} = 1 \Rightarrow v'_1 = -\frac{u_2\,F(x)}{D} = -\sec x \tan x \Rightarrow v_1 = \int -\sec x \tan x\,dx = -\sec x +$
$C_1.\ v'_2 = \frac{u_1\,F(x)}{D} = \sec x \Rightarrow v_2 = \int \sec x\,dx = \ln|\sec x + \tan x| + C_2 \Rightarrow y = C_1 \cos x + C_2 \sin x - 1 +$
$\sin x \ln|\sec x + \tan x|.\ y(0) = 1 \Rightarrow C_1 \cos 0 + C_2 \sin 0 - 1 + \sin 0 \ln|\sec 0 + \tan 0| \Rightarrow C_1 = 2 \Rightarrow$
$y = 2\cos x + C_2 \sin x - 1 + \sin x \ln|\sec x + \tan x| \Rightarrow y' = -2\sin x + C_2 \cos x + \sec x \sin x +$
$\cos x \ln|\sec x + \tan x|.\ y'(0) = 1 \Rightarrow -2\sin 0 + C_2 \cos 0 + \sec 0 \sin 0 + \cos 0 \ln|\sec 0 + \tan 0| \Rightarrow$
$C_2 = 1 \Rightarrow y = 2\cos x + \sin x - 1 + \sin x \ln|\sec x + \tan x|$

53. $y(x) + \int_0^x y(t)\,dt = x \Rightarrow \frac{dy}{dx} + y = 1 \Rightarrow \frac{dy}{dx} = 1 - y \Rightarrow \frac{dy}{1-y} = dx \Rightarrow -\ln|1 - y| = x + C \Rightarrow 1 - y = C_1\,e^{-x}$
$\Rightarrow y = C_1\,e^{-x} + 1.\ y(0) = 0 \Rightarrow 0 = C_1\,e^0 + 1 \Rightarrow C_1 = -1 \Rightarrow y = -e^{-x} + 1$

## SECTION 16.6 OSCILLATION

1. $m\frac{d^2x}{dt^2} + kx = 0 \Rightarrow \frac{d^2x}{dt^2} + \frac{k}{m}x = 0$. Let $\omega = \sqrt{\frac{k}{m}}$. Then $\frac{d^2x}{dt^2} + \omega^2 x = 0 \Rightarrow r^2 + \omega^2 = 0 \Rightarrow r = \pm\,\omega i \Rightarrow$ $x = C_1 \cos \omega t + C_2 \sin \omega t$. $x(0) = x_0 \Rightarrow C_1 \cos 0 + C_2 \sin 0 = x_0 \Rightarrow C_1 = x_0 \Rightarrow x = x_0 \cos \omega t + C_2 \sin \omega t$ $\Rightarrow x' = -x_0\omega \sin \omega t + C_2\omega \cos \omega t$. $x'(0) = v_0 \Rightarrow -x_0\omega \sin 0 + C_2\omega \cos 0 = v_0 \Rightarrow C_2 = \frac{v_0}{\omega} \Rightarrow$ $x = x_0 \cos \omega t + \frac{v_0}{\omega} \sin \omega t$. $x = C \sin(\omega t + \phi)$ where $C = \sqrt{x_0^2 + \left(\frac{v_0}{\omega}\right)^2} = \frac{\sqrt{\omega^2 x_0^2 + v_0^2}}{\omega}$ and $\phi = \tan^{-1}\left(\frac{\omega x_0}{v_0}\right) \Rightarrow x = \frac{\sqrt{\omega^2 x_0^2 + v_0^2}}{\omega} \sin\left(\omega t + \tan^{-1}\left(\frac{\omega x_0}{v_0}\right)\right)$

3. a) $L\frac{d^2i}{dt^2} + R\frac{di}{dt} + \frac{1}{C}i = \frac{dv}{dt}$. $R = 0, \frac{1}{LC} = \omega^2$, $v$ = constant $\Rightarrow L\frac{d^2i}{dt^2} + \frac{1}{C}i = 0 \Rightarrow \frac{d^2i}{dt^2} + \frac{1}{LC}i = 0 \Rightarrow$ $\frac{d^2i}{dt^2} + \omega^2 i = 0 \Rightarrow r = \pm\,\omega i \Rightarrow i = C_1 \cos \omega t + C_2 \sin \omega t$

b) $L\frac{d^2i}{dt^2} + R\frac{di}{dt} + \frac{1}{C}i = \frac{dv}{dt}$. $R = 0, \frac{1}{LC} = \omega^2$, $v = V \sin \alpha t, \alpha \neq \omega \Rightarrow \frac{d^2i}{dt^2} + \omega^2 i = \frac{V\alpha}{L} \cos \alpha t \Rightarrow$ $i_h = C_1 \cos \omega t + C_2 \sin \omega t$. $i_p = A \cos \alpha t + B \sin \alpha t \Rightarrow i'_p = -A\alpha \sin \alpha t + B\alpha \cos \alpha t \Rightarrow$ $i''_p = -A\alpha^2 \cos \alpha t - B\alpha^2 \sin \alpha t \Rightarrow -A\alpha^2 \cos \alpha t - B\alpha^2 \sin \alpha t + \omega^2\left(A \cos \alpha t + B \sin \alpha t\right) =$ $\frac{V\alpha}{L} \cos \alpha t \Rightarrow A = \frac{V\alpha}{L\left(\omega^2 - \alpha^2\right)}$, $B = 0 \Rightarrow i_p = \frac{V\alpha}{L\left(\omega^2 - \alpha^2\right)} \cos \alpha t \Rightarrow i = C_1 \cos \omega t + C_2 \sin \omega t +$ $\frac{V\alpha}{L\left(\omega^2 - \alpha^2\right)} \cos \alpha t$

c) $L\frac{d^2i}{dt^2} + R\frac{di}{dt} + \frac{1}{C}i = \frac{dv}{dt}$ $R = 0, \frac{1}{LC} = \omega^2$, $v = V \sin \omega t$, $V$ constant $\Rightarrow L\frac{d^2i}{dt^2} + \frac{1}{C}i = V\omega \cos \omega t \Rightarrow$ $\frac{d^2i}{dt^2} + \omega^2 i = \frac{V\omega}{L} \cos \omega t \Rightarrow i_h = C_1 \cos \omega t + C_2 \sin \omega t$. $i_p = At \cos \omega t + Bt \sin \omega t \Rightarrow i'_p = A \cos \omega t -$ $A\omega t \sin \omega t + B \sin \omega t + B\omega t \cos \omega t \Rightarrow y''_p = -2A\omega \sin \omega t + 2B\omega \cos \omega t - A\omega^2 t \cos \omega t -$ $B\omega^2 t \sin \omega t \Rightarrow -2A\omega \sin \omega t + 2B\omega \cos \omega t - A\omega^2 t \cos \omega t - B\omega^2 t \sin \omega t +$ $\omega^2\left(At \cos \omega t + Bt \sin \omega t\right) = \frac{V\omega}{L} \cos \omega t \Rightarrow A = 0, B = \frac{V}{2L} \Rightarrow i_p = \frac{V}{2L} t \sin \omega t \Rightarrow i = C_1 \cos \omega t +$ $C_2 \sin \omega t + \frac{V}{2L} t \sin \omega t$

d) $L\frac{d^2i}{dt^2} + R\frac{di}{dt} + \frac{1}{C}i = \frac{dv}{dt}$ $R = 50, L = 5, C = 9 \times 10^{-6}$, $v$ constant $\Rightarrow 5\frac{d^2i}{dt^2} + 50\frac{di}{dt} + \frac{1}{9} \times 10^6 i = 0 \Rightarrow$ $\frac{d^2i}{dt^2} + 10\frac{di}{dt} + \frac{1}{45} \times 10^6 i = 0 \Rightarrow r^2 + 10r + \frac{1}{45} \times 10^6 = 0 \Rightarrow r = -5 \pm 5\sqrt{-\frac{7991}{9}} \approx -5 \pm 148.99\, i \Rightarrow$ $i = e^{-5t}\left(C_1 \cos(148.99)t + C_2 \sin(148.99)t\right)$

5. $\frac{d^2\theta}{dt^2} = -\frac{2k\theta}{mr^2} \Rightarrow \frac{d^2\theta}{dt^2} + \frac{2k}{mr^2}\theta = 0 \Rightarrow r^2 + \frac{2k}{mr^2} = 0 \Rightarrow r = \pm\sqrt{\frac{2k}{mr^2}}\, i \Rightarrow \theta = C_1 \cos\sqrt{\frac{2k}{mr^2}}\, t +$
$C_2 \sin\sqrt{\frac{2k}{mr^2}}\, t.$ $t = 0 \Rightarrow \theta = \theta_0 \Rightarrow C_1 = \theta_0 \Rightarrow \theta = \theta_0 \cos\sqrt{\frac{2k}{mr^2}}\, t + C_2 \sin\sqrt{\frac{2k}{mr^2}}\, t \Rightarrow$
$\frac{d\theta}{dt} = -\theta_0\sqrt{\frac{2k}{mr^2}} \sin\sqrt{\frac{2k}{mr^2}}\, t + C_2\sqrt{\frac{2k}{mr^2}} \cos\sqrt{\frac{2k}{mr^2}}\, t.$ $\frac{d\theta}{dt} = v_0$ at $t = 0 \Rightarrow C_2\sqrt{\frac{2k}{mr^2}} = v_0 \Rightarrow$
$C_2 = v_0\sqrt{\frac{mr^2}{2k}} \Rightarrow \theta = \theta_0 \cos\sqrt{\frac{2k}{mr^2}}\, t + v_0\sqrt{\frac{mr^2}{2k}} \sin\sqrt{\frac{2k}{mr^2}}\, t$

7. a) $f(t) = A \sin \alpha t,\ \alpha \neq \sqrt{\frac{k}{m}} \Rightarrow \frac{d^2x}{dt^2} + \frac{k}{m}x = \frac{k}{m}\left(A \sin \alpha t\right) \Rightarrow r^2 + \frac{k}{m} = 0 \Rightarrow r = \pm i\sqrt{\frac{k}{m}} \Rightarrow$
$x_h = C_1 \cos\sqrt{\frac{k}{m}}\, t + C_2 \sin\sqrt{\frac{k}{m}}\, t.$ $x_p = B \cos \alpha t + C \sin \alpha t \Rightarrow x'_p = -B\alpha \sin \alpha t + C\alpha \cos \alpha t \Rightarrow$
$x''_p = -B\alpha^2 \cos \alpha t - C\alpha^2 \sin \alpha t \Rightarrow -B\alpha^2 \cos \alpha t - C\alpha^2 \sin \alpha t + \frac{k}{m}\left(B \cos \alpha t + C \sin \alpha t\right) =$
$\frac{k}{m}\left(A \sin \alpha t\right) \Rightarrow B = 0,\ C = \frac{Ak}{k - m\alpha^2} \Rightarrow x_p = \frac{Ak}{k - m\alpha^2} \sin \alpha t \Rightarrow x = \frac{Ak}{k - m\alpha^2} \sin \alpha t + C_1 \cos\sqrt{\frac{k}{m}}\, t$
$+ C_2 \sin\sqrt{\frac{k}{m}}\, t.$ $x(0) = x_0 \Rightarrow C_1 = x_0 \Rightarrow x = x_0 \cos\sqrt{\frac{k}{m}}\, t + C_2 \sin\sqrt{\frac{k}{m}}\, t + \frac{Ak}{k - m\alpha^2} \sin \alpha t \Rightarrow$
$\frac{dx}{dt} = -x_0\sqrt{\frac{k}{m}} \sin\sqrt{\frac{k}{m}}\, t + C_2\sqrt{\frac{k}{m}} \cos\sqrt{\frac{k}{m}}\, t + \frac{Ak\alpha}{k - m\alpha^2} \cos \alpha t.$ $x'(0) = 0 \Rightarrow C_2\sqrt{\frac{k}{m}} +$
$\frac{Ak\alpha}{k - m\alpha^2} = 0 \Rightarrow C_2 = -\frac{Ak\alpha}{k - m\alpha^2}\sqrt{\frac{m}{k}} \Rightarrow x = x_0 \cos\sqrt{\frac{k}{m}}\, t - \frac{A\alpha\sqrt{mk}}{k - m\alpha^2} \sin\sqrt{\frac{k}{m}}\, t + \frac{Ak}{k - m\alpha^2} \sin \alpha t$

b) $f(t) = A \sin \alpha t,\ \alpha = \sqrt{\frac{k}{m}} \Rightarrow \frac{d^2x}{dt^2} + \frac{k}{m}x = \frac{k}{m}\left(A \sin \alpha t\right) \Rightarrow r^2 + \frac{k}{m} = 0 \Rightarrow r = \pm\sqrt{\frac{k}{m}}\, i \Rightarrow$
$x_h = C_1 \cos\sqrt{\frac{k}{m}}\, t + C_2 \sin\sqrt{\frac{k}{m}}\, t.$ $x_p = Bt \cos\sqrt{\frac{k}{m}}\, t + Ct \sin\sqrt{\frac{k}{m}}\, t \Rightarrow x'_p = B \cos\sqrt{\frac{k}{m}}\, t -$
$B\sqrt{\frac{k}{m}}\, t \sin\sqrt{\frac{k}{m}}\, t + C \sin\sqrt{\frac{k}{m}}\, t + C\sqrt{\frac{k}{m}}\, t \cos\sqrt{\frac{k}{m}}\, t \Rightarrow x''_p = -2B\sqrt{\frac{k}{m}} \sin\sqrt{\frac{k}{m}}\, t +$
$2C\sqrt{\frac{k}{m}} \cos\sqrt{\frac{k}{m}}\, t - B\left(\frac{k}{m}\right)t \cos\sqrt{\frac{k}{m}}\, t - C\left(\frac{k}{m}\right)t \sin\sqrt{\frac{k}{m}}\, t \Rightarrow -2B\sqrt{\frac{k}{m}} \sin\sqrt{\frac{k}{m}}\, t +$
$2C\sqrt{\frac{k}{m}} \cos\sqrt{\frac{k}{m}}\, t - B\left(\frac{k}{m}\right)t \cos\sqrt{\frac{k}{m}}\, t - C\left(\frac{k}{m}\right)t \sin\sqrt{\frac{k}{m}}\, t + \frac{k}{m}\left(Bt \cos\sqrt{\frac{k}{m}}\, t + Ct \sin\sqrt{\frac{k}{m}}\, t\right)$
$= \frac{kA}{m} \sin \alpha t \Rightarrow B = -\frac{A\alpha}{2},\ C = 0 \Rightarrow x_p = -\frac{A\alpha}{2} t \cos \alpha t \Rightarrow x = C_1 \cos\sqrt{\frac{k}{m}}\, t + C_2 \sin\sqrt{\frac{k}{m}}\, t -$
$\frac{A\alpha}{2} t \cos \alpha t.$ $x(0) = x_0 \Rightarrow C_1 = x_0 \Rightarrow x = x_0 \cos\sqrt{\frac{k}{m}}\, t + C_2 \sin\sqrt{\frac{k}{m}}\, t - \frac{A\alpha}{2} t \cos \alpha t \Rightarrow$
$\frac{dx}{dt} = -x_0\alpha \sin \alpha t + C_2\alpha \cos \alpha t - \frac{A\alpha}{2} \cos \alpha t - \frac{A\alpha^2}{2} t \sin \alpha t.$ $x'(0) = 0 \Rightarrow C_2\alpha - \frac{A\alpha}{2} = 0 \Rightarrow$
$C_2 = \frac{A}{2} \Rightarrow x = x_0 \cos \alpha t + \frac{A}{2} \sin \alpha t - \frac{A\alpha}{2} t \cos \alpha t$

## SECTION 16.7 NUMERICAL METHODS

1.

| | $x_n$ | $y_n$ |
|---|---|---|
| $x_0$ | 0 | 1 |
| $x_1$ | 1/5 | 1.2 |
| $x_2$ | 2/5 | 1.44 |
| $x_3$ | 3/5 | 1.728 |
| $x_4$ | 4/5 | 2.0736 |
| $x_5$ | 1 | 2.48832 |

Exact Value: $y' = y \Rightarrow \frac{dy}{dx} = y \Rightarrow \frac{dy}{y} = dx \Rightarrow \ln|y| = x + C \Rightarrow e^{\ln|y|} = e^{x + C} \Rightarrow |y| = e^C e^x \Rightarrow y = C_1 e^x$. $y(0) = 1 \Rightarrow C_1 = 1 \Rightarrow y = e^x \Rightarrow y(1) = e^1 = 2.718281828...$

3.

| | $x_n$ | $y_n$ |
|---|---|---|
| $x_0$ | 0 | 1 |
| $x_1$ | 1/5 | 1.22 |
| $x_2$ | 2/5 | 1.4884 |
| $x_3$ | 3/5 | 1.815848 |
| $x_4$ | 4/5 | 2.2153346 |
| $x_5$ | 1 | 2.702708163 |

5. $y' = x^2 + y^2 \Rightarrow x_{n+1} = x_n + h \Rightarrow y_{n+1} = y_n + h\left(x_n^2 + y_n^2\right)$. $Y' = Y^2 \Rightarrow x_{n+1} = x_n + h \Rightarrow Y_{n+1} = Y_n + h(Y_n^2)$. $y_0 = h(x_0^2 + y_0^2) = Y_0 + h(Y_0)$ since $x_0 = 0 \Rightarrow y_1 = Y_1$. $y_1 + h\left(x_1^2 + y_1^2\right) > Y_1 + h(Y_1^2)$ since $x_1 > 0 \Rightarrow y_2 > Y_2$. And from here on up to $x = 1$, $y_{n+1} > Y_{n+1}$.
$Y' = Y^2 \Rightarrow \frac{dY}{dx} = Y^2 \Rightarrow \frac{1}{Y^2} dY = dx \Rightarrow -\frac{1}{Y} = x + C \Rightarrow \frac{1}{Y} = -x - C \Rightarrow Y = -\frac{1}{x + C}$. $y(0) = 1 \Rightarrow -\frac{1}{C} = 1 \Rightarrow C = -1 \Rightarrow Y = -\frac{1}{x - 1} \Rightarrow y \to \infty$ as $x \to 1^-$. Since $y_{n+1} > Y_{n+1}$, $y_{n+1} \to \infty$ as $x \to 1^-$

Note: For Exercises 7–11, the Calculus Tookit was used.

7. $y = 0.571428572$

9. $y = 0.810263855$

11. a) $y = 0.841470985$ b) $y = 0.841470983$

## PRACTICE EXERCISES

1. $e^{y-2}\,dx - e^{x+2y}\,dy = 0 \Rightarrow e^{-x}\,dx - \frac{e^{2y}}{e^{y-2}}\,dy = 0 \Rightarrow e^{-x}\,dx - e^{y+2}\,dy = 0 \Rightarrow -e^{-x} - e^{y+2} = C$. $y(0) = -2 \Rightarrow -e^0 - e^{-2+2} = C \Rightarrow C = -2 \Rightarrow e^{-x} + e^{y+2} = 2$

3. $\frac{dy}{dx} = \frac{x^2 + y^2}{2xy} \Rightarrow \frac{dy}{dx} = \frac{1 + \left(\frac{y}{x}\right)^2}{2\left(\frac{y}{x}\right)} \Rightarrow$ Homogeneous. $v = \frac{y}{x} \Rightarrow F(v) = \frac{1 + v^2}{2v} \Rightarrow \frac{dx}{x} + \frac{dv}{v - \left(\frac{1 + v^2}{2v}\right)} = 0 \Rightarrow$
$\frac{dx}{x} + \frac{2v\,dv}{v^2 - 1} = 0 \Rightarrow \ln|x| + \ln|v^2 - 1| = C \Rightarrow x(v^2 - 1) = C_1 \Rightarrow x\left(\left(\frac{y}{x}\right)^2 - 1\right) = C_1 \Rightarrow \frac{y^2}{x} - x = C_1$. $y(5) = 0 \Rightarrow 0 - 5 = C_1 \Rightarrow C_1 = -5 \Rightarrow \frac{y^2}{x} - x = -5$

5. $(x^2 + y)\,dx + \left(e^y + x\right)dy = 0 \Rightarrow \frac{\partial M}{\partial y} = 1 = \frac{\partial N}{\partial x} \Rightarrow$ Exact. $\frac{\partial f}{\partial x} = x^2 + y \Rightarrow f(x,y) = \frac{x^3}{3} + xy + k(y) \Rightarrow$
$\frac{\partial f}{\partial y} = x + k'(y) = e^y + x \Rightarrow k'(y) = e^y \Rightarrow k(y) = e^y + C \Rightarrow f(x,y) = \frac{x^3}{3} + xy + e^y + C \Rightarrow \frac{x^3}{3} + xy + e^y = C_1.$
$y(3) = 0 \Rightarrow 9 + 1 = C_1 \Rightarrow C_1 = 10 \Rightarrow \frac{x^3}{3} + xy + e^y = 10$

7. $(x + 1)\frac{dy}{dx} + 2y = x \Rightarrow P(x) = 2,\ Q(x) = x \Rightarrow \int P(x)\,dx = \int 2\,dx = 2x \Rightarrow \rho(x) = e^{2x} \Rightarrow$
$y = \frac{1}{e^{2x}}\int e^{2x}(x\,dx) = \frac{1}{e^{2x}}\left(\frac{1}{2}x\,e^{2x} - \frac{1}{4}e^{2x} + C\right) \Rightarrow y = \frac{1}{2}x - \frac{1}{4} + C\,e^{-2x}.\ y(0) = 1 \Rightarrow 1 = -\frac{1}{4} + C \Rightarrow$
$C = \frac{5}{4} \Rightarrow y = \frac{1}{2}x - \frac{1}{4} + \frac{5}{4}e^{-2x}$

9. $\frac{d^2y}{dx^2} - \left(\frac{dy}{dx}\right)^2 = 1.$ Let $p = \frac{dy}{dx} \Rightarrow \frac{dp}{dx} - p^2 = 1 \Rightarrow \frac{dp}{1 + p^2} = dx \Rightarrow \tan^{-1} p = x + C \Rightarrow p = \tan(x + C) \Rightarrow$
$\frac{dy}{dx} = \tan(x + C) \Rightarrow y = \ln|\sec(x + C)| + C_1.\ y'\left(\frac{\pi}{3}\right) = \sqrt{3} \Rightarrow \sqrt{3} = \tan\left(\frac{\pi}{3} + C\right) \Rightarrow \tan^{-1}\sqrt{3} = \frac{\pi}{3} + C \Rightarrow$
$\frac{\pi}{3} = \frac{\pi}{3} + C \Rightarrow C = 0 \Rightarrow y = \ln|\sec x| + C_1.\ y\left(\frac{\pi}{3}\right) = 0 \Rightarrow 0 = \ln\left|\sec\frac{\pi}{3}\right| + C_1 \Rightarrow C_1 = -\ln 2 \Rightarrow$
$y = \ln|\sec x| - \ln 2$

11. $\frac{d^2y}{dx^2} - 4\frac{dy}{dx} + 3y = 0 \Rightarrow r^2 - 4r + 3 = 0 \Rightarrow r_1 = 3,\ r_2 = 1 \Rightarrow y = C_1\,e^{3x} + C_2\,e^x \Rightarrow y' = 3C_1\,e^{3x} + C_2\,e^x.$
$y'(0) = -2 \Rightarrow -2 = 3C_1 + C_2.\ y(0) = 2 \Rightarrow 2 = C_1 + C_2 \Rightarrow C_1 = -2,\ C_2 = 4 \Rightarrow y = -2\,e^{3x} + 4\,e^x$

13. $\frac{d^2y}{dx^2} + 4\frac{dy}{dx} + 4y = 0 \Rightarrow r_1 = r_2 = -2 \Rightarrow y = \left(C_1x + C_2\right)e^{-2x}.\ y(0) = 0 \Rightarrow C_2 = 0 \Rightarrow y = C_1x\,e^{-2x} \Rightarrow$
$y' = C_1\left(e^{-2x} - 2x\,e^{-2x}\right).\ y'(0) = 7 \Rightarrow C_1 = 7 \Rightarrow y = 7x\,e^{-2x}$

15. $\frac{d^2y}{dx^2} + 2\frac{dy}{dx} + 2y = 0 \Rightarrow r^2 + 2r + 2 = 0 \Rightarrow r = -1 \pm i \Rightarrow y = e^{-x}\left(C_1\cos x + C_2\sin x\right).\ y(0) = 1 \Rightarrow$
$e^0\left(C_1\cos 0 + C_2\sin 0\right) = 1 \Rightarrow C_1 = 1 \Rightarrow y = e^{-x}\left(\cos x + C_2\sin x\right) \Rightarrow y' = -e^{-x}\left(\cos x + C_2\sin x\right) +$
$e^{-x}\left(-\sin x + C_2\cos x\right).\ y'(0) = 0 \Rightarrow 0 = -\left(\cos 0 + C_2\sin 0\right) + \left(-\sin 0 + C_2\cos 0\right) \Rightarrow C_2 = 1 \Rightarrow$
$y = e^{-x}(\cos x + \sin x)$

17. $\frac{d^2y}{dx^2} + 2\frac{dy}{dx} = 4x \Rightarrow r^2 + 2r = 0 \Rightarrow r_1 = 0,\ r_2 = -2 \Rightarrow y_h = C_1 + C_2\,e^{-2x}.\ y_p = Ax^2 + Bx \Rightarrow y'_p = 2Ax + B.$
$y''_p = 2A \Rightarrow 2A + 2(2Ax + B) = 4x \Rightarrow A = 1,\ B = -1 \Rightarrow y_p = x^2 - x \Rightarrow y = C_1 + C_2\,e^{-2x} + x^2 - x.$
$y(0) = 1 \Rightarrow C_1 + C_2 = 1.\ y' = -2C_2\,e^{-2x} + 2x - 1.\ y'(0) = -3 \Rightarrow -2C_2 - 1 = -3 \Rightarrow C_1 = 0,\ C_2 = 1 \Rightarrow$
$y = e^{-2x} + x^2 - x$

19. $\frac{d^2y}{dx^2} - \frac{dy}{dx} - 2y = 3\,e^{2x} \Rightarrow r^2 - r - 2 = 0 \Rightarrow r_1 = 2,\ r_2 = -1 \Rightarrow y_h = C_1\,e^{2x} + C_2\,e^{-x}$. $y_p = Ax\,e^{2x} \Rightarrow$
$y'_p = A\,e^{2x} + 2Ax\,e^{2x} \Rightarrow y''_p = 4A\,e^{2x} + 4Ax\,e^{2x} \Rightarrow 4A\,e^{2x} + 4Ax\,e^{2x} - \left(A\,e^{2x} + 2Ax\,e^{2x}\right) -$
$2\left(Ax\,e^{2x}\right) = 3\,e^{2x} \Rightarrow A = 1 \Rightarrow y_p = x\,e^{2x} \Rightarrow y = x\,e^{2x} + C_1\,e^{2x} + C_2\,e^{-x}$. $y(0) = -2 \Rightarrow C_1 + C_2 = -2$.
$y' = 2C_1\,e^{2x} - C_2\,e^{-x} + e^{2x} + 2x\,e^{2x}$. $y'(0) = 0 \Rightarrow 2C_1 - C_2 + 1 = 0 \Rightarrow C_1 = -1,\ C_2 = -1 \Rightarrow$
$y = -\,e^{2x} - e^{-x} + x\,e^{2x}$

## APPENDIX A.7 DETERMINANTS AND CRAMER'S RULE

1. $$\left|\begin{array}{ccc|cc} 2 & 3 & 1 & 2 & 3 \\ 4 & 5 & 2 & 4 & 5 \\ 1 & 2 & 3 & 1 & 2 \end{array}\right. = 30 + 6 + 8 - 5 - 8 - 36 = -5$$

3. $$\begin{vmatrix} 1 & 2 & 3 & 4 \\ 0 & 1 & 2 & 3 \\ 0 & 0 & 2 & 1 \\ 0 & 0 & 3 & 2 \end{vmatrix} = 1\begin{vmatrix} 1 & 2 & 3 \\ 0 & 2 & 1 \\ 0 & 3 & 2 \end{vmatrix} = 1\begin{vmatrix} 2 & 1 \\ 3 & 2 \end{vmatrix} = 1$$

5. a) $$\begin{vmatrix} 2 & -1 & 2 \\ 1 & 0 & 3 \\ 0 & 2 & 1 \end{vmatrix} = \begin{vmatrix} 2 & -5 & 2 \\ 1 & -6 & 3 \\ 0 & 0 & 1 \end{vmatrix} = 1\begin{vmatrix} 2 & -5 \\ 1 & -6 \end{vmatrix} = -7$$

b) $$\begin{vmatrix} 2 & -1 & 2 \\ 1 & 0 & 3 \\ 0 & 2 & 1 \end{vmatrix} = \begin{vmatrix} 2 & -1 & 2 \\ 1 & 0 & 3 \\ 4 & 0 & 5 \end{vmatrix} = -(-1)\begin{vmatrix} 1 & 3 \\ 4 & 5 \end{vmatrix} = -7$$

7. a) $$\begin{vmatrix} 1 & 1 & 0 & 0 \\ 0 & 0 & -2 & 1 \\ 0 & -1 & 0 & 7 \\ 3 & 0 & 2 & 1 \end{vmatrix} = \begin{vmatrix} 1 & 1 & 0 & 7 \\ 0 & 0 & -2 & 1 \\ 0 & -1 & 0 & 0 \\ 3 & 0 & 2 & 1 \end{vmatrix} = -(-1)\begin{vmatrix} 1 & 0 & 7 \\ 0 & -2 & 1 \\ 3 & 2 & 1 \end{vmatrix} = \begin{vmatrix} 1 & 0 & 7 \\ 0 & -2 & 1 \\ 0 & 2 & -20 \end{vmatrix} =$$

$$1\begin{vmatrix} -2 & 1 \\ 2 & -20 \end{vmatrix} = 38$$

b) $$\begin{vmatrix} 1 & 1 & 0 & 0 \\ 0 & 0 & -2 & 1 \\ 0 & -1 & 0 & 7 \\ 3 & 0 & 2 & 1 \end{vmatrix} = \begin{vmatrix} 1 & 1 & 0 & 0 \\ 0 & 0 & -2 & 1 \\ 1 & 0 & 0 & 7 \\ 3 & 0 & 2 & 1 \end{vmatrix} = -1\begin{vmatrix} 0 & -2 & 1 \\ 1 & 0 & 7 \\ 3 & 2 & 1 \end{vmatrix} = -1\begin{vmatrix} 0 & -2 & 1 \\ 1 & 0 & 7 \\ 0 & 2 & -20 \end{vmatrix} =$$

$$-1(-1)\begin{vmatrix} -2 & 1 \\ 2 & -20 \end{vmatrix} = 38$$

9. $$D = \begin{vmatrix} 1 & 8 \\ 3 & -1 \end{vmatrix} = -25. \quad x = \frac{\begin{vmatrix} 4 & 8 \\ -13 & -1 \end{vmatrix}}{-25} = \frac{100}{-25} = -4, \quad y = \frac{\begin{vmatrix} 1 & 4 \\ 3 & -13 \end{vmatrix}}{-25} = \frac{-25}{-25} = 1$$

11. $D = \begin{vmatrix} 4 & -3 \\ 3 & -2 \end{vmatrix} = 1.\ x = \dfrac{\begin{vmatrix} 6 & -3 \\ 5 & -2 \end{vmatrix}}{1} = 3,\ y = \dfrac{\begin{vmatrix} 4 & 6 \\ 3 & 5 \end{vmatrix}}{1} = 2$

13. $D = \begin{vmatrix} 2 & 1 & -1 \\ 1 & -1 & 1 \\ 2 & 2 & 1 \end{vmatrix} = \begin{vmatrix} 2 & 1 & -1 \\ 3 & 0 & 0 \\ 4 & 3 & 0 \end{vmatrix} = -1\begin{vmatrix} 3 & 0 \\ 4 & 3 \end{vmatrix} = -9.\ x = \dfrac{\begin{vmatrix} 2 & 1 & -1 \\ 7 & -1 & 1 \\ 4 & 2 & 1 \end{vmatrix}}{-9} = \dfrac{\begin{vmatrix} 2 & 1 & -1 \\ 9 & 0 & 0 \\ 6 & 3 & 0 \end{vmatrix}}{-9} =$

$\dfrac{-1\begin{vmatrix} 9 & 0 \\ 6 & 3 \end{vmatrix}}{-9} = 3,\ y = \dfrac{\begin{vmatrix} 2 & 2 & -1 \\ 1 & 7 & 1 \\ 2 & 4 & 1 \end{vmatrix}}{-9} = \dfrac{\begin{vmatrix} 2 & 2 & -1 \\ 3 & 9 & 0 \\ 4 & 6 & 0 \end{vmatrix}}{-9} = \dfrac{-1\begin{vmatrix} 3 & 9 \\ 4 & 6 \end{vmatrix}}{-9} = -2,\ z = \dfrac{\begin{vmatrix} 2 & 1 & 2 \\ 1 & -1 & 7 \\ 2 & 2 & 4 \end{vmatrix}}{-9} =$

$\dfrac{\begin{vmatrix} 3 & 0 & 9 \\ 1 & -1 & 7 \\ 4 & 0 & 18 \end{vmatrix}}{-9} = \dfrac{-1\begin{vmatrix} 3 & 9 \\ 4 & 18 \end{vmatrix}}{-9} = 2$

15. $D = \begin{vmatrix} 1 & 0 & -1 \\ 0 & 2 & -2 \\ 2 & 0 & 1 \end{vmatrix} = 2\begin{vmatrix} 1 & -1 \\ 2 & 1 \end{vmatrix} = 6.\ x = \dfrac{\begin{vmatrix} 3 & 0 & -1 \\ 2 & 2 & -2 \\ 3 & 0 & 1 \end{vmatrix}}{6} = \dfrac{2\begin{vmatrix} 3 & -1 \\ 3 & 1 \end{vmatrix}}{6} = 2,\ y = \dfrac{\begin{vmatrix} 1 & 3 & -1 \\ 0 & 2 & -2 \\ 2 & 3 & 1 \end{vmatrix}}{6} =$

$\dfrac{\begin{vmatrix} 1 & 3 & -1 \\ 0 & 2 & -2 \\ 0 & -3 & 3 \end{vmatrix}}{6} = \dfrac{1\begin{vmatrix} 2 & -2 \\ -3 & 3 \end{vmatrix}}{6} = 0,\ z = \dfrac{\begin{vmatrix} 1 & 0 & 3 \\ 0 & 2 & 2 \\ 2 & 0 & 3 \end{vmatrix}}{6} = \dfrac{2\begin{vmatrix} 1 & 3 \\ 2 & 3 \end{vmatrix}}{6} = -1$

17. $D = \begin{vmatrix} 2 & h \\ 1 & 3 \end{vmatrix} = 6 - h = 0 \Rightarrow h = 6.\ x: \begin{vmatrix} 8 & h \\ k & 3 \end{vmatrix} = 24 - hk = 24 - 6k = 0 \Rightarrow k = 4$

a) When $h = 6$, $k = 4$, there are infinitely many solutions.

b) When $h = 6$, $k \neq 4$, there are no solutions.

19. $au + bv + cw = 0 \Rightarrow v = \dfrac{-au - cw}{b},\ au' + bv' + cw' = 0 \Rightarrow v' = \dfrac{-au' - cw'}{b},\ au'' + bv'' + cw'' = 0 \Rightarrow$

$v'' = \dfrac{-au'' - cw''}{b} \Rightarrow \begin{vmatrix} u & v & w \\ u' & v' & w' \\ u'' & v'' & w'' \end{vmatrix} = \begin{vmatrix} u & \dfrac{-au - cw}{b} & w \\ u' & \dfrac{-au' - cw'}{b} & w' \\ u'' & \dfrac{-au'' - cw''}{b} & w'' \end{vmatrix} = \dfrac{1}{b}\begin{vmatrix} u & -au - cw & w \\ u' & -au' - cw' & w' \\ u'' & -au'' - cw'' & w'' \end{vmatrix} =$

19. (Continued)

$\frac{1}{b}(u(-au' - cw')w'' + u''(-au - cw)w' + w(-au'' - cw'')u' - u''(-au' - cw')w - u'(-au - cw)w'' - u(-au'' - cw'')w') = \frac{1}{b}(0) = 0$

## APPENDIX A.8 LAGRANGE MULTIPLIERS WITH TWO CONSTRAINTS

1. Let $f(x,y,z) = x^2 + y^2 + z^2$. Maximize f subject to $g_1(x,y,z) = y + 2z - 12 = 0$ and $g_2(x,y,z) = x + y - 6 = 0$ $\nabla f = 2x\,\mathbf{i} + 2y\,\mathbf{j} + 2z\,\mathbf{k}$, $\nabla g_1 = \mathbf{j} + 2\,\mathbf{k}$, $\nabla g_2 = \mathbf{i} + \mathbf{j}$. Then $2x\,\mathbf{i} + 2y\,\mathbf{j} + 2z\,\mathbf{k} = \lambda(\mathbf{j} + 2\,\mathbf{k}) + \mu(\mathbf{i} + \mathbf{j}) \Rightarrow$ $2x\,\mathbf{i} + 2y\,\mathbf{j} + 2z\,\mathbf{k} = \mu\,\mathbf{i} + (\lambda + \mu)\,\mathbf{j} + 2\lambda\,\mathbf{k} \Rightarrow 2x = \mu,\ 2y = \lambda + \mu,\ 2z = 2\lambda \Rightarrow x = \frac{\mu}{2},\ z = \lambda \Rightarrow$ $2x = 2y - z \Rightarrow x = \frac{2y - z}{2}$. Then $y + 2z - 12 = 0$ and $\frac{2y - z}{2} + y - 6 = 0 \Rightarrow 4y - z - 12 = 0 \Rightarrow z = 4 \Rightarrow$ $y = 4 \Rightarrow x = 2 \Rightarrow (2,4,4)$ is the point closest.

3. Let $g_1(x,y,z) = z - 1 = 0$ and $g_2(x,y,z) = x^2 + y^2 + z^2 - 10 = 0 \Rightarrow \nabla g_1 = \mathbf{k}$, $\nabla g_2 = 2x\,\mathbf{i} + 2y\,\mathbf{j} + 2z\,\mathbf{k}$. $\nabla f = 2xyz\,\mathbf{i} + x^2z\,\mathbf{j} + x^2y\,\mathbf{k} \Rightarrow 2xyz\,\mathbf{i} + x^2z\,\mathbf{j} + x^2y\,\mathbf{k} = \lambda(\mathbf{k}) + \mu(2x\,\mathbf{i} + 2y\,\mathbf{j} + 2z\,\mathbf{k}) \Rightarrow 2xyz = 2x\mu,\ x^2z = 2y\mu,\ x^2y = 2z + \lambda \Rightarrow xyz = x\mu \Rightarrow x = 0$ or $yz = \mu \Rightarrow \mu = y$ since $z = 1$. $x = 0 \Rightarrow 2y\mu = 0$ and $2z + \lambda = 0$ $\Rightarrow z = 1 \Rightarrow \lambda = -2 \Rightarrow y^2 - 9 = 0 \Rightarrow y = \pm 3 \Rightarrow (0,\pm 3,1)$. $\mu = y \Rightarrow x^2z = 2y^2 \Rightarrow x^2 = 2y^2$ since $z = 1$ $\Rightarrow 2y^2 + y^2 + 1 - 10 = 0 \Rightarrow 3y^2 - 9 = 0 \Rightarrow y = \pm\sqrt{3} \Rightarrow \mu = \pm\sqrt{3} \Rightarrow x^2 = 2\left(\pm\sqrt{3}\right)\left(\pm\sqrt{3}\right) = 6 \Rightarrow x =$ $\pm\sqrt{6} \Rightarrow (\pm\sqrt{6}, \pm\sqrt{3}, 1)$. $f(0,\pm 3,1) = 1$. $f(\pm\sqrt{6},\pm\sqrt{3},1) = 6\left(\pm\sqrt{3}\right) + 1 = 1 \pm 6\sqrt{3} \Rightarrow$ Maximum of f is $1 + 6\sqrt{3}$ at $(\pm\sqrt{6},\sqrt{3},1)$; minimum of f is $1 - 6\sqrt{3}$ at $(\pm\sqrt{6},-\sqrt{3},1)$

5. Let $g_1(x,y,z) = y - x = 0 \Rightarrow x = y$. Let $g_2(x,y,z) = x^2 + y^2 + z^2 - 4 = 0 \Rightarrow \nabla g_2 = 2x\,\mathbf{i} + 2y\,\mathbf{j} + 2z\,\mathbf{k}$. $\nabla g_1 = -\mathbf{i} + \mathbf{j}$; $\nabla f = y\,\mathbf{i} + x\,\mathbf{j} + 2z\,\mathbf{k} \Rightarrow y\,\mathbf{i} + x\,\mathbf{j} + 2z\,\mathbf{k} = \lambda(-\mathbf{i} + \mathbf{j}) + \mu(2x\,\mathbf{i} + 2y\,\mathbf{j} + 2z\,\mathbf{k}) \Rightarrow y = -\lambda + 2x\mu$, $x = \lambda + 2y\mu$, $2z = 2z\mu \Rightarrow z = z\mu \Rightarrow z = 0$ or $\mu = 1$. $z = 0 \Rightarrow x^2 + y^2 - 4 = 0 \Rightarrow 2x^2 - 4 = 0$ (since $x = y$) $\Rightarrow$ $x^2 = 2 \Rightarrow x = \pm\sqrt{2} \Rightarrow y = \pm\sqrt{2} \Rightarrow (\pm\sqrt{2},\pm\sqrt{2},0)$. $\mu = 1 \Rightarrow y = -\lambda + 2x$ and $x = \lambda + 2y \Rightarrow x + y =$ $2(x + y) \Rightarrow 2x = 2(2x)$ since $x = y \Rightarrow x = 0 \Rightarrow y = 0 \Rightarrow z^2 - 4 = 0 \Rightarrow z = \pm 2 \Rightarrow (0,0,\pm 2)$.
$f(0,0,\pm 2) = 4$, $f(\pm\sqrt{2},\pm\sqrt{2},0) = 2 \Rightarrow$ Maximum value of f is 4 at $(0,0,\pm 2)$; minimum value of f is 2 at $(\pm\sqrt{2},\pm\sqrt{2},0)$